D1190994

Fundamental Principles of
Polymeric Materials

SPE MONOGRAPHS

Injection Molding
Irvin I. Rubin

Nylon Plastics
Melvin I. Kohan

Introduction to Polymer Science and Technology: An SPE Textbook
Edited by Herman S. Kaufmann and Joseph J. Falcetta

Principles of Polymer Processing
Zehev Tadmor and Costas G. Gogos

Coloring of Plastics
Edited by Thomas G. Webber

The Technology of Plasticizers
J. Kern Sears and Joseph R. Darby

Plastics Polymer Science and Technology
Edited by M. D. Baijal

Fundamental Principles of Polymeric Materials
Stephen L. Rosen

Plastics vs. Corrosives
Raymond B. Seymour

Handbook of Plastics Testing Technology
Vishu Shah

Aromatic High-Strength Fibers
H. H. Yang

Giant Molecules: An SPE Textbook
Raymond B. Seymour and Charles E. Carraher

Analysis and Performance of Fiber Composites, 2nd ed.
Bhagwan D. Agarwal and Lawrence J. Broutman

Impact Modifiers for PVC: The History and Practice
John T. Lutz, Jr. and David L. Dunkelberger

The Development of Plastics Processing Machinery and Methods
Joseph Fred Chabot, Jr.

Fundamental Principles of Polymeric Materials

Second Edition

STEPHEN L. ROSEN
*Department of Chemical Engineering, University of Missouri-Rolla,
Rolla, Missouri*

A WILEY-INTERSCIENCE PUBLICATION

JOHN WILEY & SONS, INC.

New York • Chichester • Brisbane • Toronto • Singapore

This text is printed on acid-free paper.

Copyright © 1993 by John Wiley & Sons, Inc.

All rights reserved. Published simultaneously in Canada.

Reproduction or translation of any part of this work
beyond that permitted by Section 107 or 108 of the
1976 United States Copyright Act without the permission
of the copyright owner is unlawful. Requests for
permission or further information should be addressed to
the Permissions Department, John Wiley & Sons, Inc., 605 Third Avenue,
New York, NY 10158–0012

Library of Congress Cataloging in Publication Data:
Rosen, Stephen L., 1937–
    Fundamental principles of polymeric materials/Stephen L. Rosen.
—2nd ed.
        p. cm.—(SPE monographs, ISSN 0195–4288)
    "A Wiley-Interscience publication."
    Includes index.
    ISBN 0-471-57525-9 (cloth)
    1. Polymers. I. Title. II. Series.
TA455.P58R63   1993   92–10973
668.9—dc20

Printed in the United States of America

10 9 8 7 6 5

Foreword

The Society of Plastics Engineers (SPE) is pleased to sponsor and endorse this second edition of "Fundamental Principles of Polymeric Materials" authored by Stephen L. Rosen. The first edition, published over ten years ago, filled a void in the teaching literature of polymers. This second edition includes the many advances that have been made in the ensuing period, updating engineering and technology for the student.

Dr. Rosen is well known within the SPE community. He has been a longtime author and presenter of papers at meetings. Further, he has contributed his time and talents to SPE technical activities involving both the written word and identification of new technologies for incorporation into the total Society program.

SPE, through its Technical Volumes Committee, has long sponsored books on various aspects of plastics and polymers. Its involvement has ranged from identification of needed volumes to recruitment of authors. An ever-present ingredient, however, is review of the final manuscript to insure accuracy of the technical content.

This technical competence pervades all SPE activities, not only in publication of books but also in other activities such as technical conferences and educational programs. In addition, the Society publishes periodicals—*Plastics Engineering, Polymer Engineering and Science, Polymer Processing and Rheology, Journal of Vinyl Technology* and *Polymer Composites*—as well as conference proceedings and other selected publications, all of which are subject to the same rigorous technical review procedure.

The resource of some 37,000 practicing plastics engineers has made SPE the largest organization of its type worldwide. Additional information can be obtained from the Society of Plastics Engineers, 14 Fairfield Drive, Brookfield, Connecticut 06804.

Robert D. Forger

Executive Director
Society of Plastics Engineers

Technical Volumes Committee
Raymond J. Ehrig, Chairperson
Aristech Chemical Corporation

Preface

This work was written to provide an appreciation of those fundamental principles of polymer science and engineering that are currently of practical relevance. I hope the reader will obtain both a broad, unified introduction to the subject matter that will be of immediate practical value and a foundation for more advanced study.

A decade has passed since the publication of the first Wiley edition of this book. New developments in the polymer area during that decade justify an update. Having used the book in class during the period, I've thought of better ways of explaining some of the material, and these have been incorporated in this edition.

But the biggest change with this edition is the addition of end-of-chapter problems at the suggestion of some academic colleagues. This should make the book more suitable as an academic text. Most of these problems are old homework problems or exam questions. I don't know what I'm going to do for new exam questions, but I'll think of something. Any suggestions for additional problems will be gratefully accepted.

The first Wiley edition of this book in 1982 was preceded by a little paperback intended primarily as a self-study guide for practicing engineers and scientists. I sincerely hope that by adding material aimed at an academic audience I have not made the book less useful to that original audience. To this end, I have retained the worked-out problems in the chapters and added some new ones. I have tried to emphasize a qualitative understanding of the underlying principles before tackling the mathematical details, so that the former may be appreciated independently of the latter (I don't recommend trying it the other way around, however), and I have tried to include practical illustrations of the material whenever possible.

In this edition, previous material has been generally updated. In view of commercial developments over the decade, the discussion of extended-chain crystals has been increased and a section on liquid-crystal polymers has been added. The discussion of phase behavior in polymer-solvent systems has been expanded and the Flory–Huggins theory is introduced. All kinetic expressions are now written in terms of conversion (rather than monomer concentration) for greater generality and ease of application. Also, in deference to the ready availability of numerical-solution software, kinetic expressions now incorporate the possibility of a variable-volume reaction mass, and the effects of variable volume are illustrated in several examples. A section on group-transfer polymerization has been added and a quantitative treatment of Ziegler–Natta polymerization has been attempted for the first time, including three new worked-out examples. Processes based on these catalysts are presented in greater detail. The "modified Cross" model, giving viscosity as a function of both shear rate and temperature, is introduced and its utility is illustrated. A section on scaleup calculations for the laminar flow of non-Newtonian fluids has been added, including two worked-out examples. The discussion of three-dimensional stress and strain has been expanded and includes two new worked-out examples. Tobolsky's "Procedure X" for extracting discrete relaxation times and moduli from data is introduced.

Obviously, the choice of material to be covered involves subjective judgment on the part of the author. This, together with space limitations and the rapid expansion of knowledge in the field, has resulted in the omission or shallow treatment of many interesting subjects. I apologize to friends and colleagues who have suggested incorporation of their work but don't find it here. Generally, it's fine work, but too specialized for a book of this nature. The end-of-chapter references are chosen to aid the reader who wishes to pursue a subject in greater detail.

I have used the previous edition to introduce the macromolecular gospel to a variety of audiences. Parts 1, 2 and most of 3 were covered in a one-semester course with chemistry and chemical engineering seniors and graduate students at Carnegie-Mellon. At Toledo, Parts 1 and 2 were covered in a one-quarter course with chemists and chemical engineers. A second quarter covered Part 3 with additional quantitative material on processing added. The audience for this included chemical and mechanical engineers (we didn't mention chemical reactions). Finally, I covered Parts 1 and 3 in one quarter with a diverse audience of graduate engineers at the NASA–Lewis Research Labs.

A word to the student: To derive maximum benefit from the worked-out examples, make an honest effort to answer them before looking at the solutions. If you can't do one, you've missed some important points in the preceding material, and you ought to go back over it.

STEPHEN L. ROSEN

Rolla, Missouri
November 1992

Acknowledgments

I'd like to thank my students for proving time and again that the best way to learn is to teach; my teachers, B. Maxwell, L. Rahm, H. Pohl, the late A. V. Tobolsky, and, in particular Ferdinand Rodriguez, who was also my research advisor, for making the macromolecular gospel so fascinating. My industrial friends and colleagues, particularly those in the Toledo SPE section, helped keep me aware of the important "real world" problems. My Toledo academic colleagues, Ronald Fournier, Steven LeBlanc, and Sasidhar Varanasi, provided support and many helpful suggestions during the preparation of the manuscript for this edition. At Rolla, Gary Bertrand, Partho Neogi, and James Stoffer read portions of the manuscript and pointed out how little I really know about some areas. Finally, my graduate student Purnedu Rai proofread the manuscript and caught many errors that I had completely overlooked. I'm deeply grateful to them all for their help.

Contents

Fundamental Principles of
Polymeric Materials

CHAPTER 1

Introduction

Since the Second World War, polymeric materials have been the fastest-growing segments of the United States chemical industry. It has been estimated that more than a third of the chemical research dollar is spent on polymers, with a correspondingly large proportion of technical personnel working in the area.

A modern automobile contains over 300 pounds (150 kg) of plastics, and this does not include paints, the rubber in tires, or the fibers in tires and upholstery. New aircraft incorporate increasing amounts of polymers and polymer-based composites. With the need to save fuel and therefore weight, polymers will continue to replace traditional materials in the automotive and aircraft industries. Similarly, the applications of polymers in the building construction industry (piping, resilient flooring, siding, thermal and electrical insulation, paints, decorative laminates) are already impressive, and will become even more so in the future. A trip through a supermarket will quickly convince anyone of the importance of polymers in the packaging industry (bottles, films, trays). Many other examples could be cited, but to make a long story short, the use of polymers now outstrips that of metals on a mass basis.

People have objected to synthetic polymers because they're not "natural." Well, botulism is natural, but it's not particularly desirable. Seriously, if all the polyester and nylon fibers in use today were to be replaced by cotton and wool, their closest natural counterparts, calculations show that there wouldn't be enough arable land left to feed the populace, and we'd be overrun by sheep. The fact is, there simply are no practical natural substitutes for many of the synthetic polymers used in modern society.

Since nearly all modern polymers have their origins in petroleum, it has been argued that this increased reliance on polymers constitutes an unnecessary drain on energy resources. However, the raw materials for polymers account for less than two percent of total petroleum and natural gas consumption, so even the total elimination of synthetic polymers would not contribute significantly to the conservation of hydrocarbon resources. Furthermore, when *total* energy costs

1

(raw materials plus energy to manufacture and ship) are compared, the polymeric item often comes out well ahead of its traditional counterpart, e.g., glass vs. plastic beverage bottles. In addition, the manufacturing processes used to produce polymers often generate considerably less environmental pollution than the processes used to produce the traditional counterparts, e.g., polyethylene film vs. kraft paper for packaging.

Ironically, one of the most valuable properties of polymers, their chemical inertness, causes problems because polymers do not normally degrade in the environment. As a result, they contribute increasingly to litter and the consumption of scarce landfill space. Progress is being made toward the solution of these problems. Environmentally degradable polymers are being developed, although this is basically a wasteful approach and we're not yet sure of the impact of the degradation products. Burning polymer waste for its fuel value makes more sense, because the polymers retain essentially the same heating value as the raw hydrocarbons from which they were made. Still, the polymers must be collected and this approach wastes the value added in manufacturing the polymers.

The ultimate solution is recycling. If waste polymers are to be recycled, they must first be collected. Unfortunately, there are literally dozens (maybe hundreds) of different polymers in the waste mix, and mixed polymers have mechanical properties about like cheddar cheese. Thus, for anything but the least-demanding applications (e.g., parking bumpers, flower pots), the waste mix must be separated prior to recycling. To this end, automobile manufacturers are attempting to standardize on a few well-characterized plastics that can be recovered and re-used when the car is scrapped. Many objects made of the large-volume commodity plastics now have molded-in identifying marks, allowing hand sorting of the different materials.

Processes have been developed to separate the mixed plastics in the waste. The simplest of these is a sink–float scheme which takes advantage of density differences among the various plastics. Unfortunately, many plastic items are foamed, plated, or filled (mixed with nonpolymer components), which complicates density-based separations. Other separation processes are based on solubility differences between various polymers. An intermediate approach chemically degrades the waste polymer to the starting materials from which new polymer can be made.

There are five major areas of application for polymers: (1) plastics, (2) rubbers or elastomers, (3) fibers, (4) surface finishes and protective coatings, and (5) adhesives. Despite the fact that all five applications are based on polymers, and in many cases the same polymer is used in two or more, the industries grew up pretty much separately. It was only after Dr. Herman Staudinger[1,2] proposed the "macromolecular hypothesis" in the 1920s explaining the common molecular makeup of these materials (for which he won the 1953 Nobel Prize in chemistry in belated recognition of the importance of his work) that polymer science began to evolve from the independent technologies. Thus, a sound fundamental basis was established for continued technological advances. The history of polymer science is treated in detail elsewhere.[3,4]

Economic considerations alone would be sufficient to justify the impressive scientific and technological efforts expended on polymers in the past several decades. In addition, however, this class of materials possesses many interesting and useful properties that are completely different from those of the more traditional engineering materials, and that cannot be explained or handled in design situations by the traditional approaches. A description of three simple experiments should make this obvious.

Silly putty, a silicone polymer, bounces like rubber when rolled into a ball and dropped. On the other hand, if the ball is placed on a table, it will gradually spread to a puddle. The material behaves as an elastic solid under certain conditions, and as a viscous liquid under others.

If a weight is suspended from a rubber band, and the band is then heated (taking care not to burn it), the rubber band will contract appreciably. All materials other than polymers will undergo the expected thermal expansion upon heating (assuming that no phase transformation has occurred over the temperature range).

When a rotating rod is immersed in a molten polymer or a fairly concentrated polymer solution, the liquid will actually climb up the rod. This phenomenon, the Weissenberg effect, is contrary to what is observed with nonpolymer liquids, which develop a curved surface profile with a lowest point at the rod, as the material is flung outward by centrifugal force.

Although such behaviour is unusual in terms of the more familiar materials, it is a perfectly logical consequence of the *molecular structure* of polymers. This molecular structure is the key to an understanding of the science and technology of polymers, and will underlie the chapters to follow.

Figure 1.1 illustrates the questions to be considered:

1. How is the desired molecular structure obtained?
2. How do the polymer's processing (i.e., formability) properties depend on its molecular structure?
3. How do its material properties (mechanical, chemical, optical, etc.) depend on molecular structure?

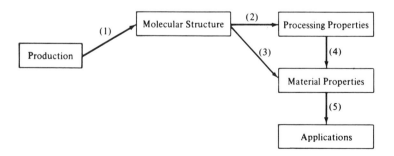

Figure 1.1   The key role of molecular structure in polymer science and technology.

4. How do material properties depend on a polymer's processing history?

5. How do its applications depend on its material properties?

The word polymer comes from the Greek for "many-membered." Strictly speaking, it could be applied to any large molecule that is formed from a relatively large number of smaller units or "mers," for example, a sodium chloride crystal, but it is most commonly (and exclusively, here) restricted to materials in which the mers are held together by covalent bonding, that is, shared electrons. For our purposes, only a few bond valences need be remembered:

$$-\overset{|}{\underset{|}{C}}- \quad \overset{|}{\underset{/\backslash}{N}} \quad -O- \quad Cl- \quad F- \quad H- \quad -\overset{|}{\underset{|}{Si}}-$$

It is always a good idea to "count the bonds" in any structure written to make sure they conform to the above.

Keep in mind also that when the covalently bound atoms differ in electronegativity, the electrons are not shared evenly between them. In gaseous HCl, for example, the electrons cluster around the electronegative chlorine atom, giving rise to a molecular dipole:

$$H \qquad :\overset{..}{\underset{}{Cl}}:$$
$$\delta^+ \qquad \delta^-$$

Electrostatic forces between such dipoles can play an important role in determining polymer properties.

A brief, concise review of organic chemistry from the polymer standpoint is available.[5] The most important constituents of living organisms, cellulose and proteins, are naturally occurring polymers, but we will confine our attention largely to synthetic polymers or to important modifications of natural polymers.

REFERENCES

1. Staudinger, H., *Ber.* **53**, 1073 (1920).
2. Staudinger, H. and J. Fritsch, *Helv. Chim. Acta* **5**, 778 (1922).
3. Morawetz, H., *Polymers: The Origins and Growth of a Science*, Wiley-Interscience, New York, 1985.
4. Stahl, G. A. *CHEMTECH*, August 1984, p. 492.
5. Richardson, P. N., and R. C. Kierstead, *SPE J.* **25**(9), 54 (1969).

PROBLEMS

1. Consider the room you're in.

   a. Identify the items in it that are made of polymers.

b. What would you make those items of if there were no polymers?

c. Why do you suppose polymers were chosen over competing materials (if any) for each particular application?

2. Repeat Problem 1 for your automobile. Don't forget to look under the hood.

3. You wish to develop a polymer to replace glass in window glazing. What properties must a polymer have for that application?

4. Vinyl chloride is the monomer from which the commercially important polymer polyvinyl chloride (PVC) is made. It has the chemical formula C_2H_3Cl. Show the structure of vinyl chloride and identify any dipoles present.

5. Acrylonitrile monomer, C_3H_3N, is an important constituent of acrylic fibers and nitrile rubber. It does not have a cyclic structure and it has only one double bond. Show the structure of acrylonitrile and identify any dipoles present.

6. Propylene, C_3H_6, is the monomer from which the fastest-growing plastic, polypropylene, is made. It contains one double bond. Show its structure and identify any dipoles present.

PART I

POLYMER FUNDAMENTALS

CHAPTER II

Types of Polymers

The large number of natural and synthetic polymers has been classified in several ways. These will be outlined below, and in the process, many terms important in polymer science and technology will be introduced.

2.1 REACTION TO TEMPERATURE

The earliest distinction between types of polymers was made long before any concrete knowledge of their molecular structure. It was a purely phenomenological distinction, based on their reaction to heating and cooling.

A Thermoplastics

It was noted that certain polymers would soften upon heating, and could then be made to flow when a stress was applied. When cooled again, they would reversibly regain their solid or rubbery nature. These polymers are known as *thermoplastics.* By analogy, ice and solder, though not polymers, behave as thermoplastics.

B Thermosets

Other polymers, although they might be heated to the point where they would soften and could be made to flow under stress *once,* would not do so reversibly; that is, heating caused them to undergo a "curing" reaction. Sometimes these materials emerge from the synthesis reaction in a cured state. Further heating of these *thermosetting* polymers ultimately leads only to degradation (as is sometimes attested to by the smell of a short-circuited electrical appliance) and not softening and flow. Again by analogy, eggs and concrete behave as thermosets. Continued heating of thermoplastics will also lead ultimately to degradation, but they will generally soften at temperatures below their degradation point.

Natural rubber is a classic example of these two categories. Introduced to Europe by Columbus, natural rubber did not achieve commercial significance for centuries; because it was a thermoplastic, articles made of it would become soft and sticky on hot days. In 1839, Charles Goodyear discovered the curing reaction with sulfur (which he called *vulcanization* in honor of the Roman god of fire) that converted the polymer to a thermoset. This allowed the rubber to maintain its useful properties to much higher temperatures, which ultimately led to its great commercial importance.

2.2   CHEMISTRY OF SYNTHESIS

Pioneering workers in the field of polymer chemistry soon observed that they could produce polymers by two familiar types of organic reactions.

A   Condensation

Polymers formed from a typical organic condensation reaction, in which a small molecule (most often water) is split out, are known, logically enough, as *condensation polymers*. The common esterification reaction of an organic acid and an organic base (alcohol) illustrates the simple "lasso chemistry" involved:

$$\text{R-O(H)} + \text{(HO)-}\overset{\displaystyle\overset{O}{\|}}{\text{C}}\text{-R}' \rightarrow \text{R-O-}\overset{\displaystyle\overset{O}{\|}}{\text{C}}\text{-R}' + H_2O$$

Alcohol + Acid   → Ester

The –OH group on the alcohol and the $\text{HO-}\overset{\displaystyle\overset{O}{\|}}{\text{C}}\text{-}$ on the acid are known as *functional groups*, those parts of a molecule that participate in a reaction. Of course, the ester formed in the preceding reaction is not a polymer because we have only hooked up two small molecules, and the reaction is finished far short of anything that might be considered "many membered."

At this point it is useful to introduce the concept of functionality. Functionality is the number of bonds a mer can form with other mers in a reaction. In condensation polymerization, it is equal to the number of functional groups on the mer.

It is obvious from the example above that each of the reactants is monofunctional, and that reactions between monofunctional mers cannot lead to polymers. But now consider what happens if each reactant is difunctional, allowing it to react at each end:

$$\text{HO-R-O(H)} + \text{(HO)-}\overset{\displaystyle\overset{O}{\|}}{\text{C}}\text{-R}'\text{-}\overset{\displaystyle\overset{O}{\|}}{\text{C}}\text{-OH} \rightarrow \text{HO-R-O-}\overset{\displaystyle\overset{O}{\|}}{\text{C}}\text{-R}'\text{-}\overset{\displaystyle\overset{O}{\|}}{\text{C}}\text{-OH} + H_2O$$

Dialcohol, diol,   Diacid   Ester
or glycol

The resulting product molecule is still difunctional. Its left end can react with another diacid molecule and its right end with another molecule of diol, and after each subsequent reaction, the growing molecule is still difunctional and capable of undergoing further growth, leading to a true polymer molecule.

In general, the *polycondensation* of x molecules of a diol with x molecules of a diacid to give a *polyester* molecule is written as

$$x \text{ HO-R-OH} + x \text{ HO-}\overset{\overset{\text{O}}{\|}}{\text{C}}\text{-R'-}\overset{\overset{\text{O}}{\|}}{\text{C}}\text{-OH} \rightarrow \text{H} \Big[\text{O-R-O-}\overset{\overset{\text{O}}{\|}}{\text{C}}\text{-R'-}\overset{\overset{\text{O}}{\|}}{\text{C}} \Big]_x \text{OH}$$

$$+ (2x - 1) \text{ H}_2\text{O}$$

Diol   Diacid   Polyester

In the polyester molecule above, the structure in brackets is the repeating unit, and is what distinguishes one polymer from another, while the $\overset{\overset{\text{O}}{\|}}{\{\text{O-C}\}}$ linkage characterizes all polyesters. The generalized organic groups R and R' may vary widely (with a consequent variation in the properties of the polymer), but as long as the repeating unit contains the $\overset{\overset{\text{O}}{\|}}{\{\text{O-C}\}}$ linkage, the polymer is a polyester.

The quantity x is the *degree of polymerization*, the number of repeating units strung together like identical beads in the polymer chain. It is sometimes also called the *chain length*, but it is a pure number, not a length.

The nomenclature above was introduced by Wallace Hume Carothers, who, with his group at Du Pont, invented neoprene and nylon in the 1930s, and was one of the founders of modern polymer science.[1]

Another functional group that is capable of taking part in a condensation is the amine ($-\text{NH}_2$) group, *one* hydrogen of which reacts with a carboxylic acid group in a manner similar to the alcoholic hydrogen to form a polyamide or nylon:

$$x \text{ H-}\overset{\overset{\text{H}}{|}}{\text{N}}\text{-R-}\overset{\overset{\text{H}}{|}}{\text{N}}\text{-H} + x \text{ HO-}\overset{\overset{\text{O}}{\|}}{\text{C}}\text{-R'-}\overset{\overset{\text{O}}{\|}}{\text{C}}\text{-OH} \rightarrow \text{H}\Big[\overset{\overset{\text{H}}{|}}{\text{N}}\text{-R-}\overset{\overset{\text{H}}{|}}{\text{N}}\text{-}\overset{\overset{\text{O}}{\|}}{\text{C}}\text{-R'-}\overset{\overset{\text{O}}{\|}}{\text{C}}\Big]_x\text{OH}$$

$$+ (2x - 1) \text{ H}_2\text{O}$$

Diamine   Diacid   Polyamide or nylon

The $\overset{\overset{\text{H O}}{| \|}}{\{\text{N-C}\}}$ linkage characterizes nylons. (Both amine hydrogens can react with two acid groups which are on adjacent carbon atoms in a benzene ring. This gives a polyimide linkage, and is illustrated in Example 4R at the end of the chapter.) The examples above serve to illustrate that reactants must be at least difunctional if a polymer is to be obtained. Molecules with higher degrees of functionality will also lead to polymers. For example, glycerine

$$
\begin{array}{ccc}
H & H & H \\
| & | & | \\
H-C-C-C-H \\
| & | & | \\
O & O & O \\
| & | & | \\
H & H & H \\
\end{array}
$$

is trifunctional in a polyesterification reaction.

It is also possible to form condensation polymers from a single monomer that contains the two complementary reactive groups in the same molecule, for example,

$$
\underset{\text{Hydroxy acid}}{HO-R-\overset{\overset{\displaystyle O}{\|}}{C}-OH} \qquad \underset{\text{Amino acid}}{\overset{\displaystyle H}{\underset{\displaystyle H}{>}}N-R-\overset{\overset{\displaystyle O}{\|}}{C}-OH}
$$

In principle, the hydroxy acid can condense directly to form a polyester and the amino acid a polyamide (proteins are poly amino acids). The reactions do not always proceed in a straightforward fashion, however. If R is large enough, say three carbon atoms or more, the difunctional monomers above may "bite their own tails," condensing to form a cyclic structure:

$$
\underset{\text{Amino acid}}{H-\overset{\overset{\displaystyle H}{|}}{N}-R-\overset{\overset{\displaystyle O}{\|}}{C}-OH} \longrightarrow \underset{\text{Lactam}}{R\overset{\overset{\displaystyle O}{\overset{\|}{C}}}{\underset{N-H}{\diagup\diagdown|}}} \quad + \quad H_2O
$$

This cyclic compound can then undergo a *ring-scission* polymerization, in which the polymer is formed without splitting out a small molecule, because the small molecule had been eliminated previously in the cyclization step:

$$
x\ R\overset{\overset{\displaystyle O}{\overset{\|}{C}}}{\underset{N-H}{\diagup\diagdown|}} \longrightarrow \underset{\text{Polyamide}}{\left[\!\!-\overset{\overset{\displaystyle H}{|}}{N}-R-\overset{\overset{\displaystyle O}{\|}}{C}\!\!-\right]_x}
$$

Despite the lack of elimination of a small molecule in the actual polymerization step, the products can be thought of as having been formed by a direct condensation from the monomer, and are usually considered condensation polymers.

Note that the characteristic nylon linkage $\begin{smallmatrix} H & O \\ | & \| \\ \{N\!-\!C\} \end{smallmatrix}$ is split up, and not immediately obvious in the repeating unit as written above. If you place a second repeating next to the one shown, it becomes evident. This illustrates the somewhat arbitrary location of the brackets, which should not obscure the fact that the polymer is a nylon.

B Addition

The second polymer-formation reaction is known as addition polymerization and its products as *addition polymers*. Addition polymerizations have two distinct characteristics:

1. No molecule is split out; hence, the repeating unit has the same chemical formula as the monomer.
2. The polymerization reaction involves the opening of a double bond.*

Monomers of the general type $\overset{|\;\;|}{\underset{|\;\;|}{C\!=\!C}}$ undergo addition polymerization:

$$x \; \overset{|\;\;|}{\underset{|\;\;|}{C\!=\!C}} \rightarrow \{ \overset{|\;\;|}{\underset{|\;\;|}{C\!-\!C}} \}_x$$

The double bond "opens up," forming bonds to other monomers at each end, so *a double bond is difunctional* according to our general definition of functionality. The question of what happens at the ends of the polymer molecule will be deferred for a discussion of polymerization mechanisms in Chapter 10.

An important subclass of the double-bond containing monomers is the vinyl monomers, $\overset{H\;\;H}{\underset{H\;\;X}{\overset{|\;\;|}{\underset{|\;\;|}{C\!=\!C}}}}$. Addition polymerization is occasionally referred to as vinyl polymerization. Table 2.1 lists some commercially important vinyl monomers.

---

* Although aromatic rings are often symbolized by ⟨///⟩, this is a poor representation of reson-ance-stabilized structures, which are completely inert to addition polymerization. They are more properly symbolized by ⟨◯⟩ to avoid confusion with the ordinary double bond. To save space, we will sometimes also use ϕ to represent the phenyl ring.

Table 2.1   Some Commercially Important Vinyl Monomers

Monomer	$-X$
Ethylene	$-H$
Styrene	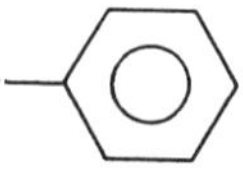
Vinyl chloride	$-Cl$
Propylene	$-CH_3$
Acrylonitrile	$-C{\equiv}N$

Example 1. Lactic acid can be dehydrated to acrylic acid according to the following reaction:

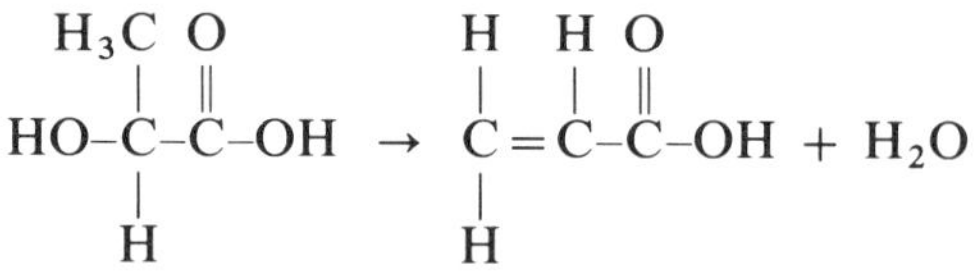

$$\underset{\text{Lactic acid}}{\overset{\displaystyle H_3C\;\;O}{HO-\overset{|}{\underset{|}{C}}-\overset{\|}{C}-OH}} \;\rightarrow\; \underset{\text{Acrylic acid}}{\overset{\displaystyle H\;\;H\;O}{C=\overset{|}{C}-\overset{\|}{C}-OH}} + H_2O$$

Each acid forms a polymer. Write the structural formulae for the repeating unit of the polymer formed from each acid.

Solution. Lactic acid is an hydroxy acid, and will undergo condensation polymerization, splitting out water from the $-OH$ and $-\overset{\displaystyle O}{\overset{\|}{C}}-OH$ groups to give a polyester:

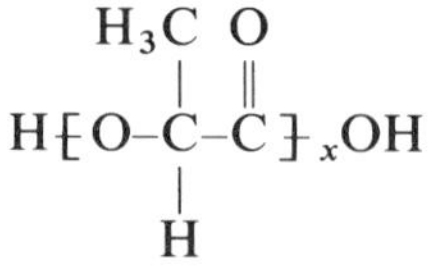

$$H{\Big[}O-\overset{\displaystyle H_3C}{\underset{\displaystyle H}{\overset{|}{\underset{|}{C}}}}-\overset{\displaystyle O}{\overset{\|}{C}}{\Big]}_x OH$$

Note again that in the above formula, the characteristic polyester linkage $\{O-\overset{\displaystyle O}{\overset{\|}{C}}\}$ is split up. Acrylic acid is a vinyl monomer, and will undergo addition polymerization to give

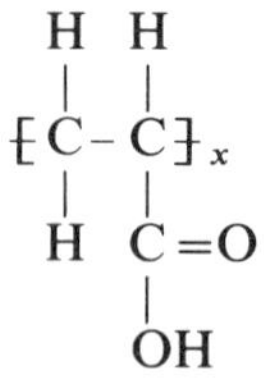

$$\{\overset{\displaystyle H}{\underset{\displaystyle H}{\overset{|}{\underset{|}{C}}}}-\overset{\displaystyle H}{\underset{\displaystyle \overset{|}{C}=O}{\overset{|}{\underset{|}{C}}}}\}_x$$
$$\underset{\displaystyle OH}{\overset{|}{}}$$

Interestingly, the ester linkage in poly(lactic acid) is easily hydrolyzed, breaking down the polymer. This for many years prevented commercial application of the polymer. Now, however, there is great interest in this polymer as a biodegradable plastic.

Dienes (the carbon atoms in which are numbered from one end)

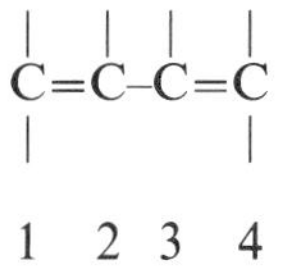

$$C=C-C=C$$

$$1 \quad 2 \quad 3 \quad 4$$

also undergo addition polymerization. Addition polymerization of a diene "uses up" one of the two double bonds per monomer, giving an *unsaturated* polymer, that is, one that contains double bonds. In this case, there is one double bond per repeating unit. Furthermore, diene polymerization can lead to several *structural isomers* in the polymer. If the monomer is symmetrical with regard to substituent groups, it can undergo 1, 2 addition and 1, 4 addition. If unsymmetrical, there is the added possibility of 3, 4 addition. (For symmetrical dienes, the 1, 2 and 3, 4 reactions are the same.) This is illustrated below for the addition polymerization of isoprene (2-methyl-1,3-butadiene):

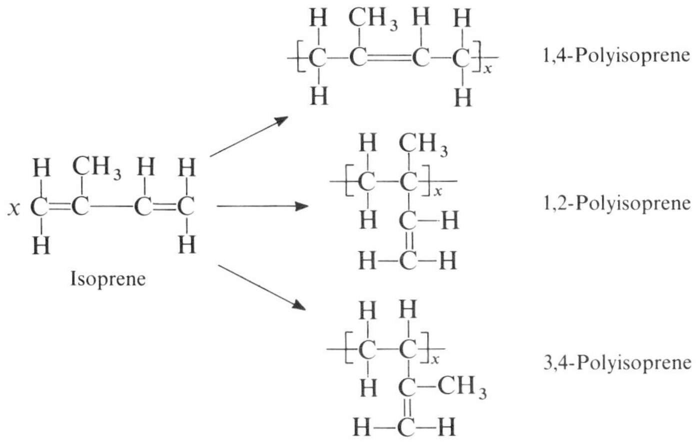

The 1, 2 and 3, 4 reactions are sometimes known as *vinyl* addition, because part of the diene monomer simply acts as an X group in a vinyl monomer.

2.3 STRUCTURE

As an appreciation for the molecular structure of polymers was gained, three major structural categories emerged. These are illustrated schematically in Fig. 2.1.

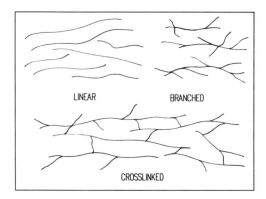

Figure 2.1 Schematic diagram of polymer structures.

A Linear

If a polymer is built from strictly difunctional monomers, the result is a *linear* polymer chain. A scale model of a typical linear polymer molecule made from 0.5-cm-diameter clothesline would be about three meters long. This isn't a bad analogy: The chains are long, flexible, essentially one-dimensional structures. The term linear can be somewhat misleading, however, because the molecules don't necessarily assume a geometrically linear conformation.

Random Copolymers

Polymers consisting of chains that contain a single repeating unit are known as *homopolymers*. If, however, the chains contain a random arrangement of two separate and distinct repeating units, the polymer is known as a *random* or *statistical copolymer*, or just plain *copolymer*. A random copolymer might be formed by the addition polymerization of a mixture of two different vinyl monomers, A and B (the degree of "randomness" depends on the relative amounts and reactivities of A and B, as we shall see later), and be represented as

AABAAABBABAAB

and called poly(A-*co*-B), where the first repeating unit listed is the one present in the greater amount. For example, a random synthetic rubber copolymer of 75% butadiene and 25% styrene would be termed poly(butadiene-*co*-styrene). Of course, ter- and higher multipolymers are possible.

It must be emphasized that the products of condensation polymerizations that require two different monomers to provide the necessary functional groups, for example, a diacid and a diamine, are not copolymers, because they contain only one repeating unit. If, however, two different diamines were to be used, leading to two distinct repeating units, the product would be a copolymer.

Example 2. Illustrate the repeating units that result when three moles of hexamethylene diamine (I) are condensed with two moles of adipic acid (II) and one mole of sebacic acid (III), and name the resulting copolymer:

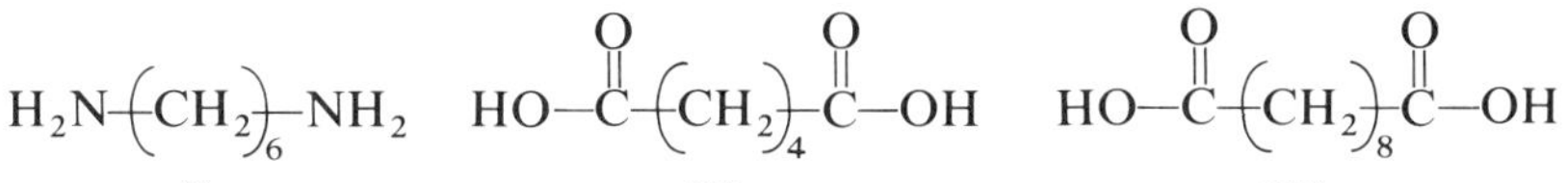

$$H_2N\left(CH_2\right)_6 NH_2 \qquad HO-\underset{\|}{\overset{O}{C}}\left(CH_2\right)_4\underset{\|}{\overset{O}{C}}-OH \qquad HO-\underset{\|}{\overset{O}{C}}\left(CH_2\right)_8\underset{\|}{\overset{O}{C}}-OH$$

(I) (II) (III)

Solution. The two repeating units are

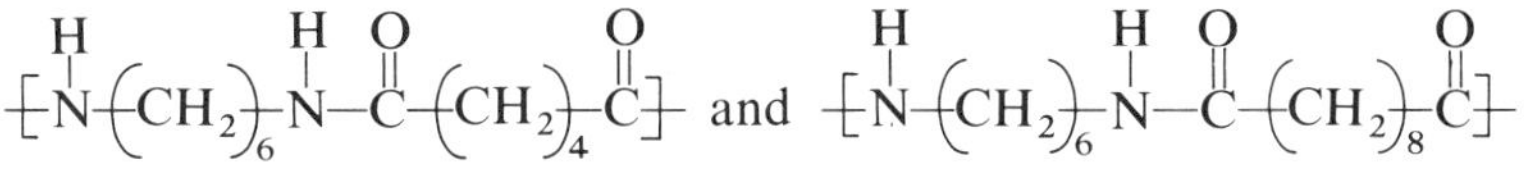

$$\left[N\left(CH_2\right)_6 N-\underset{\|}{\overset{O}{C}}\left(CH_2\right)_4\underset{\|}{\overset{O}{C}}\right] \text{ and } \left[N\left(CH_2\right)_6 N-\underset{\|}{\overset{O}{C}}\left(CH_2\right)_8\underset{\|}{\overset{O}{C}}\right]$$

From (I) and (II)   From (I) and (III)

The formal (if unwieldly) name for the copolymer containing these two repeating units is poly(hexamethylene adipamide-*co*-hexamethylene sebacamide).

Block Copolymers

Under certain conditions, linear chains can be formed that contain long contiguous blocks of two (or more) repeating units combined in the chains, a block copolymer:

AAAAAAAAAAAAAAAAAAAAAAAABBBBBBBBBBBBBBBBBB

Such a polymer would be termed poly(A–*b*–B).

B Branched

If a few points of tri (or higher) functionality are introduced (either intentionally or through side reactions) at random points along linear chains, *branched* molecules result. Branching can have a tremendous influence on the properties of polymers through steric (geometric) effects.

Graft copolymers

Under specialized conditions, branches of repeating unit B may be "grafted" to a backbone of linear A. This structure is known as a *graft copolymer*:

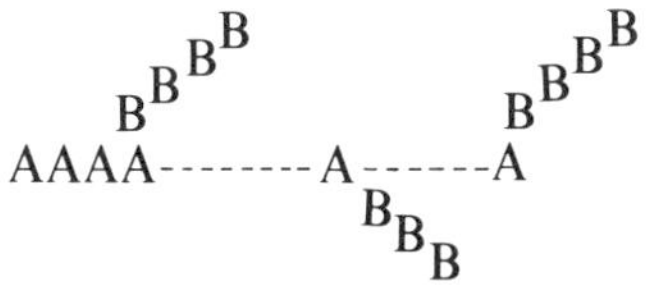

The graft copolymer above would be called poly(A–g–B), the backbone repeating unit being the one listed first.

C Crosslinked or Network

As the length and frequency of the branches on polymer chains increases, the probability that the branches will finally reach from chain to chain, connecting them together, becomes greater. When all the chains are finally connected together in three dimensions by these crosslinks, the entire polymer mass becomes one tremendous molecule, a *crosslinked* or *network* polymer. In a truly crosslinked polymer mass, all the atoms are connected to one another by covalent bonds. The polymer in a bowling ball, for example, has a molecular weight on the order of 10^{27} g/mol. Remember this the next time someone suggests that individual molecules are too small to be seen with the naked eye.

Crosslinked or network polymers may be formed in two ways: (1) by starting with reaction masses containing sufficient amounts of tri- or higher-functional monomer, or (2) by chemically creating crosslinks between previously formed linear or branched molecules ("curing"). The latter is precisely what vulcanization does to natural rubber, and this fact serves to introduce the connection between the phenomenological "reaction to temperature" classification and the more fundamental concept of molecular structure. This important connection will be clarified through a discussion of bonding in polymers.

Example 3. Show (*a*) how a linear, unsaturated polyester is produced from ethylene glycol (I) and maleic anhydride (II), and (*b*) how the linear, unsaturated polyester is crosslinked with a vinyl monomer such as styrene.

(I)   (II)

Solution. First, realize that an acid anhydride is simply a diacid with a mole of water split out from the two acid groups. (This is the only common example of acid groups condensing with one another. You can't ordinarily form polymers this way). When considering the reaction of an acid anhydride, (conceptually) hydrate it back to the diacid:

Maleic anhydride   Maleic acid

Then condense the diacid with the diol to form a polyester with one double bond per repeating unit:

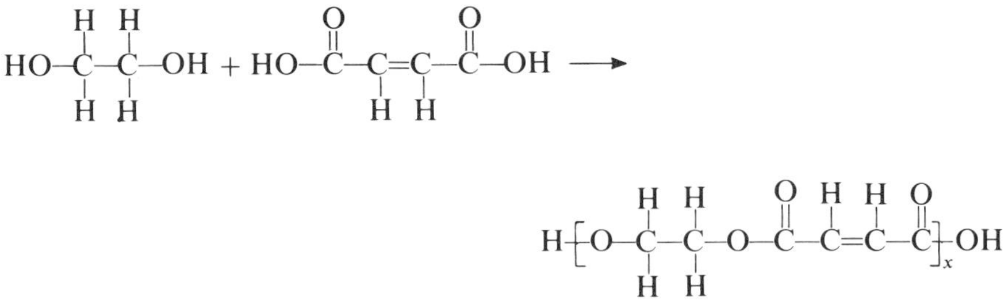

$$
\underset{\substack{|\ \ |\\ H\ H}}{HO-\overset{\substack{H\ H\\ |\ \ |}}{C-C}-OH} + HO-\overset{\overset{O}{\|}}{C}-\underset{\substack{|\ \ |\\ H\ H}}{C=C}-\overset{\overset{O}{\|}}{C}-OH \longrightarrow
$$

$$
H\!\left[\!O-\underset{\substack{|\ \ |\\ H\ H}}{\overset{\substack{H\ H\\ |\ \ |}}{C-C}}-O-\overset{\overset{O}{\|}}{C}-\underset{\substack{|\ \ |\\ H\ H}}{C=C}-\overset{\overset{O}{\|}}{C}\!\right]_{\!x}\!\!-OH
$$

The double bond in the maleic acid is inert toward polycondensation. Note that the degree of unsaturation (average number of double bonds per repeating unit) could be varied from zero to one by employing mixtures of a saturated diacid, e.g., phthalic anhydride (III) with the maleic to form copolyesters with saturated and unsaturated repeating units:

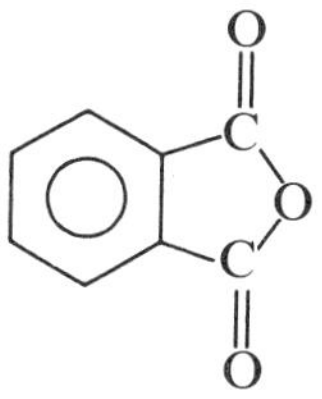

(III)

Commercially, the degree of polymerization x is maintained low (say 8–10) (by techniques considered in Chapter 9) so that the product is a viscous liquid.

The linear, unsaturated polyester is then diluted with a liquid vinyl monomer, most often styrene. Before use, a chemical that promotes addition polymerization (Chapter 10) is added, causing the vinyl monomer to undergo addition copolymerization with the double bonds in the polyester to form a highly crosslinked, rigid network:

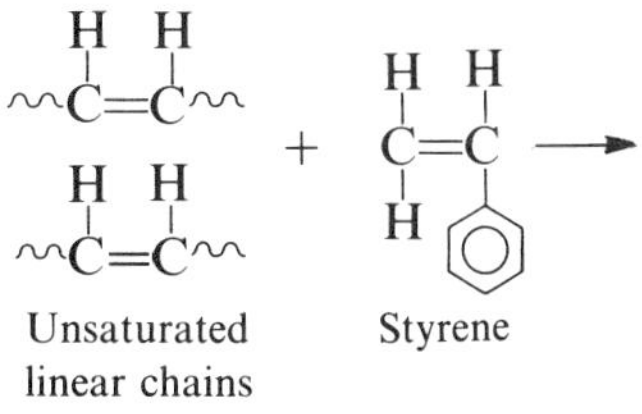

Unsaturated   Styrene
linear chains

Network structure

The liquid polyester–styrene mixture is often used to impregnate fiberglass and is cured to form boat hulls, auto (Corvette) bodies, and other so-called fiberglass objects (really fiberglass-reinforced polyester).

Example 4. Show the structural formulas of the repeating units for each of the following polymers and classify them according to structure and chemistry of formation. All the polymers are commercially important. Most follow the rules outlined above, but some don't, and have been included here to illustrate their structures, chemistries of formation, and characteristic linkages.

A. Polystyrene
B. Polyethylene
C. Poly(butylene terephthalate) (PBT)
D. Poly(ethylene terephthalate) (PET) (Dacron,® Mylar®)
E. Nylon 6/6 (The numbers designate carbon atoms in the diamine/diacid.)
F. Nylon 6 (The number designates carbon atoms in the monomer.)
G. Glyptal (glycerol + phthalic anhydride)
H. Poly(diallyl phthalate)
I. Melamine-formaldehyde (Melmac,® Formica®)
J. Polytetrafluoroethylene (Teflon TFE®)
K. Poly(phenylene oxide) (PPO) (*Hint*: polymerized in presence of O_2)
L. Polypropylene
M. Acetal (polyformaldehyde or polyoxymethylene) (Celcon®, Delrin®)
N. Polycarbonate (Calibre®, Lexan®, Makrolon®)
O. Epoxy or phenoxy
P. Poly(dimethyl siloxane) (silicone rubber) (*Hint*: polymerized in presence of H_2O)

Q. Polyurethane

R. Polyimide

S. Polysulfone (Udel®)

Starting monomers are shown below.

A   H—C=CH$_2$

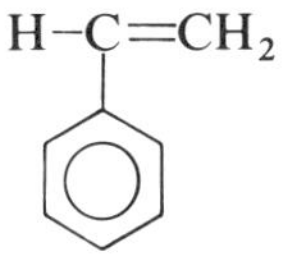

Styrene

B   H$_2$C=CH$_2$

Ethylene

C   HO—(CH$_2$)$_4$—OH

Butylene glycol
(1,4 butane diol)

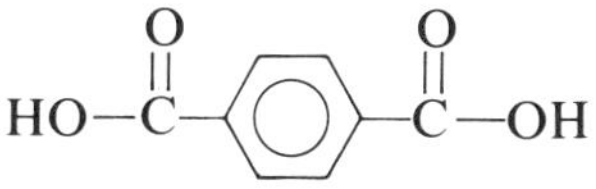

Terephthalic acid

D   H$_3$C—O—C—(ring)—C—O—CH$_3$

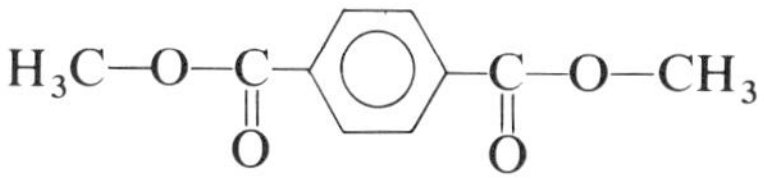

Dimethyl terephthalate

HO—(CH$_2$)$_2$—OH

Ethylene glycol

E   H$_2$N—(CH$_2$)$_6$—NH$_2$

Hexamethylene diamine

HO—C—(CH$_2$)$_4$—C—OH

Adipic acid

F

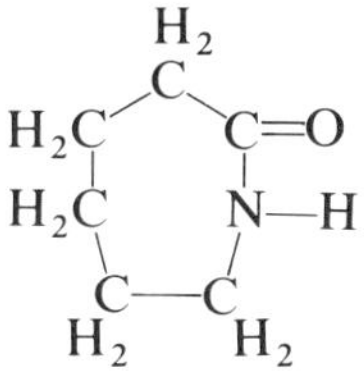

ε-Caprolactam

G

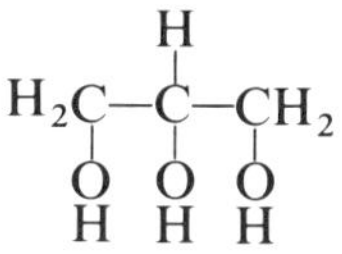

Glycerol

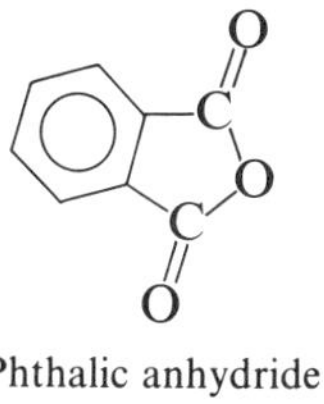

Phthalic anhydride

H

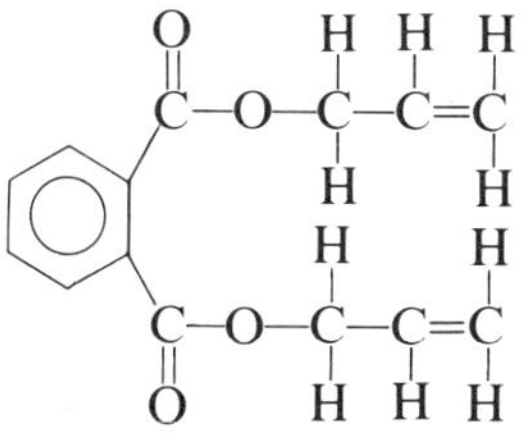

Diallyl phthalate

I

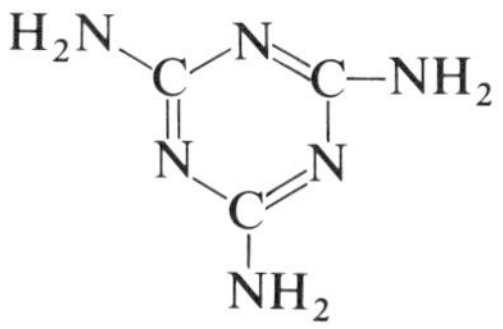

Melamine

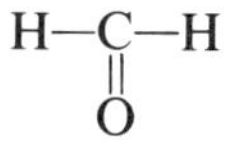

H—C—H

Formaldehyde

J $F_2C{=}CF_2$
Tetrafluoroethylene

K   HO⬡ CH_3 (2,6-Dimethyl phenol)

CH_3

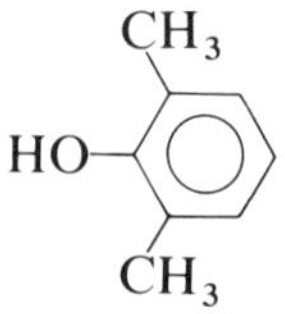

2,6-Dimethyl phenol

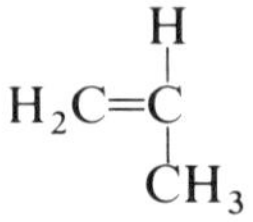

L $H_2C{=}C\overset{H}{\underset{CH_3}{|}}$
Propylene

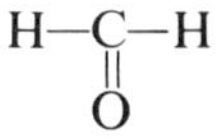

M $H{-}\overset{}{\underset{O}{C}}{-}H$ or $H_2C\overset{O}{\diagdown}CH_2$
Formaldehyde

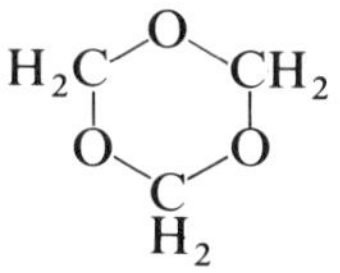

Trioxane

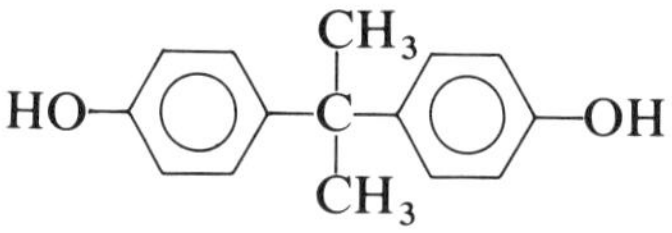

N   HO⬡$\overset{CH_3}{\underset{CH_3}{C}}$⬡OH
Bisphenol-A

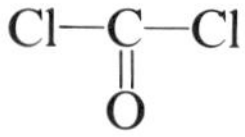

$Cl{-}\overset{}{\underset{O}{C}}{-}Cl$
Phosgene

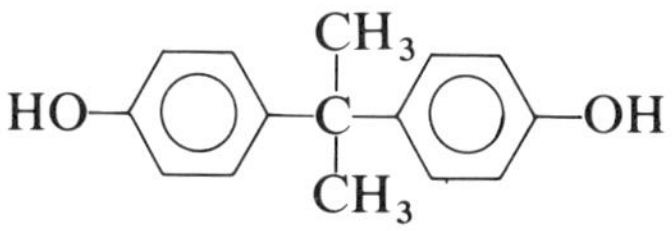

O   HO⬡$\overset{CH_3}{\underset{CH_3}{C}}$⬡OH
Bisphenol-A

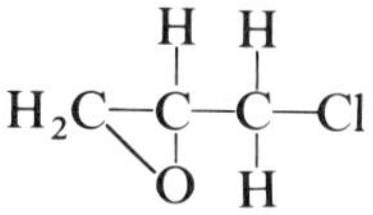

$H_2C\overset{}{\diagdown}\overset{H}{\underset{O}{C}}\overset{H}{\underset{H}{C}}{-}Cl$
Epichlorohydrin

P $Cl{-}\overset{CH_3}{\underset{CH_3}{Si}}{-}Cl$
Dimethyl dichlorosilane

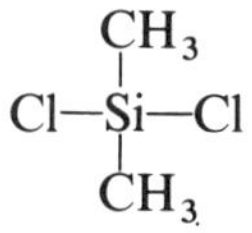

Q   HOROH
Diol or glycol

$O{=}C{=}N{-}R'{-}N{=}C{=}O$
Diisocyanate

R 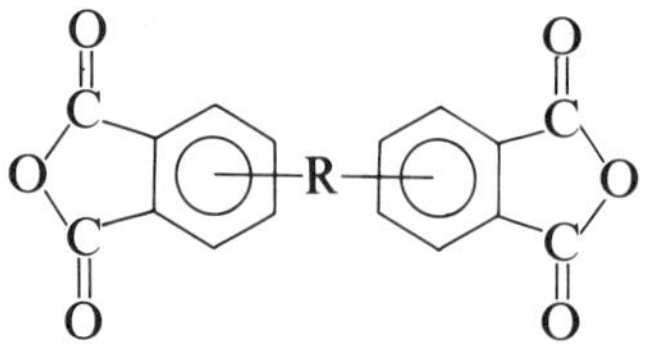   Dianhydride

$H_2NR'NH_2$
Diamine

S   NaO⬡$\overset{CH_3}{\underset{CH_3}{C}}$⬡ONa

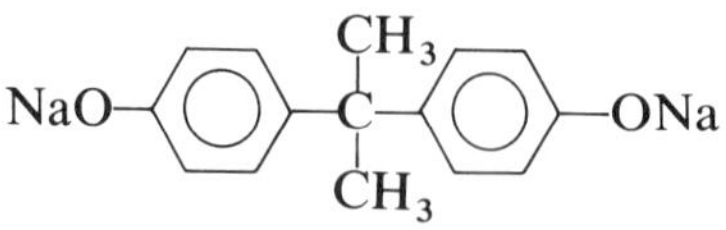

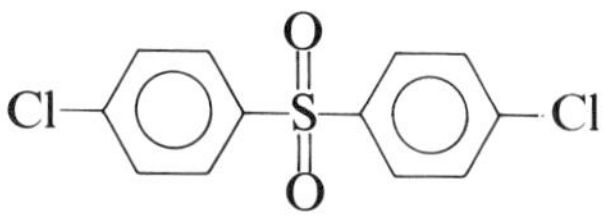

Cl⬡$\overset{O}{\underset{O}{S}}$⬡$Cl$

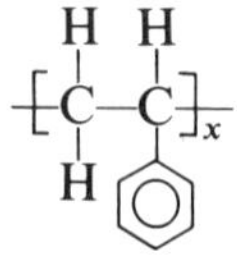

Solution

A $\{CH_2{-}CH(C_6H_5)\}_x$ Linear addition

B $\{CH_2{-}CH_2\}_x$ Linear or branched from side reactions, addition

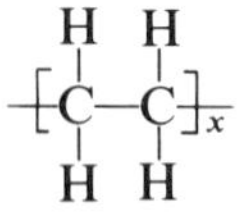

C $H\left[O-(CH_2)_4-O-\overset{O}{\underset{}{C}}-\bigcirc-\overset{O}{\underset{}{C}}\right]_x OH$ (H$_2$O out)   Linear, condensation

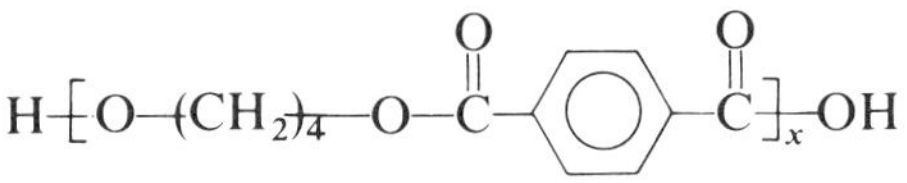

D $H_3C\left[O-\overset{}{\underset{O}{C}}-\bigcirc-\overset{}{\underset{O}{C}}-O-(CH_2)_2\right]_x OH$ CH$_3$OH   Can also be made from the diacid (see above) splitting out H$_2$O.

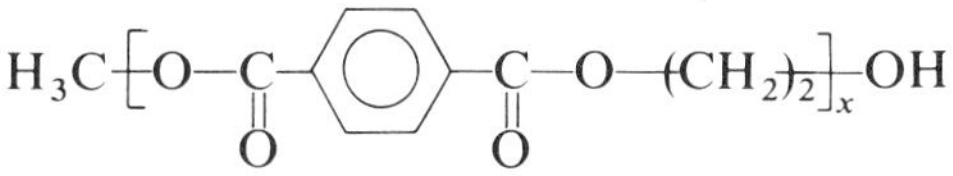

Split out

Linear, condensation

E $H\left[\overset{H}{\underset{}{N}}(CH_2)_6-\overset{H}{\underset{}{N}}-\overset{O}{\underset{}{C}}-(CH_2)_4-\overset{O}{\underset{}{C}}\right]_x OH$ (H$_2$O out) linear, condensation

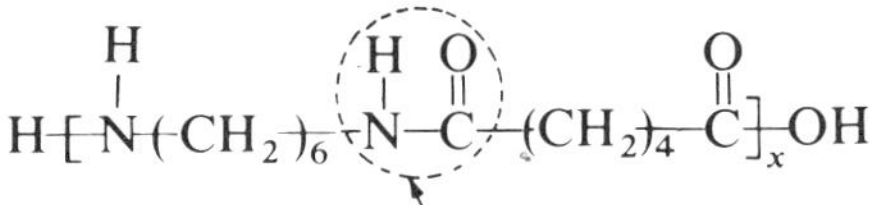

Characteristic nylon linkage

F $\left[\overset{O}{\underset{}{C}}-\overset{H}{\underset{}{N}}-(CH_2)_5\right]_x$ (nothing out)   Monomer made from $HO\overset{O}{\underset{}{C}}-(CH_2)_5-\overset{H}{\underset{}{N}}-H$

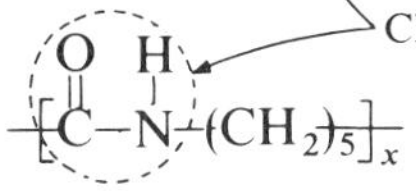

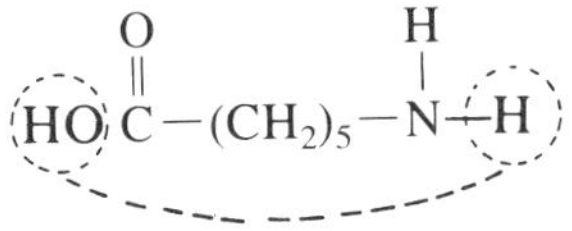

Linear condensation

G $HO-\overset{H}{\underset{H}{C}}-\overset{H}{\underset{O}{C}}-\overset{H}{\underset{H}{C}}-O-\overset{O}{\underset{}{C}}-\bigcirc-\overset{O}{\underset{}{C}}-O-$

$\overset{}{\underset{}{C}}=O$

$\overset{}{\underset{}{C}}=O$

$\overset{}{\underset{}{O}}$

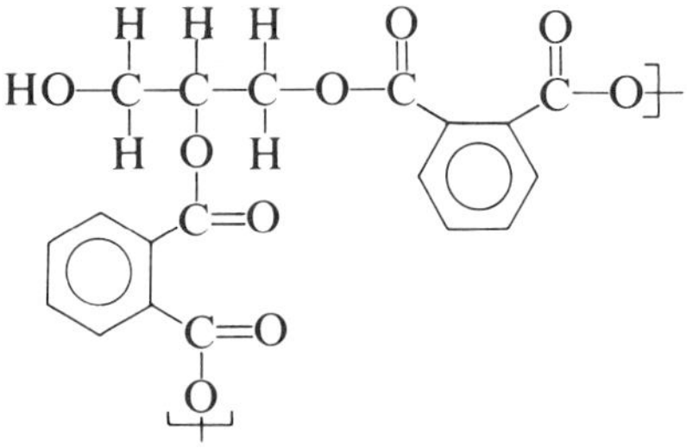

Condensation, structure depends on ratio of reactants (Chapter IX), but usually will be highly crosslinked.

H $\left[\overset{H}{\underset{}{C}}-\overset{H}{\underset{H}{C}}\right]$

CH_2

O

$O=C$

$\bigcirc$

$O=C$

O

CH_2

$\left[\overset{H}{\underset{H}{C}}-\overset{H}{\underset{H}{C}}\right]$

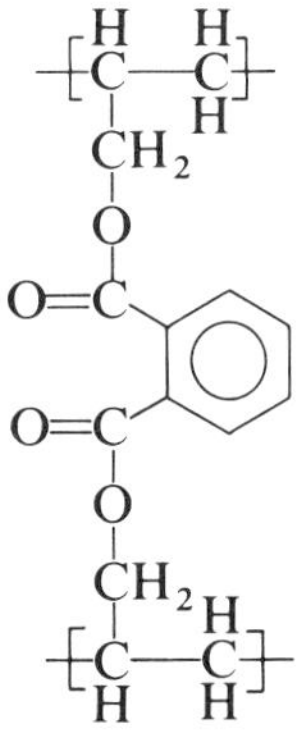

Network, addition

Monomer is 4-functional with two double bonds.

I 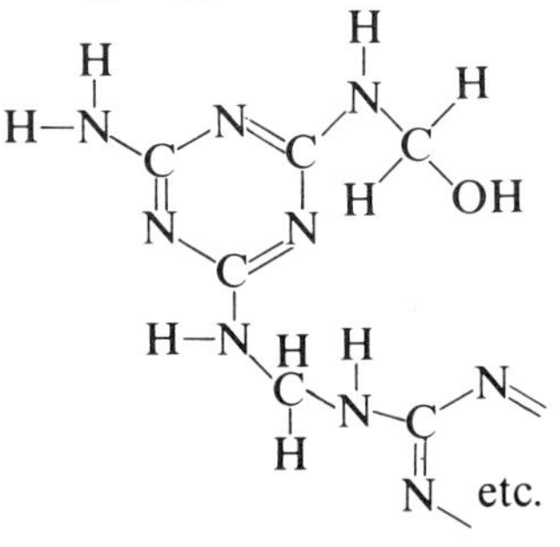   Network, condensation

Similar structures result when formaldehyde is condensed with urea or phenol.

etc.

J $\left[\begin{array}{c} F\ F \\ C-C \\ F\ F \end{array}\right]_x$ Linear, addition

K $\left[O-\underset{CH_3}{\overset{CH_3}{\bigcirc}}\right]_x$ Linear, condensation   "Oxidative coupling" H_2O out

L $\left[\begin{array}{c} H\ H \\ C-C \\ H\ CH_3 \end{array}\right]_x$ Linear, addition

M $\left[\begin{array}{c} H \\ C-O \\ H \end{array}\right]_x$ Linear, addition   Only other double bond that forms addition polymers.

N $H\left[O-\bigcirc-\underset{CH_3}{\overset{CH_3}{C}}-\bigcirc-O-\underset{O}{\overset{}{C}}\right]_x Cl$ (HCl out) Linear, condensation $\left(O-\overset{O}{\overset{\|}{C}}-O\right)$ Characteristic carbonate linkage

O $H\left[O-\bigcirc-\underset{CH_3}{\overset{CH_3}{C}}-\bigcirc-O-\underset{H}{\overset{H}{C}}-\underset{O}{\overset{H}{C}}-\underset{H}{\overset{H}{C}}\right]_x$ *Note*: Get both condensation
$-OH + Cl \rightarrow HCl$
and epoxide ring scission.

$$-OH + H_2C-\underset{O}{\overset{H\ H}{C-C}}-Cl + HO- \rightarrow -O-\underset{H\ H\ H}{\overset{H\ O\ H}{C-C-C}}-O- + HCl$$

Linear, condensation

for $x < 8$, liquid epoxy ⎫ Generally crosslinked later with diamines or
$8 < x < 20$, solid epoxy ⎭ acid anhydrides through $-OH$ and terminal

$$-\underset{O}{\overset{H}{C}-CH_2} \text{ groups.}$$

$x \approx 100$, linear, "phenoxy" plastic

P $\left[\begin{array}{c} CH_3 \\ Si-O \\ CH_3 \end{array}\right]_x$ (HCl out) Linear, condensation   Commercial materials contain some

$$Cl-\underset{Cl}{\overset{CH_3}{Si}}-Cl$$

which is 3-functional and allows crosslinking. Also acetate may be substituted for Cl.

Characteristic urethane linkage

Q $\left[O-R-O-\underset{}{\overset{O}{\overset{\|}{C}}}-\underset{}{\overset{H}{N}}-R'-\underset{}{\overset{H}{N}}-\underset{}{\overset{O}{\overset{\|}{C}}}\right]_x$ R and R' vary widely—often R is already a low-x polymer.

Linear, condensation nothing out, but diisocyanate can be considered.

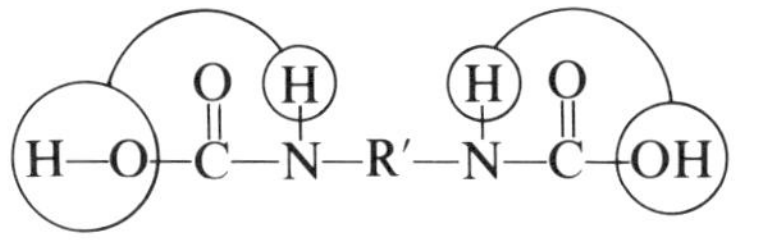

minus 2H$_2$O

can be crosslinked through amine groups, or by using higher functional alcohols or isocyanates.

R 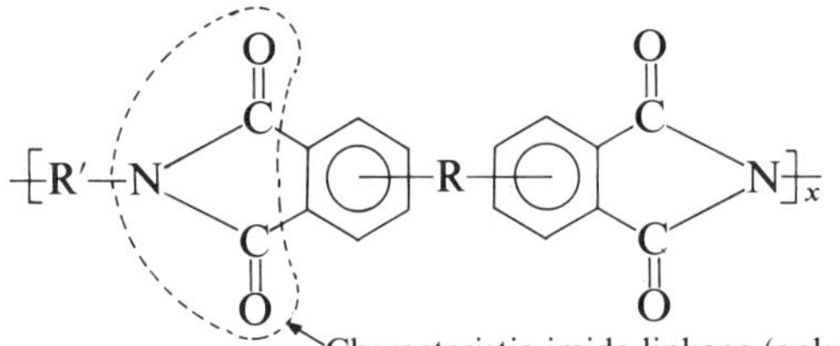   Linear, condensation

Characteristic imide linkage (only example where both amine H's react).

S 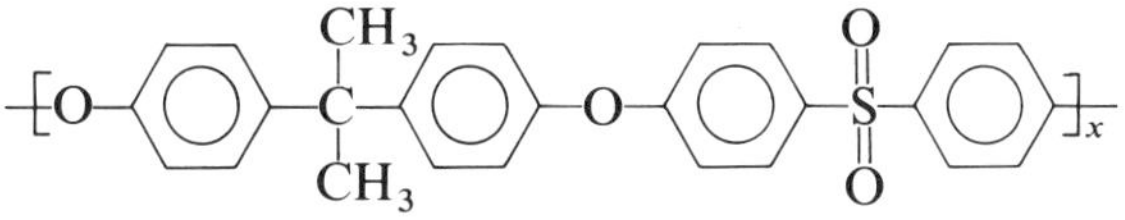   (NaCl out)

REFERENCE

1. Hounshell, D. A., and J. K. Smith, Jr., *Invention & Technology*, Fall 1988, p. 40.

PROBLEMS

1. A 40-ft "fiberglass" boat hull weighs 3000 lb. Sixty percent of that is a fully cured polyester resin (the remainder is the glass reinforcing fibers, pigment, fillers, etc.). What is the molecular weight of the polymer?

2. Pyromellitic dianhydride, PMDA (below), is useful in a variety of polymerization reactions:

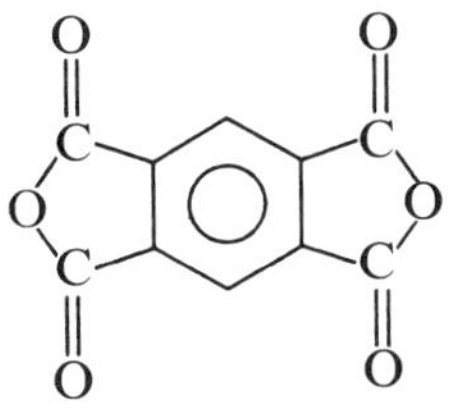

   a. What is its functionality in polyesterification reactions? Structurally, what type of polymer would you expect to get by reacting PMDA with a glycol?

   b. Show the repeating unit of the polymer formed by reacting PMDA with hexamethylene diamine (Example 4E).

3. The thermotropic liquid crystal polymer Xydar® is reported to be made by condensing terephthalic acid (HOOC–ϕ–COOH), p,p'-dihydroxybiphenyl (HO–ϕ–ϕ–OH) and p-hydroxybenzoic acid (HO–ϕ–COOH). What repeating unit(s) would you expect to find in this polymer?

4. These are to be considered additions to Example 4:

   T. Polyurea—starting monomers: diamine (see part E) and diisocyanate (part Q).

   U. Polyarylate—starting monomers: terephthalic acid (part C) and bisphenol-A (part N).

   V. Polyphenylene sulfide (PPS) (Ryton®, Supec®)—starting monomes: para-dichlorobenzene and Na_2S.

   W. Poly(pivalolactone)—starting monomer:

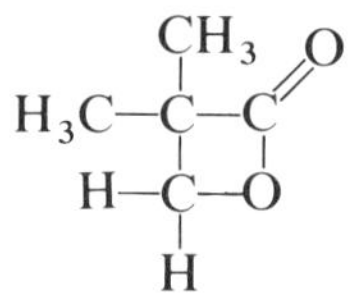

$$H_3C-\underset{\underset{H}{\overset{|}{\underset{|}{C}}-O}}{\overset{CH_3}{\overset{|}{C}}}-C\overset{O}{\diagup}$$

5. A crosslinked polymer is made by polymerizing a reaction mass that contains 98 mol% styrene, $H_2C=CH\phi$, and 2 mol% divinyl benzene (DVB), $H_2C=CH-\phi-HC=CH_2$. Assuming that the reaction goes to completion, what is the averge number of polystyrene repeating units between the DVB crosslinks?

6. A crosslinked polyester is made by reacting (assume completely) 2 mol of glycerine (Example 4G) with 3 mol of terephthalic acid (Example 4C). What is the molecular weight of the resulting polymer?

7. Assume that you have supplies of the reagents below. Consider only the reactions discussed in the chapter and assume that they can be carried out completely and irreversibly.

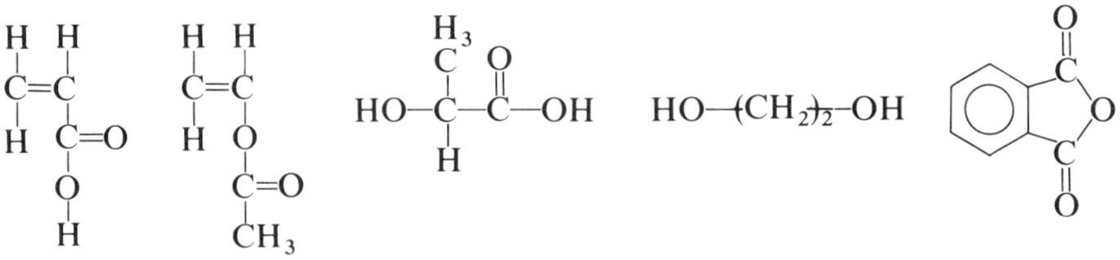

(I)   (II)   (III)   (IV)   (V)

   a. Describe (show reagents and reaction steps) how you would make a graft copolymer with the branches distributed randomly along the main chain, and with no possibility of crosslinking. Show the main-chain and branch repeating units.

b. Describe how you would make a block copolymer $[A]_x[B]_y[A]_z$ (x, y, and z are not necessarily the same). Show the A and B repeating units.

8. Given supplies of isocyanoethyl methacrylate (IEM),

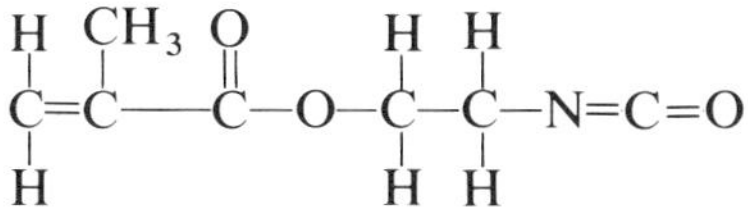

$$
\begin{array}{ccccc}
\text{H} & \text{CH}_3 & \text{O} & & \text{H} \quad \text{H} \\
| & | & \| & & | \quad | \\
\text{C}=\text{C}&-&\text{C}-\text{O}-\text{C}-\text{C}-\text{N}=\text{C}=\text{O} \\
| & & & & | \quad | \\
\text{H} & & & & \text{H} \quad \text{H}
\end{array}
$$

and a glycol, HO–R–OH, and using only reactions discussed in the chapter, do the following:

a. Show the repeating unit(s) of all the linear homopolymers that could be produced.

b. Illustrate, showing the reactions involved, two different methods of producing a crosslinked polymer, again using only the IEM and the glycol.

9. Given supplies of acrylic acid (I), adipic acid (II), and propylene glycol (III)

$$H_2C=CHCOOH \qquad HOOC(CH_2)_4COOH \qquad HO(CH_2)_3OH$$

(I)   (II)   (III)

a. Show the repeating unit(s) of all the linear homopolymers that could be produced.

b. Describe three different procedures by which crosslinked polymers could be produced (again using only I, II, and III).

CHAPTER III

Bonding in Polymers

3.1 TYPES OF BONDS

Various types of bonds hold together the atoms in polymeric materials, unlike in metals, for example, where only one type of bond exists. These types are (1) primary covalent, (2) hydrogen bond, (3) dipole interaction, (4) van der Waals and, (5) ionic, examples of which are shown in Fig. 3.1.[1] Hydrogen bonds, dipole interactions, van der Waals bonds, and ionic bonds are known collectively as secondary bonds. The distinctions are not always clear-cut; that is, hydrogen bonds may be considered as the extreme of dipole interactions.

3.2 BOND DISTANCES AND STRENGTHS

Regardless of the type of bond, the potential energy of the interacting atoms as a function of the separation between them is represented qualitatively by the potential function sketched in Fig. 3.2. As the interacting centers are brought together from large separation, an increasingly great attraction tends to draw them together (negative potential energy). Beyond the separation r_m, as the atoms are brought closer together, their electronic "atmospheres" begin to interact, and a powerful repulsion is set up. At r_m, the system is at a potential energy minimum, its most probable or equilibrium separation, r_m being the equilibrium bond distance. The "depth" of the potential well ε is the energy required to break the bond, separating the atoms completely.

Table 3.1 lists the approximate bond strengths and interatomic distances of the bonds encountered in polymeric materials. The important fact to notice here is how much stronger the primary covalent bonds are than the others. As the material's temperature is raised and its thermal energy (kT) is thereby increased, the primary covalent bonds will be the last to dissociate, when the available thermal energy exceeds their dissociation energy.

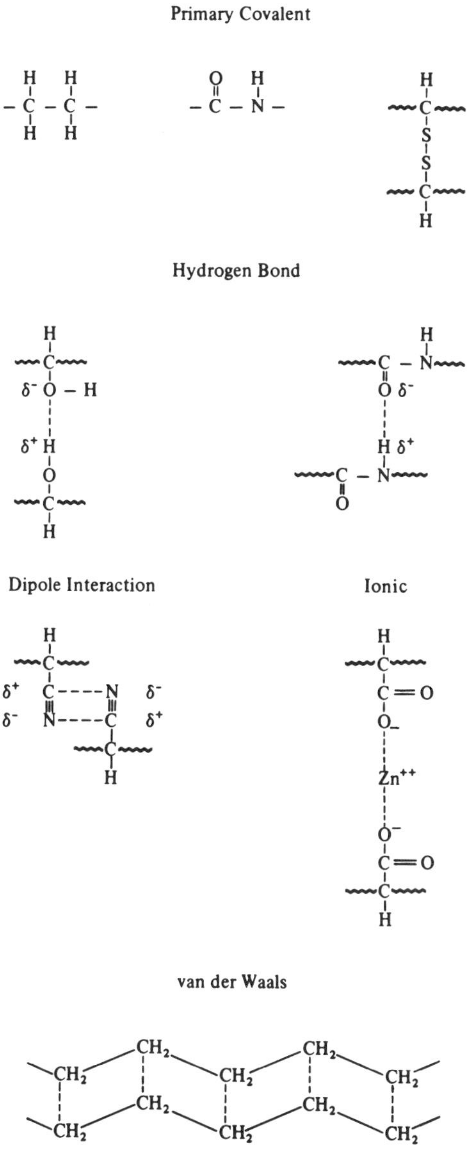

Primary Covalent

Hydrogen Bond

Dipole Interaction   Ionic

van der Waals

Figure 3.1   Bonding in polymer systems.

3.3   BONDING AND RESPONSE TO TEMPERATURE

In linear and branched polymers, only the secondary bonds hold the individual polymer chains together (neglecting temporary mechanical entanglements). Thus, as the temperature is raised, a point will be reached where the forces

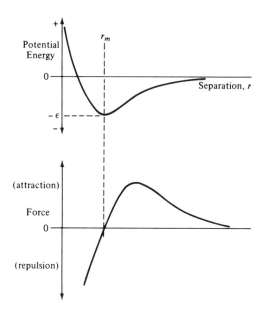

Figure 3.2   Interatomic potential energy and force.

Table 3.1   Bond Parameters[1,2]

Bond Type	Interatomic Distance r_m (nm)	Dissociation Energy ε (kcal/mol)
Primary covalent	0.1–0.2	50–200
Hydrogen bond	0.2–0.3	3–7
Dipole interaction	0.2–0.3	1.5–3
van der Waals bond	0.3–0.5	0.5–2
Ionic bond	0.2–0.3	10–20

holding the chains together become insignificant, and the chains are then free to slide past one another, that is, to *flow* upon the application of stress. Therefore, *linear* and *branched* polymers are generally *thermoplastic*. The crosslinks in a crosslinked polymer, on the other hand, are held together by the same primary covalent bonds as are the main chains. When the thermal energy exceeds the dissociation energy of the primary covalent bonds, both main-chain and cross-link bonds fail randomly, and the polymer degrades. Hence, *crosslinked polymers are thermosetting*.

There are some exceptions to these generalizations. It is occasionally possible for secondary bonds to make up for in quantity what they lack in quality. Polyacrylonitrile

$$
\begin{array}{cc}
\mathrm{H} & \mathrm{H} \\
\mid & \mid \\
\left[\!\!\!\begin{array}{cc} \mathrm{C}\!\!\!\!\!&\!\!\!\!\!\mathrm{C} \end{array}\!\!\!\right]_x \\
\mid & \mid \\
\mathrm{H} & \mathrm{C}\!\equiv\!\mathrm{N}
\end{array}
$$

for example, is capable of strong dipole interactions at the nitrile groups on every other carbon atom along the chains. If these secondary bonds could be broken one by one, that is, "unzipped," polyacrylonitrile would behave as a typical thermoplastic. This is impossible, of course, and by the time enough of the secondary bonds have been dissociated to free the chains and allow flow, the dissociation energy of some carbon–carbon main-chain bonds has been exceeded and the material degraded. Extreme stiffness of the polymer chain also contributes to this sort of behavior. Cellulose has a bulky, complex repeating unit that contains three hydroxyl groups. Though linear, its chains are therefore stiff and strongly hydrogen bonded, and it is not thermoplastic. If the hydroxyls are reacted with acids, such as nitric, acetic, and butyric, the resulting cellulose esters are typical thermoplastics largely because of the reduced hydrogen bonding:

$$
\underset{\text{Cellulose}}{[\mathrm{R(OH)_3}]_x} + 3x\ \underset{\text{Acetic acid}}{\mathrm{HO}\!-\!\overset{\displaystyle \mathrm{O}}{\overset{\|}{\mathrm{C}}}\!-\!\mathrm{CH_3}} \rightarrow \underset{\text{Cellulose acetate}}{[\mathrm{R(O}\!-\!\overset{\displaystyle \mathrm{O}}{\overset{\|}{\mathrm{C}}}\!-\!\mathrm{CH_3})_3]_x} + 3x\ \mathrm{H_2O}
$$

Polytetrafluoroethylene (Teflon TFE) is another example, due to the close packing and extensive secondary bonding of the main chains.

3.4   ACTION OF SOLVENTS

The action of solvents on polymers is in many ways similar to that of heat. Appropriate solvents, that is, those that can form strong secondary bonds with the polymer chains, can penetrate, replace the interchain secondary bonds, and thereby pull apart and dissolve linear and branched polymers. The polymer–solvent secondary bonds cannot overcome primary valence crosslinks, however, so crosslinked polymers are not soluble, although they may swell considerably. (Try soaking a rubber band in toluene overnight.) The amount of swelling is, in fact, a convenient measure of the extent of crosslinking. A lightly crosslinked polymer such as a rubber band will swell tremendously, while one with extensive crosslinking, for example, an ebonite ("hard rubber") bowling ball, will not swell noticeably at all.

REFERENCES

1. Platzer, N., *Ind. Eng. Chem.* **61**(5), 10, 1969.
2. Miller, M. L., *The Structure of Polymers*. Reinhold, New York, 1966.

CHAPTER IV

Stereoisomerism

It is obvious that the chemical nature of a polymer is of considerable importance in determining the polymer's properties. Of comparable significance is the way the atoms are arranged geometrically within the individual polymer chains.

4.1 INTRODUCTION

The carbon atom is normally (exclusively, for our purposes) tetravalent. In compounds such as methane (CH_4) and carbon tetrachloride (CCl_4) the four identical substituents surround it in a symmetrical tetrahedral geometry. If the substituent atoms are not identical, the symmetry is destroyed, but the general tetrahedral pattern is maintained. This is still true for each carbon atom in the interior of a linear polymer chain, where two of the substituents are chains. If a polyethylene chain were to be stretched out, for example, the carbon atoms in the chain backbone would lie in a zigzag fashion in a plane, with the hydrogen substituents on either side of the plane (Fig. 4.1). In the case of polyethylene, in which all the substituents are the same, this is the only arrangement possible. With vinyl polymers, however, there are several possibilities.

4.2 STEREOISOMERISM IN VINYL POLYMERS

Before beginning a discussion of isomerism in vinyl polymers, it must be pointed out that the monomers ($H_2C=CHX$) polymerize almost exclusively in a head-to-tail fashion, placing the X groups on every other carbon atom along the chain. The reasons for this are the steric interference of similar substituent groups and the electrostatic repulsion of groups with similar polarities. There are then three possible ways in which the X group may be arranged with respect

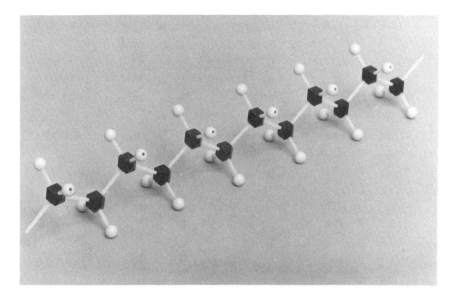

Figure 4.1 The geometry of a polyethylene chain.

to the carbon–carbon backbone plane (see Fig. 4.2), that is, three *stereoisomers*:

1. *Atactic*. A random arrangement of the X groups. Its lack of regularity has important consequences.
2. *Isotactic*. The structure in which all the X groups are lined up on the same side of the backbone plane.
3. *Syndiotactic*. Alternating placement of the X group on either side of the plane.

The terms above were coined by Dr. Giulio Natta, who shared the 1964 Nobel Chemistry Prize for his work in the area.

Although useful for descriptive purposes, the planar zigzag arrangement of the main-chain carbon atoms is not always the one preferred by nature; that is, it is not necessarily the minimum free-energy configuration. In the case of poly-ethylene it is, but for isotactic and syndiotactic polypropylene, for example, steric interference of the –CH$_3$ groups dictates other arrangements. For these materials, the preferred (minimum-energy) configurations are quite regular, while the atactic is irregular (Fig. 4.3).

Atactic polypropylene has a consistency somewhat like used chewing gum, whereas the stereoregular forms are hard, rigid plastics. The reason why this regularity (or lack of it) has such a profound effect on mechanical properties is discussed in the next chapter.

The type of stereoregularity described above is a direct result of the dissym-metry of vinyl monomers. It is established in the polymerization reaction, and

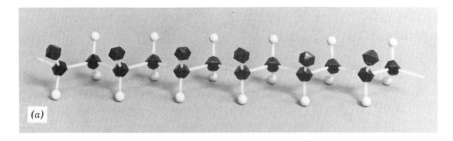

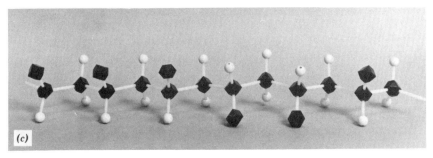

Figure 4.2 Stereoisomerism in vinyl polymers: (*a*) isotactic; (*b*) syndiotactic; (*c*) atactic.

no amount of twisting and turning the chain about its bonds can convert one stereoisomer into another (molecular models are a real help, here).

The situation is even more complex for monomers of the form HXC=CHX'. This is discussed by Natta,[1] but is currently of no commercial importance.

Example 1. Both isotactic and atactic polymers of propylene oxide,

$$H-\underset{\underset{O}{\diagdown\diagup}}{\overset{H\ H}{\underset{|\ \ |}{C-C}}}-CH_3$$

have been prepared by ring scission polymerization.

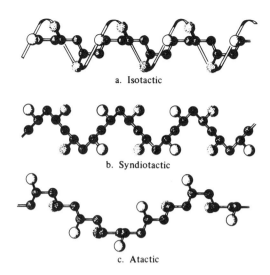

a. Isotactic

b. Syndiotactic

c. Atactic

Figure 4.3 Configurations of polypropylene chains. The large balls represent methyl (–CH$_3$) groups, and hydrogen atoms are not shown. (From Natta.'[1] Copyright © 1961 by Scientific American, Inc. All rights reserved.)

a. Write the general structural formula for the polymer.

b. Indicate how the atactic, isotactic, and syndiotactic structures differ.

Solution

a.

$$
\begin{array}{c}
\quad H\ \ H \\
\quad |\ \ \ | \\
\text{---}\!\!\left[O\text{---}C\text{---}C\right]_{\!x} \\
\quad |\ \ \ | \\
\quad H\ \ CH_3
\end{array}
$$

b.

$$
\begin{array}{c}
\ \ H\ \ H \ \ \ \ \ \ H\ \ H \ \ \ \ \ \ H\ \ CH_3 \\
\ \ |\ \ \ | \ \ \ \ \ \ \ \ |\ \ \ | \ \ \ \ \ \ \ \ |\ \ \ | \\
\text{---}O\text{---}C\text{---}C\text{---}O\text{---}C\text{---}C\text{---}O\text{---}C\text{---}C\text{---} \quad \text{Atactic} \\
\ \ |\ \ \ | \ \ \ \ \ \ \ \ |\ \ \ | \ \ \ \ \ \ \ \ |\ \ \ | \\
\ \ H\ \ CH_3 \ \ \ \ H\ \ CH_3 \ \ \ \ H\ \ H
\end{array}
$$

$$
\begin{array}{c}
\ \ H\ \ H \ \ \ \ \ \ H\ \ H \ \ \ \ \ \ H\ \ H \\
\ \ |\ \ \ | \ \ \ \ \ \ \ \ |\ \ \ | \ \ \ \ \ \ \ \ |\ \ \ | \\
\text{---}O\text{---}C\text{---}C\text{---}O\text{---}C\text{---}C\text{---}O\text{---}C\text{---}C\text{---} \quad \text{Isotactic} \\
\ \ |\ \ \ | \ \ \ \ \ \ \ \ |\ \ \ | \ \ \ \ \ \ \ \ |\ \ \ | \\
\ \ H\ \ CH_3 \ \ \ \ H\ \ CH_3 \ \ \ \ H\ \ CH_3
\end{array}
$$

$$
\begin{array}{c}
\ \ H\ \ H \ \ \ \ \ \ H\ \ CH_3 \ \ \ \ H\ \ H \\
\ \ |\ \ \ | \ \ \ \ \ \ \ \ |\ \ \ | \ \ \ \ \ \ \ \ |\ \ \ | \\
\text{---}O\text{---}C\text{---}C\text{---}O\text{---}C\text{---}C\text{---}O\text{---}C\text{---}C\text{---} \quad \text{Syndiotactic} \\
\ \ |\ \ \ | \ \ \ \ \ \ \ \ |\ \ \ | \ \ \ \ \ \ \ \ |\ \ \ | \\
\ \ H\ \ CH_3 \ \ \ \ H\ \ H \ \ \ \ \ \ H\ \ CH_3
\end{array}
$$

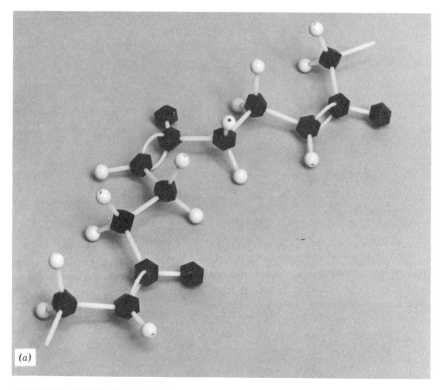

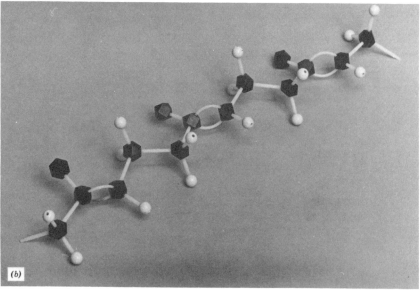

Figure 4.4   Stereoisomers of 1,4-polyisoprene: (*a*) *cis*, (*b*) *trans*.

4.3 STEREOISOMERISM IN DIENE POLYMERS

Another type of stereoisomerism arises in the case of poly-1,4-dienes because of the impossibility of rotation about a double bond. The substituent groups on the double-bonded carbons may be either on the same side of the chain (*cis*) or on opposite sides (*trans*), as shown in Fig. 4.4 for 1,4-polyisoprene. The chains of *cis*-1,4-polyisoprene assume a tortured, irregular configuration because of the steric interference of the substituents adjacent to the double bonds. This stereoisomer is familiar as natural rubber, as in rubber bands. *trans*-1,4-Polyisoprene chains assume a regular structure. The polymer is known as gutta-percha, a tough but not elastic material, long used as a golf-ball cover.

Note that stereoisomerism in poly-1,4-dienes does not depend on dissymmetry of the repeating unit. The same isomers are possible with butadiene, in which all carbon substituents are hydrogen.

Example 2. Identify all the possible structural and stereoisomers that can result from the addition polymerization of butadiene,

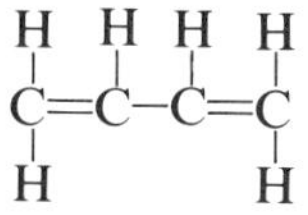

$$
\begin{array}{cccc}
H & H & H & H \\
| & | & | & | \\
C{=}C & {-}C & {=}C \\
| & & & | \\
H & & & H \\
\end{array}
$$

Solution. Recall from Chapter 2 that butadiene may undergo 1,4 addition or 1,2 addition. The 1,4 polymer can have *cis* and *trans* isomers like the 1,4-polyisoprene in Fig. 4.4. The 1,2 polymer

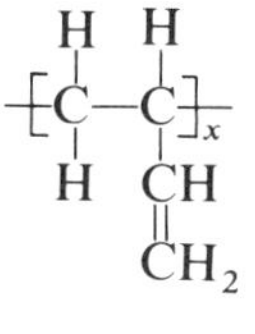

$$
\begin{array}{cc}
H & H \\
| & | \\
{+}C{-}C{+}_x \\
| & | \\
H & CH \\
& \| \\
& CH_2 \\
\end{array}
$$

can have atactic, isotactic, and syndiotactic stereoisomers like any vinyl polymer.

Example 3. Identify all the possible structural and stereoisomers that can result from the addition polymerization of chloroprene,

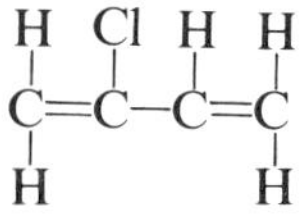

$$
\begin{array}{cccc}
H & Cl & H & H \\
| & | & | & | \\
C{=}C & {-}C & {=}C \\
| & & & | \\
H & & & H \\
\end{array}
$$

(known commercially as Neoprene).

Solution. As in Example 2, the 1,4 polymer can have *cis* and *trans* stereoisomers. This unsymmetrical diene monomer can also undergo both 1,2 addition and 3,4 addition:

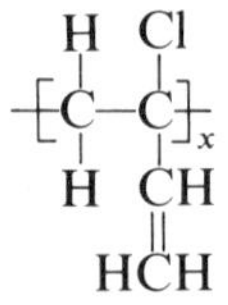

Each of these structural isomers may have atactic, isotactic, or syndiotactic stereoisomers. Thus, in principle at least, there are *eight* different isomers of polychloroprene possible.

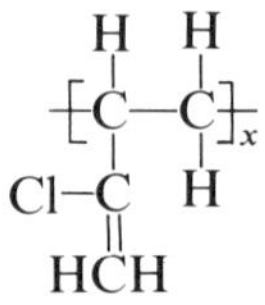

REFERENCE

1. Natta, G., *Sci. Am.* **205**(2), 33 (1961).

PROBLEMS

1. Given the monomers shown below

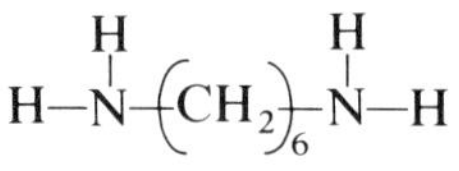

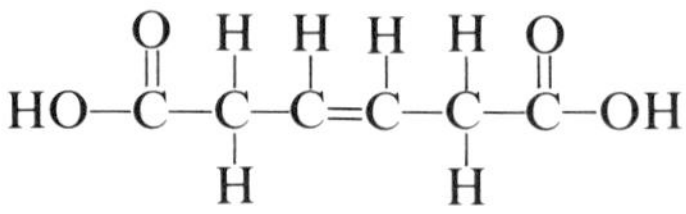

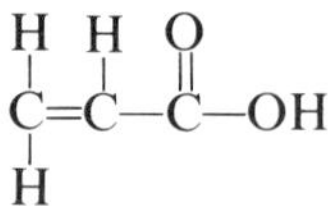

  a. Show the repeating units of all the *linear homopolymers* that could be produced.

  b. Which of the polymers, if any, could have *cis* and *trans* isomers? Illustrate. How would this differ from the isomerism that arises in 1,4-polydienes?

  c. Which of the polymers, if any, could have isotactic and syndiotactic isomers? Illustrate.

CHAPTER V

Polymer Morphology

It was pointed out in the previous chapter that the geometric arrangement of the atoms in polymer chains can exert a significant influence on the properties of the bulk polymer. To appreciate why this is so, the subject of *polymer morphology*, the structural arrangement of the chains in the polymer, is introduced here.

5.1 REQUIREMENTS FOR CRYSTALLINITY

Although the precise nature of crystallinity in polymers is still under investigation, a number of facts have long been known about the requirements for polymer crystallinity. First, an ordered, regular chain structure is necessary to allow the chains to pack into an ordered, regular, three-dimensional crystal lattice. Thus, stereoregular polymers are more likely to be crystalline than those that have irregular chain structures. Irregularly spaced, protruding side groups hinder crystallinity. Second, no matter how regular the chains, the secondary forces holding the chains together in the crystal lattice must be strong enough to overcome the disordering effect of thermal energy; so hydrogen bonding or strong dipole interactions promote crystallinity, and other things being equal, raise the crystalline melting point.

X-ray studies show that there are numerous polymers that do not meet the above criteria and show no traces of crystallinity; that is, they are completely *amorphous*. In contrast to the regular, ordered arrangement in a crystal lattice, the chains in an amorphous polymer mass assume a more-or-less random, twisted, entangled, "balled-up" configuration. A common analogy is a bowl of cooked spaghetti. A better analogy, in view of the constant thermal motion of the chain segments, is a bowl of wriggling snakes. The latter analogy forms the basis of a currently popular quantitative model of polymer behavior known as reptation (for reptiles) theory.

At the other extreme, despite intensive efforts, no one has succeeded in producing a completely crystalline polymer. The crystalline content may in certain cases be pushed up to about 98%, but at least a few percent noncrystalline material always remains. In the case of metals, crystal defect concentrations are normally on the order of parts per million, so they can be considered perfectly crystalline in comparison with even the most highly crystalline polymers. Interestingly, amorphous metals are a recent and highly promising development.

References 1–5 contain extensive discussions of the techniques used to study crystallinity in polymers as well as reviews of research results.

5.2   THE FRINGED MICELLE MODEL

The first attempt to explain the observed properties of crystalline (the word should be prefaced by "semi," but rarely is) polymers was the *fringed micelle* model (Fig. 5.1). This model pictures crystalline regions known as fringed micelles or *crystallites* interspersed in an amorphous matrix. The crystallites, whose dimensions are on the order of tens of nanometers, are small volumes in which portions of the chains are regularly aligned parallel to one another, tightly packed into a crystal lattice. The individual chains, however, are many times longer than the dimensions of a crystallite, so they pass from one crystallite to another through amorphous areas.

This model explains nicely the coexistence of crystalline and amorphous material in polymers, and also explains the increase in crystallinity that is observed when fibers are drawn (stretched). Stretching the polymer orients the chains in the direction of the stress, increasing the alignment in the amorphous areas and producing greater degrees of crystallinity (Fig. 5.1*b*). Since the chains

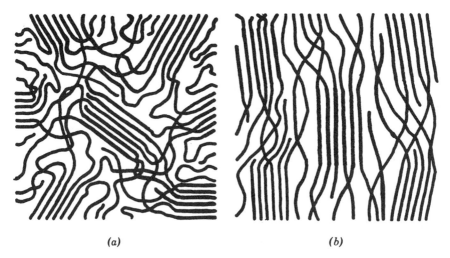

(a)   (b)

Figure 5.1   The fringed micelle model: (*a*) unoriented; (*b*) chains oriented by applied stress.

pass randomly from one crystallite to another, it is easy to see why perfect crystallinity can never be achieved. This also explains why the effects of crystallinity on properties are in many ways similar to those of cross-linking, because, like crosslinks, the crystallites tie the individual chains together. Unlike crosslinks, though, the crystallites will generally melt before the polymer degrades and solvents that form strong secondary bonds with the chains can dissolve them.

The fringed micelle model has been superseded as new information has led to the development of new models. Nevertheless, it still does a good job of qualitatively predicting the effects of crystallinity on the mechanical properties of polymers.

Example 1. Polyvinyl alcohol is made by the hydrolysis of polyvinyl acetate because the vinyl alcohol monomer is unstable.

$$
\left[\begin{array}{c} \text{H} \quad \text{H} \\ \text{C}\!-\!\text{C} \\ \text{H} \quad \text{O} \\ \quad\;\; | \\ \quad\;\; \text{C}\!=\!\text{O} \\ \quad\;\; | \\ \quad\;\; \text{CH}_3 \end{array}\right]_x + \text{H}_2\text{O} \longrightarrow \left[\begin{array}{c} \text{H} \quad \text{H} \\ \text{C}\!-\!\text{C} \\ \text{H} \quad \text{OH} \end{array}\right]_x + \text{H}_3\text{C}\!-\!\overset{\text{O}}{\underset{}{\text{C}}}\!-\!\text{OH}
$$

Polyvinyl acetate — Polyvinyl alcohol — Acetic acid

The extent of reaction may be controlled to yield polymers with anywhere from 0 to 100% of the original acetate groups hydrolyzed. Pure polyvinyl acetate (0% hydrolyzed) is insoluble in water. It has been observed, however, that as the extent of hydrolysis is increased, the polymers become more water soluble, up to about 87% hydrolysis, after which further hydrolysis decreases water solubility at room temperature. Explain briefly.

Solution. The normal polyvinyl acetate is atactic, and the irregular arrangement of the acetate side groups renders it completely amorphous. Water cannot form strong enough secondary bonds with the chains to dissolve it. As the acetate groups are replaced by hydroxyls, sites are introduced that can form hydrogen bonds with water, thereby increasing the solubility. At high degrees of substitution, replacement of the bulky acetate groups with the compact hydroxyls allows the chains to pack into a crystal lattice. The hydroxyl groups provide hydrogen-bonding sites between the chains that help hold them in the lattice, and thus solubility is reduced.

Example 2. Explain the following experiment: A weight is tied to the end of a polyvinyl alcohol fiber. The weight and part of the fiber are dunked in a beaker of boiling water. As long as the weight remains suspended, the situation is stable, but when the weight is rested on the bottom of the beaker, the fiber dissolves.

Solution. As long as the weight is suspended from the fiber, the stress maintains the alignment of the chains in a crystal lattice against the disordering effects of thermal energy and solvent (water) penetration. When the weight is rested on the bottom, the stress is removed, allowing the polymer to dissolve.

5.3 LAMELLAR CRYSTALS[1-7]

The first direct observations of the nature of polymer crystallinity resulted from the growth of single crystals from dilute solution. Either by cooling or evaporation of solvent, thin, pyramidal or plate-like polymer crystals (lamellae) were precipitated from dilute solutions (Fig. 5.2). These crystals were on the order of 10 μm along a side and only about 0.01 μm thick. This was fine, except that X-ray measurements showed that the polymer chains were aligned perpendicular to the large flat faces of the crystals, and it was known that the extended length of the individual chains was on the order of 0.1 μm. How could a chain fit into a crystal one-tenth its length? The only answer is that the chain must fold back on itself, as shown in Fig. 5.2. Two competing models of chain folding are illustrated.

This folded-chain model has been well substantiated for single polymer crystals. The lamellae are about 50 to 60 carbon atoms thick, with about five carbon atoms in a direct reentry fold. The atoms in a fold, whether direct or indirect reentry, can never be part of a crystal lattice.

It is now well established that similar lamellar crystallites exist in bulk polymer samples crystallized from the melt, although the lamellae may be up to 1 μm thick. Recent results support the presence of a third, interfacial region between the crystalline lamellae and the amorphous phase. This interfacial phase can make up some 10 to 20% of the material. Furthermore, there doesn't seem to be much, if any, direct-reentry folding of chains in bulk-crystallized lamellae. This is illustrated in Fig. 5.3. Orientation of the lamellae along with additional orientation and crystallization in the interlamellar amorphous re-

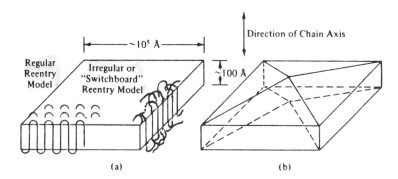

Figure 5.2 Polymer single crystals: (*a*) flat lamellae; (*b*) pyramidal lamellae. Two concepts of chain reentry are illustrated.

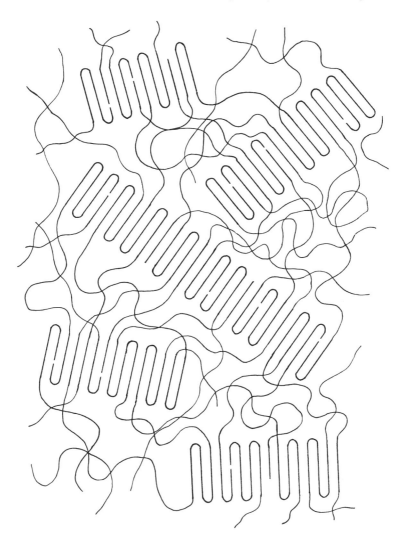

Figure 5.3 Compromise model showing folded-chain lamellae tied together by inter-lamellar amorphous chains.

gions, as in the fringed-micelle model, is usually invoked to explain the increase in the degree of crystallinity with drawing.

5.4 THE EFFECT OF CRYSTALLINITY ON MECHANICAL PROPERTIES

For simplicity, let's assume that only crystalline and amorphous phases are present. With a given polymer, the properties of each phase remain the same, but the relative amounts of the phases can vary, and this can strongly influence the bulk properties, particularly the mechanical properties. This is illustrated below for polyethylene.

Since the polymer chains are packed together more efficiently and tightly in the crystalline areas than in the amorphous, the crystallites will have a higher density. Thus, low-density ($0.92\,g/cm^3$) polyethylene is estimated to be about 43% crystalline, while high-density ($0.97\,g/cm^3$) polyethylene is about 76% crystalline. Density is, in fact, a convenient measure of the degree of crystallinity. Because the volumes of the crystalline and amorphous phases are additive, density and degree of crystallinity are related by

$$\frac{1}{\rho} = \frac{w_c}{\rho_c} + \frac{w_a}{\rho_a} = \frac{w_c}{\rho_c} + \frac{(1 - w_c)}{\rho_a} \tag{5.1}$$

where the w's are weight fractions and the subscripts c and a refer to the crystalline and amorphous phases, respectively.

In the case of polyethylene, the differences in degree of crystallinity arise largely from branching that occurs during polymerization (although it is also influenced by molecular weight and the rate of cooling). The branch points sterically hinder packing into a crystal lattice in their immediate vicinity, and thus lower the degree of crystallinity.

Low-density polyethylene (LDPE) has traditionally been made by a high-pressure process (25 000–50 000 psi) and high-density polyethylene (HDPE) by a low-pressure process (~100 psi). Thus, LDPE is sometimes referred to as high-pressure polyethylene and HDPE as low-pressure polyethylene.

If that weren't confusing enough, low-density polyethylenes are now also made by low-pressure processes similar to those used for HDPE. These materials have most unfortunately (and inaccurately) been termed linear, low-density polyethylene (LLDPE). If they were truly linear, they wouldn't be low-density.

Traditional (high-pressure) LDPE is a homopolymer. Its long, "branched branches" arise from a side reaction during polymerization. LLDPE, however, has short branches that are introduced by random copolymerization with minor amounts (say 8–10%) of one or more α-olefins (vinyl monomers with hydrocarbon X groups) (1-butene, 1-hexene, 1-octene, and 4-methyl-1-pentene are commonly used). This same approach is extended (by adding more comonomer) to make VLDPE (very-low . . .) or ULDPE (ultra-low . . .) ($\rho < 0.915\,g/cm^3$). The nature of the branching affects some properties to a certain extent. LLDPEs form stronger, tougher films than LDPEs of equivalent density, for example. The various polyethylenes are summarized in Table 5.1 and their molecular architectures are sketched in Fig. 5.4. Equation 5.1 must be used with caution where the density is varied with a comonomer, as in LLDPE, because ρ_a will, in general, vary with copolymer composition.

Since the polymer chains are more closely packed in the crystalline areas than in the amorphous, there are more of them available per unit area to support a stress. Also, since they are in close and regular contact over relatively long

Table 5.1 The Influence of Crystallinity on Some of the Properties of Polyethylene[a]

Commercial product	Low density	Medium density	High density
Density range, g/cm^3	0.910–0.925	0.926–0.940	0.941–0.965
Approximate % crystallinity	42–53	54–63	64–80
Branching, equivalent CH$_3$ groups/1000 carbon atoms	15–30	5–15	1–5
Crystalline melting point, °C	110–120	120–130	130–136
Hardness, Shore D	41–46	50–60	60–70
Tensile modulus, psi	$0.14–0.38 \times 10^5$	$0.25–0.55 \times 10^5$	$0.6–1.8 \times 10^5$
(N/m^2)	$(0.97–2.6 \times 10^8)$	$(1.7–3.8 \times 10^8)$	$(4.1–12.4 \times 10^8)$
Tensile strength, psi	600–2300	1200–3500	3100–5500
(N/m^2)	$(0.41–1.6 \times 10^7)$	$(0.83–2.4 \times 10^7)$	$(2.1–3.8 \times 10^7)$
Flexural modulus, psi	$0.08–0.6 \times 10^5$	$0.6–1.15 \times 10^5$	$1.0–2.6 \times 10^5$
(N/m^2)	$(0.34–4.1 \times 10^8)$	$(4.1–7.9 \times 10^8)$	$(6.9–18 \times 10^8)$

[a] It must be kept in mind that mechanical properties are influenced by factors other than the degree of crystallinity (molecular weight, in particular).

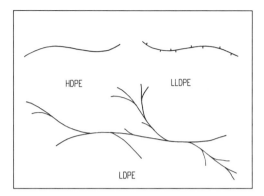

Figure 5.4   Molecular architecture of various polyethylenes. LDPE, low-density polyethylene; LLDPE, linear, low-density polyethylene; HDPE, high-density polyethylene.

distances in the crystallites, the secondary forces holding them together are cumulatively greater than in the amorphous regions. Thus, crystallinity can significantly increase the strength and rigidity of a polymer. For this reason, the stereoregular polypropylenes, which can and do crystallize, are fairly hard, rigid plastics, while the irregular, atactic polymer is amorphous, soft, and sticky. And, other things being equal, the greater the ratio of crystalline to amorphous phase, the stronger, harder, more rigid, and less easily deformable the polymer will be, as is illustrated in Table 5.1.

5.5   THE EFFECT OF CRYSTALLINITY ON OPTICAL PROPERTIES

The optical properties of polymers are also influenced by crystallinity. When light passes between two phases with different refractive indices, some of it is scattered at the interface if the dimensions of the discontinuities are comparable to or greater than the wavelength of visible light (0.4–0.7 μm). Thus, a block of ice is transparent, but snow appears white because light must pass alternately from air to ice crystals many times. Similarly, in a crystalline polymer, the usually denser crystalline areas have a higher refractive index than the amorphous areas, so crystalline polymers are either opaque or translucent because light is scattered as it passes from one phase to the other. So, in general, *transparent polymers are completely amorphous.* An interesting exception arises in the case of isotactic poly(4-methyl-1-pentene). The refractive indices (and densities) of the amorphous and crystalline phases are almost identical. Thus, as far as light is concerned, it's a homogeneous material, and is the one transparent polymer that is known to be highly crystalline. Also, as the dimensions of the dispersed-phase particles become smaller than the wavelength of visible light, scattering decreases, so a polymer with very small crystallites and a low degree of crystallinity might appear nearly transparent, particularly in thin sections.

The converse is not necessarily true; lack of transparency in a polymer may be due to crystallinity, but it can also be caused by an added second phase, such as a filler. If the polymer is known to be a pure homopolymer, however, translucency is a sure sign of crystallinity. Thus, commercial homopolystyrene, which is atactic and therefore completely amorphous because of the irregular arrangement of the bulky phenyl side groups, is perfectly transparent. Ironically, it is often called "crystal" polystyrene because of its crystal clarity. Isotactic polystyrene has been synthesized in the laboratory and does crystallize. It has the white, translucent appearance typical of polyethylene. Foamed polystyrene (cups, packaging supports) is white because light must pass between atactic polystyrene and gas bubbles. High-impact polystyrene consists of a dispersion of 1–10 μm polybutadiene rubber particles in a continuous phase of atactic polystyrene. Thus, while each phase is completely amorphous and individually transparent, they have different refractive indices, so the composite scatters light and appears white and translucent.

Example 3. Explain the following facts:

a. Polyethylene and polypropylene produced with stereospecific catalysts are each fairly rigid, translucent plastics, while a 65–35 copolymer of the two, produced in exactly the same manner, is a soft, transparent rubber.

b. A plastic is similar in appearance and mechanical properties to the polyethylene and polypropylene described in (***a***), but it consists of 65% ethylene and 35% propylene units. The two components of this plastic cannot be separated by any physical or chemical means without degrading the polymer.

Solution. ***a.*** The polyethylene produced with these catalysts is linear and thus highly crystalline. The polypropylene is isotactic and also highly crystalline. The crystallinity confers mechanical strength and translucency. The 65–35 copolymer, ethylene–propylene rubber (EPR) is a *random* copolymer, so the CH_3 groups from the propylene monomer are arranged at irregular intervals along the chain, preventing packing in a regular crystal lattice, giving an amorphous, rubbery polymer.

b. Since the components cannot be separated, they must be chemically bound within the chains. The properties indicate a crystalline polymer, so the CH_3 groups from the propylene cannot be spaced irregularly, as in the random copolymer in (***a***). Thus, these materials must be *block* copolymers of ethylene and stereoregular polypropylene, poly(ethylene-*b*-propylene). The long blocks of ethylene pack into a polyethylene lattice and the propylene blocks into a polypropylene lattice.

5.6 EXTENDED-CHAIN CRYSTALS

Polymers crystallized from a quiescent melt (such as those in Table 5.1) will have, in general, a random orientation of the chains at the macroscopic level. While the chains in any crystallite are oriented in a particular direction, the multiple crystallites are randomly oriented. It has long been known that if the chain axes could share a common orientation at the macroscopic level, the material would have superior mechanical properties in the orientation direction, since the chains would be arranged most efficiently to support a stress. In fact, fiber drawing does just this to a certain extent, which is why it is practiced.

Polymers crystallized while being subjected to an extensional flow (drawing), which tends to disentangle the chains and align them in the direction of flow, form fibrillar structures. These are believed to be extended-chain crystals, in which the chains are aligned parallel to one another over great distances, with little or no chain folding.[8] When fibers of such materials are further drawn, a super-strong fiber results. Very high molecular weight polymers are favored for this purpose. The longer chains align more readily, and there are fewer crystalline defects due to chain ends. Also, formation of the fibers from relatively dilute solution allows easier disentanglement and better alignment of the chains.

For linear polyethylene fibers made in this fashion, a tensile modulus of 44 GPa (6.4×10^6 psi) (steel is about 206 GPa (30×10^6 psi) but has a density 7.6 times as great!) and a tensile strength of 1.8 GPa (2.6×10^5 psi) have been claimed.[9] Compare these figures with those for ordinary, quiescently crystallized HDPE in Table 5.1.

Extended-chain crystals have also been identified as tying together lamellae in bulk-crystallized polymers and forming the core ("skewer") of the interesting "shish kebab" structures grown from dilute solutions subjected to shearing.[10]

5.7   LIQUID-CRYSTAL POLYMERS[11]

As noted above, if polymer chains can be aligned in the melt prior to crystallization, high degrees of extended-chain crystallinity may be formed, imparting remarkable mechanical properties, at least in the chain direction. Another way of doing it is by starting with molecules that show a degree of order in the liquid phase. Such liquid-crystalline materials have been known for years, but until fairly recently were limited to relatively small molecules. Now, however, several types of polymers are known which exhibit liquid-crystalline order, either *lyotropic* (order in solution) or *thermotropic* (order in the melt).[12]

Kevlar® is an aromatic polyamid ("aramid") with the repeating unit

It is spun into fibers from a lyotropic liquid-crystalline solution in concentrated H_2SO_4. The solution is extruded through small holes into a bath that leaches out the acid, forming the fibers ("wet spinning"). Because the molecules are oriented prior to crystallization, the fibers maintain a high degree of

Table 5.2   Comparison of Liquid-Crystalline Kevlar® 49 Fibers with Nylon 6/6 Fibers[13]

	Kevlar® 49	Nylon 6/6
Tensile strength, GPa (psi)	2.95 (4.3×10^5)	1.28 (1.8×10^5)
Tensile modulus, GPa (psi)	130 (19×10^6)	6.2 (3.0×10^5)
Elongation to break, %	2.3	19

extended-chain crystalline order in the fiber direction, imparting remarkable strength. Table 5.2 compares some properties of these fibers with those of ordinary nylon 6/6 fibers. With a density of only 1.44 g/cm^3, they compare very favorably with other reinforcing fibers (glass, graphite, etc.) on strength/weight basis. As a result, Kevlar® fibers are used in bullet-proof vests and other ballistic armor, as a tire cord, and as a reinforcement in high-strength composites.

Several thermotropic aromatic copolyesters have also been commercialized. Vectra A® is reported to have the repeating units

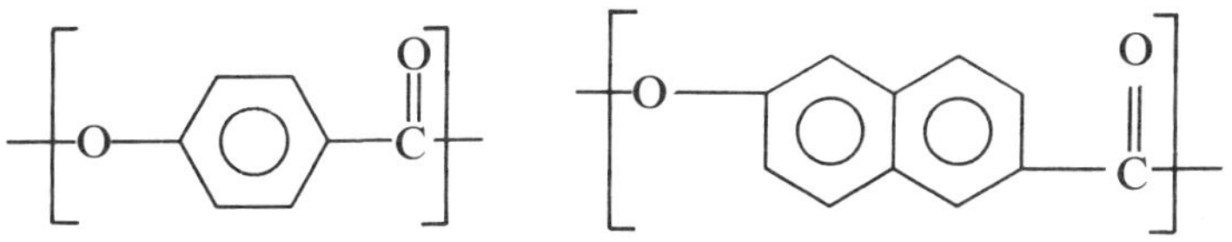

and Xydar® the repeating units

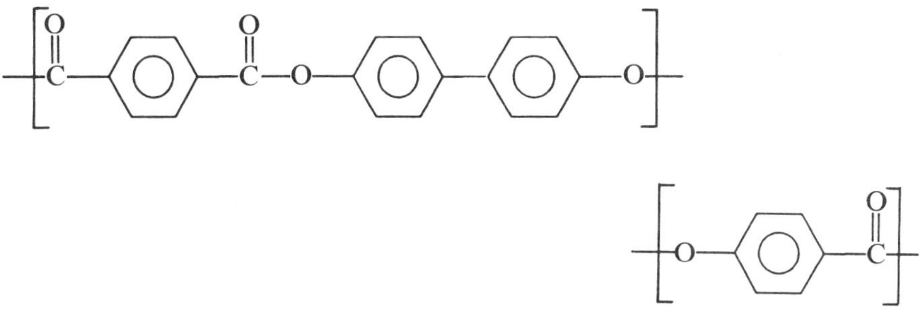

These thermotropic liquid-crystalline polymers have high melting points but can be melt processed like other thermoplastics. The macroscopic orientation of the extended-chain crystals depends on the orientation imparted by flow during processing (molding, extrusion, etc.). Because of the fibrous nature of the extended-chain crystals, these plastics behave as "self-reinforced composites," with excellent mechanical properties, at least in the chain direction. This is illustrated in Table 5.3 for molded specimens of a liquid-crystalline copolyester of ethylene glycol, terephthalic acid, and *p*-hydroxybenzoic acid.[14] In the direction parallel to flow, the properties listed in Table 5.3 compare favorably with

Table 5.3   Properties of a Molded Liquid-Crystalline Copolyester[14]

	∥ to Flow	⊥ to Flow
Tensile strength, MPa	107	28.4
Elongation, %	8	10
Flexural modulus, GPa	11.8	1.67
Linear coefficient of expansion, $°C^{-1}$	0	4.5×10^{-5}

ordinary crystalline thermoplastics (nylons, polyesters) reinforced with up to 30% glass fibers.

It is obvious from the repeating units above that liquid-crystalline behaviour is promoted by highly aromatic chains. The phenyl rings inhibit rotation, resulting in stiff, rigid, extended chains, and because they are nearly planar, they stack next to one another, promoting long-range order in both the liquid and solid states.

5.8   SPHERULITES

Not only are polymer chains often arranged to form crystallites, but these crystallites often aggregate into supermolecular structures known as *spherulites*. Spherulites are in some ways similar to the grain structure in metals. They grow radially from a point of nucleation until other spherulites are encountered. Thus, the size of the spherulites can be controlled by the number of nuclei present, with more nuclei resulting in more but smaller spherulites. They are typically about 0.01 mm in diameter and have a Maltese cross appearance between crossed polaroids. Figure 5.4 shows how the polymer chains are thought to be arranged in the spherulites.

Large spherulites contribute to brittleness in polymers and also scatter a lot of light. To minimize brittleness and enhance transparency, nucleating agents

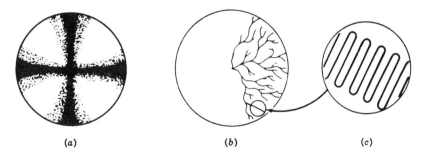

(a)   (b)   (c)

Figure 5.5   Spherulites: (a) appearance between crossed polarizers; (b) branching of lamellae; (c) orientation of chains in lamellae.

plasticiser reduces T_g

are often added or the polymer is shock cooled (which increases the nucleation rate) to promote smaller spherulites.

REFERENCES

1. Schultz, J. M., *Polymer Material Science*, Prentice-Hall, Englewood Cliffs, NJ, 1974.
2. Wunderlich, B., *Macromolecular Physics*, Vol. 1, *Crystal Structure, Morphology, Defects*, Academic, New York, 1973.
3. Tadukoru, H. *Structure of Crystalline Polymers*, Wiley-Interscience, New York, 1979.
4. Mandlekern, L., Chapter 4 in *Physical Properties of Polymers*, American Chemical Society, Washington, DC, 1984.
5. Bovey, F. A., Chapter 5 in *Macromolecules: An Introduction to Polymer Science*, F. A. Bovey and F. H. Winslow (eds.), Academic, New York, 1979.
6. Geil, P. H., *Polymer Single Crystals*, Interscience, New York, 1963.
7. Oppenlander, G. C., *Science* **159**, 1311 (1968).
8. Southern, J. H., and R. S. Porter, *J. Appl. Polymer Sci.* **14**, 2305 (1970); *J. Macromol. Sci. Phys.* **B4**, 541 (1970).
9. Kavesh, S., and D. C. Prevorsek, *U.S. Patent* 4,413,110.
10. Pennings, A. J., *J. Polymer Sci. (Symposia)* **59**, 55 (1977).
11. Weiss, R. A., and C. K. Ober (eds.), *Liquid-Crystalline Polymers*, American Chemical Society, Washington, DC, 1990.
12. Blumstein, A. (ed.), *Liquid Crystalline Order in Polymers*, Academic, New York, 1978.
13. Du Pont product literature.
14. Menges, G., and G. Hahn, *Modern Plastics*, October 1981, p. 56.

PROBLEMS

1. Derive Eq. 1.

2. X-ray diffraction measurements show that polyethylene crystallizes in a body-centered orthorhombic unit cell with lattice parameters $a = 0.7417$ nm, $b = 0.4945$ nm, and $c = 0.2547$ nm at $25°C$.[3] The a, b, and c axes are orthogonal. The chain axes run in the c direction and there are two repeating units per unit cell (one running up the center and one-quarter in each of the four corners). Calculate the crystal density of polyethylene.

3. The density of amorphous polyethylene is estimated to be 0.855 g/cm^3 at $25°C$ by extrapolating values from above the melting point. Use this value and your answer to Problem 2 to estimate the degree of crystallinity of a 0.93-g/cm^3 polyethylene and the density of a 72% crystalline polyethylene.

4. Suppose that you know the theoretical enthalpy of melting for a 100% crystalline polymer. Show how you could obtain the degree of crystallinity for an actual sample by measuring its enthalpy of melting.

5. The entry in Table 5.1 "Branching, equivalent CH_3 groups/1000 C atoms" is determined by calculating that number for a poly(ethylene-*co*-propylene) with the same density. How many CH_3 groups are there per 1000 main-chain C atoms in a 65(weight)% ethylene–35% propylene copolymer? This copolymer, ethylene–propylene rubber (EPR) is completely amorphous.

6. Traditional introductory materials science texts, concerned mostly with metals and ceramics, devote considerable space to crystal structures and crystal defects. These topics have been studied for polymers, but they are hardly mentioned here. Why are they much less important in polymers?

7. Equation 5.1 relates the density of a (semi)crystalline polymer to its composition and the densities of the individual phases. Develop two simple models that relate the tensile modulus of a crystalline polymer E to the moduli of the individual phases E_c and E_a. The densities of each phase are known, as are the % crystallinity. Base each model on a *layered* structure of the two phases (OK, so this is pretty naive). In one, the layers are assumed to be perpendicular to the applied stress, and in the other, parallel.

8. Both Kevlar® (Section 5.7) and extended-chain linear polyethylene (Section 5.6) fibers contain a high degree of extended-chain crystallinity, yet the former is somewhat stronger than the latter. Why?

9. Nylon 6/6 (Chapter 2, Example 4E) is opaque when pure. However, if isophthalic acid (a diacid in which the two acid groups are *meta* on a phenyl ring) is substituted for the adipic acid, the resulting polymer is transparent. Explain.

10. Two perfectly transparent polymers are physically mixed. The resulting mixture is also perfectly transparent. What does this mean? *Hint*: There is more than one possibility.

CHAPTER VI

Characterization of Molecular Weight

6.1 INTRODUCTION

With the exception of a few naturally occurring polymers, all polymers consist of molecules with a *distribution* of chain lengths. It is therefore necessary to characterize the entire distribution quantitatively, or at least to define and measure average chain lengths or molecular weights for these materials, because many important properties of the polymer depend on these quantities. Extensive reviews are available[1,2] concerning the effects of molecular weight and molecular weight distribution on the mechanical properties of polymers.

The concept of an *average* molecular weight causes some initial difficulty because we're used to thinking in terms of ordinary low molecular weight compounds in which the molecules are identical and there is a single, well-defined molecular weight for the compound. Where the molecules in a sample vary in size, however, the results depend on how you count.

In the case of pure, low molecular weight compounds, the molecular weight is defined as

$$M = \frac{W}{N} \tag{6.1}$$

where W = total sample weight
          N = number of moles in the sample

6.2 AVERAGE MOLECULAR WEIGHTS

Where a distribution of molecular weights exists, a *number-average* molecular weight $\bar{M}_n$ may be defined in an analogous fashion to (6.1):

$$\bar{M}_n = \frac{W}{N} = \frac{\sum\limits_{x=1}^{\infty} n_x M_x}{\sum\limits_{x=1}^{\infty} n_x} = \frac{n_1 M_1}{\Sigma n_x} + \frac{n_2 M_2}{\Sigma n_x} + \cdots = \sum_{x=1}^{\infty} \left(\frac{n_x}{N}\right) M_x \quad (6.2)$$

where W = total sample weight = $\sum\limits_{x=1}^{\infty} w_x = \sum\limits_{x=1}^{\infty} n_x M_x$

w_x = total *weight* of x-mer

N = total *number of moles* in the sample (of all sizes) = $\sum\limits_{x=1}^{\infty} n_x$

n_x = *number of moles* of x-mer

M_x = molecular weight of x-mer

(n_x/N) = mole fraction of x-mer

Any analytical technique that determines the *number* of moles present in a sample of known weight, regardless of their size, will give the number-average molecular weight.

Rather than count the number of molecules of each size present in a sample, it is possible to define an average in terms of the *weights* of molecules present at each size level. This is the weight-average molecular weight $\bar{M}_w$. A good way to illustrate the differences between the two averages is to consider the analogy of a mixture of various-sized ball bearings rolling down a trough into which successively larger slots have been cut (Fig. 6.1). The smallest ball bearings fall into a compartment beneath the first slot, the next larger size into a compartment beneath the second slot, and so on, the last compartment holding the

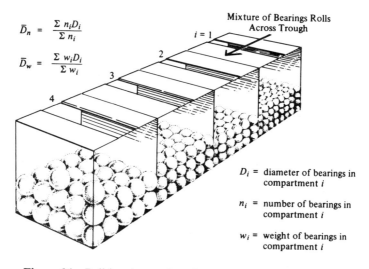

$$\bar{D}_n = \frac{\Sigma\, n_i D_i}{\Sigma\, n_i}$$

$$\bar{D}_w = \frac{\Sigma\, w_i D_i}{\Sigma\, w_i}$$

Mixture of Bearings Rolls Across Trough

D_i = diameter of bearings in compartment i

n_i = number of bearings in compartment i

w_i = weight of bearings in compartment i

Figure 6.1 Ball bearing analogy for average molecular weights.

largest ball bearings. (We are assuming that there are a few discrete bearing sizes and therefore that each compartment holds bearings of a single diameter.) The subscript i serves to identify the compartments in order of increasing slot size.

The number-average ball bearing diameter $\bar{D}_n$ is obtained by *counting* the numbers of ball bearings in each compartment:

$$\bar{D}_n = \frac{\Sigma n_i D_i}{\Sigma n_i}$$

where D_i = diameter of bearings in compartment i
n_i = number of bearings in compartment i

and is analogous to the number-average molecular weight. An equally valid average diameter, the *weight-average* $\bar{D}_w$ is obtained by *weighing* the ball bearings in each compartment

$$\bar{D}_w = \frac{\Sigma w_i D_i}{\Sigma w_i}$$

where w_i = weight of all ball bearings in the ith compartment
D_i = diameter of the bearings in compartment i

The analogous *weight-average* molecular weight $\bar{M}_w$ is then

$$\bar{M}_w = \frac{\Sigma w_x M_x}{\Sigma w_x} = \frac{w_1 M_1}{\Sigma w_x} + \frac{w_2 M_2}{\Sigma w_x} + \cdots = \Sigma \left(\frac{w_x}{W}\right) M_x = \frac{\Sigma n_x M_x^2}{\Sigma n_x M_x} \tag{6.3}$$

where w_x = weight of x-mer in sample = $n_x M_x$
w_x/W = weight fraction of x-mer in sample

Analytical procedures that, in effect, determine the weight of molecules at a given size level result in the weight-average molecular weight.

The number-average molecular weight is the *first moment* of the molecular weight distribution, analogous to the center of gravity (the first moment of the mass distribution) in mechanics. The weight-average molecular weight, the *second moment* of the distribution, corresponds to the radius of gyration in mechanics. Higher moments, for example, $\bar{M}_z$, the third moment, may be defined and are used occasionally.

It is sometimes more convenient to represent the size of polymer molecules in terms of the degree of polymerization or chain length x, rather than molecular weight. They are simply related by

$$\bar{M}_z = \frac{\Sigma n_x M_x^3}{\Sigma n_x M_x^2} \qquad M_x = mx \tag{6.4a}$$

$$\bar{M}_n = m\bar{x}_n \tag{6.4b}$$

$$\bar{M}_w = m\bar{x}_w \tag{6.4c}$$

where m = molecular weight of a repeating unit

$\bar{x}_n$ = number-average degree of polymerization or chain length

$\bar{x}_w$ = weight-average degree of polymerization or chain length

(These relations neglect the difference between the end groups on the molecule and the repeating units. This is perfectly justifiable in most cases since the end groups are an insignificant portion of a typical large polymer molecule.) In terms of chain lengths, Eqs. 6.2 and 6.3 become

$$\bar{x}_n = \frac{\Sigma n_x x}{\Sigma n_x} = \Sigma \left(\frac{n_x}{N}\right) x \tag{6.2a}$$

$$\bar{x}_w = \frac{\Sigma n_x x^2}{\Sigma n_x x} = \Sigma \left(\frac{w_x}{W}\right) x \tag{6.3a}$$

It may be shown that $\bar{M}_w \geq \bar{M}_n (\bar{x}_w \geq \bar{x}_n)$. The averages are equal only for a *monodisperse* (all molecules the same size) polymer. The ratio $\bar{M}_w/\bar{M}_n = \bar{x}_w/\bar{x}_n$ is known as the *polydispersity index* PI, and is a measure of the breadth of the molecular weight distribution. Values range from about 1.02 for carefully fractionated or anionic addition polymers, to over 50 for some commercial polymers.

Example 1. Measurements on two essentially monodisperse fractions of a linear polymer, A and B, yield molecular weights of 100 000 and 400 000, respectively. Mixture 1 is prepared from one part by weight of A and two parts by weight of B. Mixture 2 contains two parts by weight of A and one of B. Determine the weight- and number-average molecular weights of mixtures 1 and 2.

Solution. For mixture 1

$$n_A = \frac{1}{100\,000} = 1 \times 10^{-5}$$

$$n_B = \frac{2}{400\,000} = 0.5 \times 10^{-5}$$

$$\bar{M}_n = \frac{\Sigma n_i M_i}{\Sigma n_i} = \frac{(1 \times 10^{-5})(10^5) + (0.5 \times 10^{-5})(4 \times 10^5)}{1 \times 10^{-5} + 0.5 \times 10^{-5}} = 2.0 \times 10^5$$

$$\bar{M}_w = \Sigma \left(\frac{w_i}{W}\right) M_i = \tfrac{1}{3}(1 \times 10^5) + \tfrac{2}{3}(4 \times 10^5) = 3 \times 10^5$$

For mixture 2

$$n_A = \frac{2}{100\,000} = 2 \times 10^{-5}$$

$$n_B = \frac{1}{400\,000} = 0.25 \times 10^{-5}$$

$$\bar{M}_n = \frac{(2 \times 10^{-5})(10^5) + (0.25 \times 10^{-5})(4 \times 10^5)}{2 \times 10^{-5} + 0.25 \times 10^{-5}} = 1.33 \times 10^5$$

$$\bar{M}_w = \tfrac{2}{3}(1 \times 10^5) + \tfrac{1}{3}(4 \times 10^5) = 2 \times 10^5$$

Example 2. Two *polydisperse* samples are mixed in equal weights. Sample A has $\bar{M}_n = 100\,000$ and $\bar{M}_w = 200\,000$. Sample B has $\bar{M}_n = 200\,000$ and $\bar{M}_w = 400\,000$. What are $\bar{M}_n$ and $\bar{M}_w$ of the mixture?

Solution. First, let's derive general expressions for calculating the averages of mixtures:

$$\bar{M}_n \equiv \frac{W}{N} = \frac{\sum\limits_i W_i}{\sum\limits_i N_i}$$

where the subscript i refers to the various polydisperse components of the mixture. Now, for a given component,

$$N_i = \frac{W_i}{\bar{M}_{ni}}$$

$$\bar{M}_n(\text{mixture}) = \frac{\sum\limits_i W_i}{\sum\limits_i (W_i/\bar{M}_{ni})} \tag{6.5}$$

$$\bar{M}_w = \frac{\sum w_x M_x}{W} = \frac{\sum\limits_i \left(\sum\limits_x w_x M_x\right)_i}{\sum\limits_i W_i}$$

$$\bar{M}_{wi} = \frac{\left(\sum\limits_x w_x M_x\right)_i}{W_i}$$

$$\bar{M}_w(\text{mixture}) = \frac{\sum\limits_i (\bar{M}_{wi} W_i)}{\sum\limits_i W_i} = \sum\limits_i \left(\frac{W_i}{\sum\limits_i W_i}\right) \bar{M}_{wi} \tag{6.6}$$

where $(W_i/\sum W_i)$ is the weight fraction of component i in the mixture. In this case, let $W_A = 1$ g and $W_B = 1$ g. Then

$$\bar{M}_n = \frac{W_A + W_B}{(W_A/\bar{M}_{n_A}) + (W_B/\bar{M}_{n_B})} = \frac{1 + 1}{(1/10^5) + (1/2 \times 10^5)} = 133\,000$$

$$\bar{M}_w = \left(\frac{W_A}{W_A + W_B}\right)\bar{M}_{w_A} + \left(\frac{W_B}{W_A + W_B}\right)\bar{M}_{w_B}$$

$$= (\tfrac{1}{2})2 \times 10^5 + (\tfrac{1}{2})4 \times 10^5 = 300\,000$$

Note that even though the polydispersity index of each component of the mixture is 2.0, the PI of the mixture is greater, 2.25.

6.3 DETERMINATION OF AVERAGE MOLECULAR WEIGHTS

In this section we consider common techniques for measuring average molecular weights with the object of gaining a qualitative understanding of their operation and an appreciation for their advantages and limitations. Additional theoretical and experimental details are available elsewhere.[3]

In general, techniques for the determination of average molecular weights fall into two categories: *absolute* and *relative*. In the former, measured quantities are theoretically related to the average molecular weight; in the latter, a quantity is measured that is in some way related to molecular weight, but the exact relation must be established by calibration with one of the absolute methods.

A Absolute Methods

End-Group Analysis

If the chemical nature of the end groups on the polymer chains is known, standard analytical techniques may sometimes be employed to determine the concentration of the end groups and thereby of the polymer molecules, giving directly the number-average molecular weight. For example, in a linear polyester formed from a stoichiometrically equivalent batch, there are, on the average, one unreacted acid group and one unreacted –OH group per molecule. These groups may sometimes be analyzed by appopriate titration. If an addition polymer is known to terminate by disproportionation (see Chapter 10), there will be one double bond per every two molecules, which may be detectable quantitatively by halogenation or by infrared measurements. Other possibilities include the use of a radioactively tagged initiator which remains in the chain ends.

These methods have one drawback in addition to the necessity of knowing the nature of the end groups. As the molecular weight increases, the concentration of end groups (number per unit volume) decreases, and the measurement sensitivity drops off rapidly. For this reason, these methods are generally limited to the range $\bar{M}_n < 10\,000$.

Colligative Property Measurements

When a solute is added to a solvent, it causes a change in the activity and chemical potential (partial molal Gibbs free energy) of the solvent. The magnitude of the change is directly related to the solute concentration. For example, when pure water is boiled, the chemical potentials of the liquid and the vapor in equilibrium with it are the same. If, now, some salt is added to the water, it lowers the chemical potential of the liquid water. To reestablish equilibrium with the pure water vapor above the salt solution, the temperature of the system must be raised, causing a boiling-point elevation. In a similar fashion, the addition of ethylene glycol antifreeze depresses the freezing point of water.

Freezing-point depression, boiling-point elevation, and a third technique, osmotic pressure, may be used to determine the number of moles of polymer per unit volume of solution and thereby establish the number-average molecular weight. The following are the relevant thermodynamic equations for the three techniques:

$$\lim_{c \to 0} \frac{\Delta T_b}{c} = \frac{RT^2}{\rho \Delta H_v \bar{M}_n} \qquad \text{(boiling-point elevation)} \qquad (6.7)$$

$$\lim_{c \to 0} \frac{\Delta T_f}{c} = \frac{-RT^2}{\rho \Delta H_f \bar{M}_n} \qquad \text{(freezing-point depression)} \qquad (6.8)$$

$$\lim_{c \to 0} \frac{\pi}{c} = \frac{RT}{\bar{M}_n} \qquad \text{(osmotic pressure)} \qquad (6.9)$$

where $\quad c$ = solute (polymer) concentration, mass/volume
$\qquad T$ = absolute temperature
$\qquad R$ = gas constant
$\qquad \Delta H_v$ = solvent enthalpy of vaporization
$\qquad \Delta H_f$ = solvent enthalpy of fusion
$\qquad \Delta T_b$ = boiling-point elevation
$\qquad \Delta T_f$ = freezing-point depression
$\qquad \pi$ = osmotic pressure
$\qquad \rho$ = density _solvent_

It is important to note that the thermodynamic equations apply only for ideal solutions, a condition that can be reached only in the limit of infinite dilution of the solute.

Freezing-point depression and boiling-point elevation require precise measurements of very small temperature differences. Although they are used

$$\left(\frac{\pi}{c} \right)^{1/2} = \left(\frac{RT}{M_n} \right)^{1/2} \left\{ 1 + \frac{A_2 M_n c}{2} \right\}$$

second virial coefficient

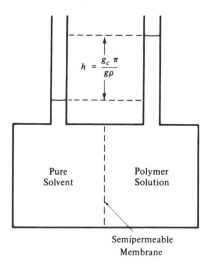

$$h = \frac{g_c\,\pi}{g\rho}$$

Pure
Solvent

Polymer
Solution

Semipermeable
Membrane

Figure 6.2   Schematic diagram of an osmometer.

occasionally, the difficulties involved have prevented their widespread application. Osmotic pressure, on the other hand, is the most common absolute method of determining $\bar{M}_n$. A schematic diagram of an osmometer is shown in Fig. 6.2. The solution and solvent chambers are separated by a "semipermeable" membrane, one that ideally allows passage of solvent molecules but not solute molecules. The solvent flows through the membrane to dilute the solution. This is a natural consequence of the tendency of the system to increase its entropy, which is accomplished by the dilution of the solution. This dilution continues until the tendency toward further dilution is counterbalanced by the increased pressure in the solution chamber. At this point, the chemical potential of the solvent is the same in both chambers and the pressure difference between the chambers is the osmotic pressure, π. By making measurements at several concentrations, plotting (π/c) vs. c and extrapolating to zero concentration, the number-average molecular weight is established through Eq. 6.9.

High-speed, automated membrane osmometers monitor the fluid volume in one of the chambers and externally *apply* the osmotic pressure to the solution chamber to *prevent* flow. Since no flow through the membrane is necessary, they can reduce measurement time from days to hours or even minutes.

One of the major difficulties with membrane osmometry is finding suitable semipermeable membranes. Ordinary cellophane or modifications of it are commonly used. Unfortunately, if the membrane is sufficiently tight to prevent the passage of low molecular weight chains, the rate of solvent passage is slow, and it takes longer to reach equilibrium. In practice, all membranes allow some of the low molecular weight polymer in a distribution to sneak through. Also, as the average molecular weight increases, π decreases, so the measurement precision also decreases. These factors usually limit the applicability of the technique to $50\,000 < \bar{M}_n < 1\,000\,000.$

A related colligative technique is vapor-pressure osmometry. Two thermistors are placed in a carefully thermostated chamber that contains a pure-solvent reservoir so that the atmosphere is saturated with solvent vapor. A drop of solvent is placed on one thermistor, and a drop of polymer solution on the other. Because of the solvent's lower chemical potential in the solution, solvent vapor condenses on the solution drop, giving up its heat of condensation, warming the solution drop relative to the pure solvent drop. In principle, the equilibrium ΔT is thermodynamically related to the molar solution concentration, thereby allowing calculation of $\bar{M}_n$. In practice, heat losses (mainly along the thermistor leads) require that the instrument be calibrated for precise results, really making it a relative technique. On a routine basis, commercial instruments are probably limited to maximum $\bar{M}_n$ values of $40\,000$–$50\,000$.[4]

The techniques discussed to this point establish the number of molecules present per unit volume of solution, regardless of their size or shape. Other methods measure quantities that are related to the average *mass* of the molecules in solution, thereby giving the weight-average molecular weight. One of the more common of these is *light scattering*, which is based on the fact that the intensity of light scattered by a polymer molecule is proportional, among other things, to the square of its mass. A light-scattering photometer measures the intensity of scattered light as a function of the scattering angle. Measurements are made at several concentrations. By a double extrapolation to zero angle and zero concentration (*Zimm plot*), and with a knowledge of the dependence of the solution refractive index vs. concentration, $\bar{M}_w$ is established. This technique also provides information on the solvent–solute interaction and on the configuration of the polymer molecules in solution, since the quantitative nature of the scattering depends also on the size of the particles. Light scattering is generally applicable over the range $10\,000 < \bar{M}_w < 10\,000\,000$. The weight-average molecular weight can also be obtained with an ultracentrifuge, which distributes the molecules according to their mass in a centrifugal force field.

B   Relative Methods

The molecular weight determination techniques discussed to this point allow direct calculation of the average molecular weight from experimentally measured quantities through known theoretical relations. Sometimes, however, these relations are not known, and although something is measured that is known to be related to molecular weight, one of the absolute methods above must be used to calibrate the technique.

A case in point is solution viscosity. It has long been known that relatively small amounts of dissolved polymer could cause tremendous increases in viscosity, and it is logical to assume that, other things being equal, larger molecules will impede flow more than smaller ones and give a higher solution viscosity. The solution viscosity will also depend on the solvent viscosity, temperature, solute concentration, the particular polymer and solvent, because the interactions between the polymer and solvent influence the conformation of the polymer molecules in solution, and entanglements between the polymer molecules. In functional form, we may write

$$\eta = \eta(\eta_s, T, \text{polymer, solvent, } c, \text{entanglements, } M)$$

where η = solution viscosity
η_s = solvent viscosity

We can get rid of the effect of solvent viscosity by calculating the fractional increase in viscosity caused by the solute, the *specific viscosity* η_{sp}:

$$\eta_{sp} = \frac{\eta - \eta_s}{\eta_s} = \frac{\eta}{\eta_s} - 1 = \eta_r - 1 \qquad (6.10)$$

where $\eta_r = \eta/\eta_s$ is known as the _relative viscosity_. Similarly, we can normalize for concentration by dividing the specific viscosity by concentration to get the *reduced viscosity* η_{red}:

$$\eta_{red} = \frac{\eta_{sp}}{c} = \frac{(\eta/\eta_s) - 1}{c} \qquad (6.11)$$

To get rid of the influence of entanglements on viscosity, we extrapolate the reduced viscosity to zero concentration to get the *intrinsic viscosity* $[\eta]$:

$$[\eta] = \lim_{c \to 0} \frac{(\eta/\eta_s) - 1}{c} \qquad (6.12)$$

The intrinsic viscosity, then, should be a function of the molecular weight of the polymer in solution, the polymer–solvent system, and the temperature. If measurements are made at constant temperature using a specified solvent for a particular polymer, it should be quantitatively related to the polymer's molecular weight.

Huggins proposed a relation between reduced viscosity and concentration for dilute polymer solutions ($\eta_r < 2$):

$$\frac{\eta_{sp}}{c} = [\eta] + k'[\eta]^2 c \qquad \text{(Huggins equation)} \qquad (6.13a)$$

Interestingly enough, k' turns out to be approximately equal to 0.4 for a variety of polymer–solvent systems, providing a convenient means of estimating dilute solution viscosity vs. concentration if the intrinsic viscosity is known. By expanding the natural logarithm in a power series, it may be shown that an equivalent form of the Huggins equation is

$$\eta_{inh} = \frac{\ln \eta_r}{c} = [\eta] + k''[\eta]^2 c \qquad Kraemy \qquad (6.13b)$$

where η_{inh} = *inherent viscosity*
$\quad\quad k'' = k' - 0.5$

From (6.13b) it is seen that an alternative definition of intrinsic viscosity is

$$[\eta] = \lim_{c \to 0} \eta_{inh} = \lim_{c \to 0} \left[\frac{\ln(\eta/\eta_s)}{c} \right] \qquad (6.12a)$$

Plots of the reduced and inherent viscosities are linear with concentration, at least at low concentrations, in accord with (6.13a and b), and have a common intercept, the intrinsic viscosity (Fig. 6.3). Exceptions occur with polyelectrolytes, where the degree of ionization and therefore the chemical nature of the polymer changes with concentration.

Note that the intrinsic viscosity has dimensions of reciprocal concentration. For some strange reason, concentrations were traditionally given in grams/deciliter (= 100 mL), although g/mL is now becoming more common. In fact, the relative, specific, reduced, intrinsic, and inherent viscosities are not true viscosities, and don't have dimensions of viscosity. More appropriate terminology has been proposed, but has not been widely adopted. Table 6.1 summarizes the various quantities defined and typical units.

Now that the intrinsic viscosity has been established, how is it related to molecular weight? Studies of the intrinsic viscosity of essentially monodisperse polymer fractions whose molecular weights have been established by one of the

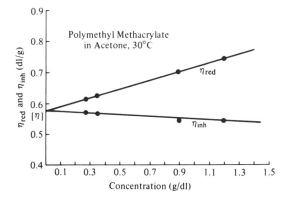

Figure 6.3   Determination of intrinsic viscosity (Example 3).

Table 6.1 Solution Viscosity Terminology[5]

Quantity	Common units	Common name	Recommended name
η	Centipoise	Solution viscosity	Solution viscosity
η_s	Centipoise	Solvent viscosity	Solvent viscosity
$\eta_r = \eta/\eta_s$	Dimensionless	Relative viscosity	Viscosity ratio
$\eta_{sp} = (\eta - \eta_s)/\eta_s = \eta_r - 1$	Dimensionless	Specific viscosity	—
$\eta_{red} = \eta_{sp}/c = \eta_r - 1/c$	Deciliters/gram	Reduced viscosity	Viscosity number
$\eta_{inh} = \ln \eta_r/c$	Deciliters/gram	Inherent viscosity	Logarithmic viscosity number
$[\eta] = \lim_{c \to 0} \eta_{red} = \lim_{c \to 0} \eta_{inh}$	Deciliters/gram	Intrinsic viscosity	Limiting viscosity number

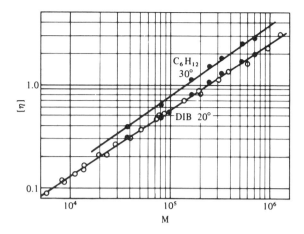

Figure 6.4 Intrinsic viscosity–molecular weight relations for polyisobutylene in cyclohexane at 30°C and diisobutylene at 20°C. Reprinted from P. J. Flory, *Principles of Polymer Chemistry.* Copyright 1953 by Cornell University. Used by permission of Cornell University Press.

absolute methods indicate a rather simple relation (Fig. 6.4):

$$[\eta]_x = K(M_x)^a \qquad (0.5 < a < 1) \tag{6.14}$$

where the subscript x refers to a monodisperse sample of a particular molecular weight. Equation 6.14 is known as the Mark–Houwink–Sakurada (MHS) relation.

What about an unfractionated, polydisperse sample? Experimentally, the measured intrinsic viscosity of a mixture of monodisperse fractions is a weight average:

$$[\eta] = \frac{\Sigma [\eta]_x w_x}{\Sigma w_x} = \Sigma \left(\frac{w_x}{W} \right) [\eta]_x \tag{6.15}$$

A *viscosity-average* molecular weight $\bar{M}_v$ is defined in terms of this measured intrinsic viscosity as

$$\bar{M}_v = \left(\frac{[\eta]}{K} \right)^{1/a} = \left\{ \frac{\Sigma M_x^a w_x}{W} \right\}^{1/a} = \left\{ \frac{\Sigma M_x^a n_x M_x}{\Sigma n_x M_x} \right\}^{1/a} = \left\{ \frac{\Sigma n_x M_x^{(1+a)}}{\Sigma n_x M_x} \right\}^{1/a} \tag{6.16}$$

With $0.5 < a < 1.0$, $\bar{M}_n < \bar{M}_v < \bar{M}_w$, but $\bar{M}_v$ is closer to $\bar{M}_w$ than $\bar{M}_n$. If the molecular weight distribution of a series of samples does not differ too much, that is, if the ratios of the various averages remain nearly the same, approximate equations of the type $[\eta] = K'(\bar{M}_w)^{a'}$ may be applicable.

Example 3. The following data were obtained for a sample of polymethyl methacrylate in acetone at 30°C:

η_r	c (g/100 mL)
1.170	0.275
1.215	0.344
1.629	0.896
1.892	1.199

For polymethyl methacrylate in acetone at 30°C, $[\eta] = 5.83 \times 10^{-5}(\bar{M}_v)^{0.72}$. Determine $[\eta]$ and $\bar{M}_v$ for the sample and k', the constant in the Huggins equation.

Solution. For the data above, calculations give

η_{sp}	η_{red} (dL/g)	$\ln \eta_r$	η_{inh} (dL/g)
0.170	0.618	0.157	0.571
0.215	0.625	0.195	0.567
0.629	0.702	0.488	0.545
0.892	0.744	0.638	0.532

In Fig. 6.3, η_{red} and η_{inh} are plotted against concentration and extrapolated to a common intercept at zero concentration, the intrinsic viscosity, $[\eta] = 0.580$ dL/g:

$$\bar{M}_v = \left(\frac{[\eta]}{5.83 \times 10^{-5}}\right)^{1/0.72} = \left(\frac{0.580}{5.83 \times 10^{-5}}\right)^{1.39} = 357\,000$$

The regression slope of the upper line is 0.137. From (6.13a) $k' = $ slope/$[\eta]^2 = 0.408$ (dimensionless).

Example 4. By assuming that the fractions in Example 1 are polymethyl methacrylate, calculate $\bar{M}_v$ for mixtures 1 and 2 in acetone at 30°C and compare with $\bar{M}_n$ and $\bar{M}_w$ (a is given in Example 3).

Solution. From (6.16), for mixture 1

$$\bar{M}_v = \left\{\sum \left(\frac{w_x}{W}\right) M_x^a\right\}^{1/a} = \{\tfrac{1}{3}(1 \times 10^5)^{0.72}$$

$$+ \tfrac{2}{3}(4 \times 10^5)^{0.72}\}^{1/0.72} = 288\,000 \qquad \begin{array}{l} \bar{M}_n = 200\,000 \\ \bar{M}_w = 300\,000 \end{array}$$

for mixture 2

$$\bar{M}_v = \{\tfrac{2}{3}(1 \times 10^5)^{0.72}$$
$$+ \tfrac{1}{3}(4 \times 10^5)^{0.72}\}^{1/0.72} = 187\,000 \qquad \begin{array}{l} \bar{M}_n = 133\,000 \\ \bar{M}_w = 200\,000 \end{array}$$

Viscosities for molecular weight determination are usually measured in glass capillary viscometers, in which the solution flows through a capillary under its own head. Two common types, the Ostwald and Ubbelohde, are sketched in Fig. 6.5. (Since polymer solutions are non-Newtonian, intrinsic viscosity must be defined, strictly speaking, in terms of the zero-shear or lower-Newtonian viscosity (Chapter XV). This is rarely a problem, because the low shear rates in the usual glassware viscometers give just that. Occasionally, however, extrapolation to zero-shear conditions is required.)

Flow times t are related to the viscosity of the solution by an equation of the form

$$\frac{\eta}{\rho} = v = at + \frac{b}{t} = \alpha \rho \left(t - \frac{\beta}{\alpha t} \right) \quad (6.17)$$

where a and b are instrument constants, ρ is the solution density, and v is the kinematic viscosity. The last term, the kinetic energy correction, is generally negligible for flow times of over a minute, and since the densities of the dilute polymer solutions differ little from that of the solvent:

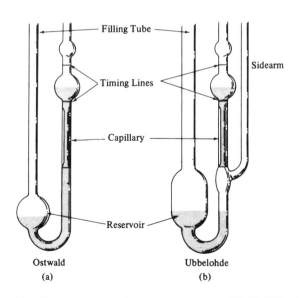

Figure 6.5 Dilute-solution viscometers: (a) Ostwald; (b) Ubbelohde.

$$\frac{\eta}{\eta_s} \approx \frac{t}{t_s} \tag{6.18}$$

The Ubbelohde viscometer has the distinct advantage that the driving fluid head is independent of the amount of solution in it; hence, dilution can be carried out right in the instrument.

The equipment necessary for intrinsic viscosity determination is inexpensive, and the measurements straightforward and rapid. The MHS constants K and a in (6.14) and (6.16) are extensively tabulated for a wide variety of polymer-solvent systems and temperatures.[6]

6.4   MOLECULAR WEIGHT DISTRIBUTIONS

A typical synthetic polymer might consist of a mixture of molecules with degrees of polymerization x ranging from one to perhaps millions. The complete molecular weight distribution specifies the mole (number) or mass (weight) fraction of molecules at each size level in a sample. (Actually, moles or masses could be specified, but they vary linearly with sample size, i.e., are extensive quantities, while the fractions are intensive, independent of sample size, and are therefore preferable.)

Distributions are often presented in the form of a plot of mole (n_x/N) or mass (w_x/W) fraction of x-mer vs. either x or M_x. Since x and M_x differ by a constant factor, the molecular weight of the repeating unit m (Eq. 6.4a), it makes little difference which is used. Because x can assume only integral values, a true distribution must consist of a series of spikes, one at each integral value of x, or separated by m molecular weight units if plotted against M_x. The height of the spike represents the mole or mass fraction of that particular x-mer. No analytical technique is capable of resolving the individual x-mers, so distributions are drawn (and represented mathematically) as continuous curves, the locus of the spike tops. This is sketched in Fig. 6.6.

The averages may be calculated from either the mole- or mass-fraction distributions. For continuous distributions, the summations in Eqs. 6.2 and 6.3 must be replaced by the corresponding integrals (the subscript x has been dropped for clarity and because of the continuous nature of the distributions).

From the mole- (number-) fraction distribution (n/N),

$$\overline{M}_n = \text{first moment} = \frac{\int_0^\infty nM\,dM}{\int_0^\infty n\,dM} = \frac{\int_0^\infty (n/N)M\,dM}{\int_0^\infty (n/N)\,dM} \tag{6.19}$$

Since $\overline{M}_n = m\bar{x}_n$ and $M = mx$

$$\bar{x}_n = \int_0^\infty \left(\frac{n}{N}\right) x\,dx \tag{6.20}$$

$$N(x) = \frac{n_x}{N}$$

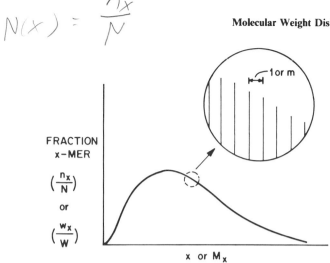

Figure 6.6 Molecular weight distributions illustrating the actual discrete distribution and the usual continuous approximation to it.

Note that

$$\int_0^\infty \left(\frac{n}{N}\right) dx = 1 \tag{6.21}$$

that is, the mole fractions must sum to 1.

$$\bar{M}_w = \text{2nd moment} = \frac{\int_0^\infty nM^2\,dM}{\int_0^\infty nM\,dM} = \frac{\int_0^\infty (n/N)M^2\,dM}{\int_0^\infty (n/N)M\,dM} \tag{6.22}$$

Since $\bar{M}_w = m\bar{x}_w$,

$$\bar{x}_w = \frac{\int_0^\infty (n/N)x^2\,dx}{\int_0^\infty (n/N)x\,dx} = \frac{1}{\bar{x}_n}\int_0^\infty \left(\frac{n}{N}\right)x^2\,dx \tag{6.23}$$

From the mass- (weight-) fraction distribution (w/W):

$$\bar{M}_n = \frac{\int_0^\infty nM\,dM}{\int_0^\infty n\,dM} = \frac{\int_0^\infty w\,dM}{\int_0^\infty (w/M)\,dM} = \frac{\int_0^\infty (w/W)\,dM}{\int_0^\infty (1/M)(w/W)\,dM} \tag{6.24}$$

$$\bar{x}_n = \frac{\int_0^\infty (w/W)\,dx}{\int_0^\infty (1/x)(w/W)\,dx} \tag{6.25}$$

$$\bar{M}_w = \frac{\int_0^\infty nM^2\,dM}{\int_0^\infty nM\,dM} = \frac{\int_0^\infty (nM)M\,dM}{\int_0^\infty (nM)\,dM} = \frac{\int_0^\infty wM\,dM}{\int_0^\infty w\,dM} = \frac{\int_0^\infty (w/W)M\,dM}{\int_0^\infty (w/W)\,dM} \tag{6.26}$$

$$\bar{x}_w = \int_0^\infty \left(\frac{w}{W} \right) x \, dx \tag{6.27}$$

Note that

$$\int_0^\infty \left(\frac{w}{W} \right) dx = 1 \tag{6.28}$$

that is, the mass fractions must sum to 1.

Given the mole- (number-) fraction distribution, the mass- (weight-) fraction distribution may be calculated, and vice versa. Since $w = nM$ and, by definition of $\bar{M}_n$ (Eq. 6.2), $W = N\bar{M}_n$,

$$\frac{w}{W} = \frac{M}{\bar{M}_n} \left(\frac{n}{N} \right) = \frac{x}{\bar{x}_n} \left(\frac{n}{N} \right) \tag{6.29}$$

Example 5. One analytic representation of a distribution of chain lengths that finds some practical application (as we will see later) is

$$\frac{n}{N} = \frac{1}{C} e^{-(x/C)}$$

where C is a constant.

    a. Determine $\bar{x}_n$ for this distribution.

    b. Determine $\bar{x}_w$ and the polydispersity index $\bar{x}_w/\bar{x}_n$ for this distribution.

    c. Obtain the expression for the weight-fraction distribution, (w/W).

    d. Sketch the number- and weight-fraction distributions in the form $(n/N)\bar{x}_n$ and $(w/W)\bar{x}_n$ vs. $x/\bar{x}_n$.

$$\textit{Solution.} \quad \textbf{\textit{a.}} \quad \bar{x}_n = \int_0^\infty \left(\frac{n}{N} \right) x \, dx = \frac{1}{C} \int_0^\infty x e^{-x/C} \, dx$$

Fortunately for most of us, the above definite integral is tabulated in standard references. The result is

$$\bar{x}_n = C$$

b.

$$\bar{x}_w = \frac{1}{\bar{x}_n} \int_0^\infty \left(\frac{n}{N} \right) x^2 \, dx = \frac{1}{\bar{x}_n^2} \int_0^\infty x^2 e^{-x/\bar{x}_n} \, dx$$

This integral is also tabulated, and gives the simple result

$$\bar{x}_w = 2\bar{x}_n$$

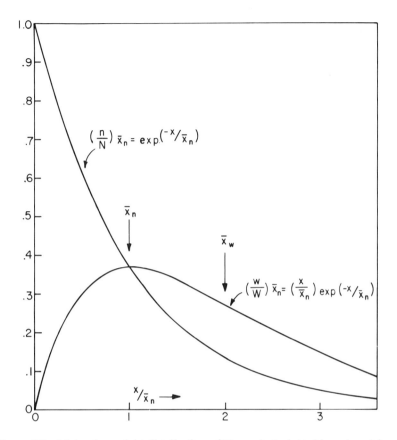

Figure 6.7 Molecular weight distribution of Example 5 plotted in reduced form.

$$\left(\frac{w}{W}\right)(x) = \frac{x^2}{333\frac{1}{3}}\left(.002(1-.001x)\right)$$
$$= 6 \times 10^{-6} x^2 - 6 \times 10^{-9} x^3$$

that is, $PI = \bar{x}_w/\bar{x}_n = 2$ for this distribution.

c.

$$\frac{w}{W} = \left(\frac{n}{N}\right)\frac{x}{\bar{x}_n} = \frac{x}{\bar{x}_n^2}e^{-x/\bar{x}_n}$$

d. The distributions are shown in Fig. 6.7. Note that while the number of molecules decreases exponentially with x, the weight of a molecule increases linearly with x. The weight fraction therefore goes through a maximum in this case.

This distribution is derived for a particular kind of polymerization reaction in Section 10.7.

6.5   SIZE-EXCLUSION CHROMATOGRAPHY (SEC)[7,*]

Until fairly recently, the approximate molecular weight distribution of a polymer could be determined only by laborious fractionation of the sample according to molecular weight followed by determination of the molecular weights of the individual fractions with one of the techniques previously discussed. The bases for these fractionation techniques are discussed in Chapter VII. Suffice it to say that they tend to be difficult and time-consuming, and so, in general, are avoided whenever possible.

Now, however, size-exclusion chromatography has been firmly established as a means of rapidly determining molecular weight averages and distributions. SEC makes use of a column, or series of columns, packed with particles of a porous substrate. The term gel in gel-permeation chromatography refers to a crosslinked polymer that is swollen by the solvent used. This is perhaps the most common type of substrate, but others, for example, porous glass beads, are used. The column is maintained at a constant temperature, and solvent is passed through it at a constant rate. At the start of a run, a small amount of polymer solution is injected just ahead of the column. The solvent flow carries the polymer through the column. The smaller molecules in the sample have easy access to the substrate pores and diffuse in and out of the pores, following a circuitous route as they progress through the column. The large molecules simply can't fit into the pores, and are swept more or less directly through the interstices in the packing. Thus, a separation is obtained, the largest molecules being washed through the column first, followed by successively smaller ones.

A concentration-sensitive detector is placed at the outlet of the column. The most common detector is a differential refractometer, which measures the difference in refractive index between pure solvent and the polymer solution leaving the column, a sensitive measure of polymer concentration. UV or IR detectors can also be used if the polymer has some group that absorbs the radiation. Regardless of the type of detector, it is essential that it measure some quantity Q that is proportional only to the *mass* concentration of polymer at the column outlet (Q must be independent of M):

$$Q = kc \tag{6.30}$$

where Q = detector readout
k = proportionality constant
c = mass concentration of polymer (g/cm^3)

Thus, a SEC curve consists of a plot of Q (usually in arbitrary scale divisions) vs. v, the volume of solvent that has passed through the detector since sample injection, called *elution* or *retention volume*. The chart may be "blipped" every $5\ cm^3$ of volume.

---

* This technique is also commonly known as gel-permeation chromatography (GPC).

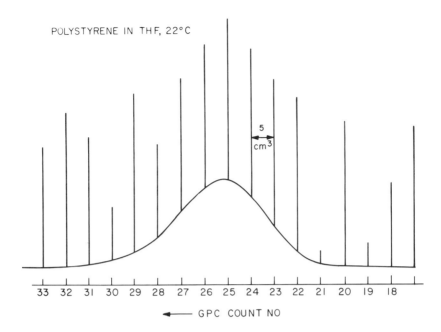

Figure 6.8 SEC curve for polystyrene. Tetrahydrofuran solvent, 22°C.

A SEC curve is shown in Fig. 6.8. Such a curve is often immediately useful for quality-control purposes. For example, if a curve is available for a material that performs acceptably in a particular application, the SEC curves for subsequent batches may be checked against it, revealing qualitatively variations in molecular weight distribution from the standard. This may be enough to take appropriate corrective action. SEC is also useful in detecting low molecular weight additives such as plasticizers and stabilizers. These materials show up as peaks at the low-M (high v) end of the spectrum.

Conventional SEC is a relative method. To provide quantitative results, the relation between M and v must be established by calibration with monodisperse polymer standards. A calibration curve is shown in Fig. 6.9. Typically, when the data are plotted in the form $\log M$ vs. v, the curve is linear over much of the range, but sometimes turns up sharply at low v (high M).

Example 6. Discuss the physical significance of an upturn in the calibration curve at low v.

Solution. The point at which the curve begins to shoot up is set by the largest pore size in the column. All molecules that are too large to fit in those pores will pass through the column at the same rate, giving the infinite slope; that is, the technique cannot discriminate between molecules above that size. If the samples to be analyzed contain significant material above that size, it would be advisable

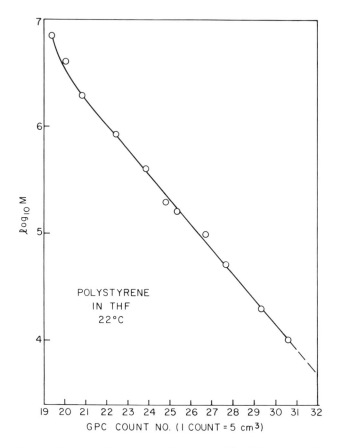

Figure 6.9 SEC calibration. Narrow-polydispersity ($\bar{M}_w/\bar{M}_n$) < 1.1 polystyrene samples in tetrahydrofuran at 22°C.

to tack on another column (or replace the existing one) with a substrate containing larger pores to extend the calibration.

The calibration curve, strictly speaking, applies only to the particular polymer, solvent, temperature, flow rate, and column for which it was established. Change any one, and the calibration is no longer valid. Most SEC calibrations are obtained with polystyrene, because the necessary monodisperse standards are readily available at reasonable cost. What do you do when you want to analyze another polymer? One approach was suggested by Grubisic et al.[8] When they plotted $\log([\eta]M)$ (the product $[\eta]M$ is proportional to the hydrodynamic volume of the molecules) vs. v, they obtained a single, universal calibration curve for a variety of different polymers. Thus, if you measure intrinsic viscosity along with SEC elution volume for your polystyrene standards (or calculate $[\eta]$ for them with (6.14) and literature values for K and a), you can plot the universal calibration curve for your column.

To back out a calibration curve for a different polymer, you would need to know K and a under the new conditions:

$$[\eta]_0 M_0 = [\eta] M = KM^{(a+1)} \tag{6.31}$$

$$M = \left(\frac{[\eta]_0 M_0}{K}\right)^{1/(a+1)} \tag{6.32}$$

where the subscript 0 above refers to the calibration conditions, i.e., values from the universal calibration.

Balke et al.[9] have proposed a calibration method that requires only a single *polydisperse* sample of known $\bar{M}_n$ and $\bar{M}_w$. The method *assumes* that the relation between $\log M$ and v is linear, and therefore it can be characterized by two parameters, a slope and an intercept. Given the SEC curve and the calibration, $\bar{M}_n$ and $\bar{M}_w$ can be calculated (using techniques outlined below). With the SEC data on the standard, a computer program adjusts the two calibration parameters until the known $\bar{M}_n$ and $\bar{M}_w$ are obtained, in effect using the two known average molecular weights to solve for the two unknown calibration parameters. If the calibration curve is not linear (and you would have no way of knowing), serious errors can result.

Nowadays, most SEC packages include the software necessary to calculate molecular weight averages and the complete distribution from the data and the calibration, so one need only push the appropriate buttons. Nevertheless, the equations are outlined below to provide an understanding of what the software is doing and what its limitations might be.

Molecular weight averages may be approximated directly from the SEC curve and calibration by breaking the curve into arbitrary volume increments Δv. Usually, Δv is taken as $5\,\mathrm{cm}^3$, the volume between "counts" on a preautomatic machine, but smaller Δv's improve accuracy. The number of moles of polymer in a volume increment Δv is n_i:

$$n_i = \frac{c_i \Delta v}{M_i} \tag{6.33}$$

where c_i = polymer mass concentration in the ith volume increment

       M_i = molecular weight of polymer in the ith volume increment (assumed essentially constant over small Δv)

Because $c_i = Q_i/k$ (6.30),

$$n_i = \frac{Q_i \Delta v}{kM_i} \tag{6.34}$$

If the volume increment Δv is constant, inserting (6.34) into (6.2) and (6.3) gives

$$\bar{M}_n = \frac{\sum\limits_i n_i M_i}{\sum\limits_i n_i} = \frac{\sum\limits_i Q_i}{\sum\limits_i (Q_i/M_i)} \tag{6.35}$$

$$\bar{M}_w = \frac{\sum\limits_i n_i M_i^2}{\sum\limits_i n_i M_i} = \frac{\sum\limits_i Q_i M_i}{\sum\limits_i Q_i} \tag{6.36}$$

Note that the proportionality constant k between concentration and SEC readout cancels out, and therefore need not be known. Q_i is normally read from the SEC curve as the *distance above the baseline* (in any convenient units) at each count. The baseline presumably represents the detector output with pure solvent. Establishing a good baseline is not always a trivial procedure, and the results can be quite sensitive to its location. M_i is read from the calibration curve at the v corresponding to the particular count. For greater accuracy, if necessary, the integral analogs of (6.35) and (6.36) may be used, with the integrals evaluated numerically:

$$\bar{M}_n = \frac{\int_0^\infty Q \, dv}{\int_0^\infty (Q/M) \, dv} \tag{6.37}$$

$$\bar{M}_w = \frac{\int_0^\infty QM \, dv}{\int_0^\infty Q \, dv} \tag{6.38}$$

In most cases, the averages as calculated above and the qualitative information about the distribution provided by the SEC curve are all that are necessary. However, the technique is capable of providing the true distributions, if desired. The necessary calculations are outlined below:

$$\text{moles of polymer in volume increment } dv = \frac{c \, dv}{M} = \frac{Q \, dv \, (\text{g/cm}^3)(\text{cm}^3)}{k \, M \quad (\text{g/g mol})} \tag{6.39}$$

$$\text{moles of polymer/unit of } M = \frac{c}{M} \frac{dv}{dM} = \frac{Q \, dv}{kM \, dM} \tag{6.40}$$

$$n = \text{moles of polymer in a molecular weight interval } m = \frac{mc \, dv}{M \, dM} = \frac{Qm \, dv}{kM \, dM}$$

$$= \frac{mc}{2.303 \, M^2} \left(\frac{dv}{d \log M} \right) = \frac{Qm}{2.303 \, kM^2} \left(\frac{dv}{d \log M} \right) \tag{6.41}$$

Note:

$$\frac{dv}{d \log M} = \frac{1}{\text{slope}} \quad \text{of calibration}$$

The preceding calculations assign all the moles of polymer in a range m of the continuous GPC curve to a single spike (see Section 6.4) to calculate the correct height of the distribution:

$$N = \int_0^\infty \frac{c}{M} dv = \frac{1}{k} \int_0^\infty \frac{Q}{M} dv \tag{6.42}$$

$$\frac{n}{N} = \frac{(Qm/2.303\,M^2)(dv/d\log M)}{\int_0^\infty (Q/M)\,dv} \tag{6.43}$$

w = weight of polymer in a mol wt interval $m = nM$

$$= \frac{mc}{2.303M}\left(\frac{dv}{d\log M}\right) = \frac{Qm}{2.303\,kM}\left(\frac{dv}{d\log M}\right) \tag{6.44}$$

$$W = \text{total sample wt} = \int_0^\infty c\,dv = \frac{1}{k}\int_0^\infty Q\,dv$$

$$= \frac{\text{area under GPC curve}}{k} \tag{6.45}$$

$$\frac{w}{W} = \frac{(Qm/2.303\,M)(dv/d\log M)}{\int_0^\infty Q\,dv} \tag{6.46}$$

Again, k cancels out. Once the distributions are known, the exact averages may be calculated by the techniques in Section 6.4. For example, insertion of (6.43) into (6.19) and (6.22) gives (6.37) and (6.38).

A multiangle laser light scattering photometer (DAWN®) is now available which can be added to the usual concentration detector. This combination performs an on-line light scattering determination of the M of the polymer in the eluant stream. This makes SEC an absolute technique, eliminating the need for calibration. Similarly, a differential viscometer is commercially available which performs an on-line intrinsic viscosity measurement on the column effluent.[10] Recall that the ordinate of the universal calibration is $[\eta]M$. This device requires simple calibration (in effect locating the universal calibration curve for the particular column, temperature, solvent, etc.) to provide absolute molecular weight distributions and also provides information on the hydrodynamic volume of the molecules in solution and therefore on branching, because a branched molecule in solution has a smaller hydrodynamic volume than a linear molecule of the same molecular weight.

REFERENCES

1. Martin, J. R., J. F. Johnson, and A. R. Cooper, *J. Macromol. Sci. Rev.* **C8**(1), 57 (1972).

2. Nunes, R. W., J. R. Martin, and J. F. Johnson, *Polym. Eng. Sci.* **22**(4), 193 (1982).

3. Collins, E. A., et al., *Experiments in Polymer Science*, Wiley, New York, 1973.

4. Burge, D. E., **9**(7), 41 (1977).

5. Billmeyer, F., *Textbook of Polymer Science*, 2d ed., Wiley, New York, 1971, Chapter 3.

6. Kurata, M., and Y. Tsunashima, Viscosity–molecular weight relationships and unperturbed dimensions of linear chain molecules, Chapter VII/1 in *Polymer Handbook*, 3d ed., J. Brandrup and E. H. Immergut (eds.), Wiley, New York, 1975.

7. Provder, T. (ed.), *Size Exclusion Chromatography: Methodology and Characterization of Polymers and Related Materials*, ACS Symposium Series No. 245, American Chemical Society, Washington, DC, 1984.

8. Grubisic, Z., et al., *J. Polym. Sci.* **B-5**, 753 (1967).

9. Balke, S. T., et al., *Ind. Eng. Chem., Prod. Res. Dev.* **8**, 54 (1969).

10. Yau, W. W., and S. W. Rementer, *J. Liquid Chromatogr.* **13** (4), 626 (1990); Wang, P. J., and B. S. Glassbrenner, *ibid*, **11**(16), 3321 (1988).

PROBLEMS

1. The weights of the (extremely) offensive linemen of the MIT (Monongahela Institute of Technology) football team are:

Cussler	split end	180 lb
Miller	left tackle	270
Westerberg	left guard	256
Brenner	center	285
Anderson	right guard	260
Jain	right tackle	305
Prieve	tight end	250

Calculate the number- and weight-average line weights.

2. Consider a copolymer with repeating units A and B in which the mole fraction of A is y_A. The molecular weights of the repeating units are m_A and m_B. Obtain the analog of Eq. 6.4a for the copolymer, in which x represents the total number of repeating units of both kinds.

3. A *discrete* distribution of chain lengths has the same number of moles of each species over the range $1 \leq x \leq a$ and nothing outside that range. Obtain expressions for $\bar{x}_n$, $\bar{x}_w$ and the polydispersity index for this distribution. *Hints*: Start with Eqs. 6.2a and 6.3a. You may have to look up some series sums.

4. Now consider the *continuous* distribution of chain lengths in which the number of moles of each species is constant over the range $0 \leq x \leq a$ and zero outside that range.

   a. Obtain expressions for $\bar{x}_n$, $\bar{x}_w$, and the polydispersity index.

   b. What is the constant value of (n/N) between 0 and a?

c. Obtain an expression for the weight-fraction distribution, $(w/W)(x)$ and sketch it.

d. Show that your answers to problem 3 above agree with these when $a \gg 1$.

5. Calculate $\overline{M}_n$, $\overline{M}_w$, and the polydispersity index for the polystyrene sample in Fig. 6.8 using Eqs. 6.35 and 6.36 and the calibration curve in Fig. 6.9.

6. Suppose the SEC data in Fig. 6.8 were obtained from a polymethyl methacrylate (PMMA) sample using acetone as a solvent at 30°C. Using the polystyrene calibration curve, Fig. 6.9, obtain $\overline{M}_n$, $\overline{M}_w$ and the polydispersity index for the PMMA sample. Necessary data are as follows:

	K (mL/g)	a
PS in THF, 22°C	11×10^{-3}	0.725
PMMA in acetone, 30°C	5.83×10^{-3}	0.720

7. Several commercial polymers are now produced with bimodal molecular weight distributions to achieve certain processing and product properties. They are sometimes made by mixing two samples of the same polymer with high and low average molecular weights. Such a material gives a SEC curve with two separate and distinct peaks. How would you determine the percentages of the high and low molecular weight components in the mixture?

8. Crud Chemicals is using GPC to analyze polystyrene samples plasticized with a low molecular weight mixed hydrocarbon oil, $\overline{M}_n = 400$. Their calibration curve is identical to the one in Fig. 6.9 and from past experience they know the unplasticized polymer gives a GPC curve just like Fig. 6.8.

a. Sketch what you think a GPC curve for a plasticized material would look like.

b. If you know that $k_{oil} = 0.70\,k_{PS}$ (in Eq. 6.30), could you use the GPC curve to determine the weight fraction plasticizer in the sample? If so, clearly describe how. If not, what additional information would you need?

9. Given the hypothetical distribution of chain lengths

$$\frac{n}{N} = C(1 - 10^{-3}x) \qquad 0 \le x \le 10^3$$

a. Determine the value of the constant C.

b. Calculate $\bar{x}_n$, $\bar{x}_w$, and the polydispersity index for this distribution.

c. Obtain the expression for the weight-fraction distribution, $(w/W)(x)$, and sketch it.

10. Some authorities prefer to represent molecular weight distributions in cumulative (integral) form rather than the differential form used here. Cumulative distributions are the number or weight fraction of material in the sample with chain length $\leq x$ (or molecular weight $\leq M$).

For the distribution in Example 5, obtain expressions for the cumulative mole and weight distributions in terms of $(x/\bar{x}_n)$ and sketch them.

11. Consider the continuous distribution of chain lengths in Example 5. It is shown there that $\bar{x}_w/\bar{x}_n = 2.0$ for this distribution. Obtain an expression for $\bar{x}_v/\bar{x}_n$ for this distribution in terms of the MHS constant a in Eqs. 6.14 and 6.16. Given that $0.5 \leq a \leq 1$, what is the possible range of $\bar{x}_v/\bar{x}_n$ for this distribution?

   Hints: 1. Express Eq. 6.16 in integral form.

   2. All necessary definite integrals are tabulated in standard references.

   3. Express your result in terms of the gamma (Γ) function.

12. A polystyrene has the distribution of chain lengths in Example 5. A light-scattering measurement gives a molecular weight of 208 000. Calculate the *mass* of material with $x = 100$ in a 100-g sample of this polystyrene.

13. The people at Crud Chemicals eagerly await the first SEC analysis of the polystyrene from their new polymerization process. There is a hushed silence as the first sample is injected into the SEC. Suddenly, the chart pen begins to move, and to everyone's amazement, the following SEC curve emerges:

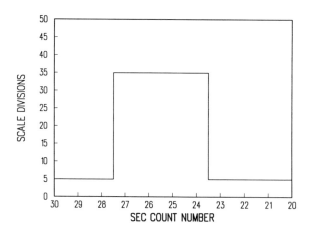

Not knowing what to make of this, they call you in as a consultant. Their SEC operates at 22°C with THF solvent. Its calibration curve is identical to Fig. 6.9.

a. Sketch the number-fraction molecular weight distribution for their sample, including quantitative molecular weight extremes.

b. Calculate $\bar{M}_n$ and $\bar{M}_w$ for their sample. You can do this approximately using the discrete equations but really impress them and do it exactly by treating the continuous distribution.

14. Consider the polymer mixture formed in Example 2. Each component of the mixture follows the distribution of chain lengths given in Example 5. For simplicity, the molecular weight of a repeating unit may be taken as $m = 100$. What are the mole and weight fractions of material with $x = 10$ (n_{10}/N) and (w_{10}/W) in the mixture?

15. A polymer sample consists of a mixture of N components, each of which has the distribution of chain lengths given in Example 5. The number-average chain lengths of the individual components of the mixture are $\bar{x}_{ni}$, and the weight fractions of each component in the mixture are w_i (a slight change in notation here for the sake of simplicity). Obtain expressions for the number- and weight-fraction distributions of chain lengths in the mixture, $(n/N)_{mix}(x)$ and $(w/W)_{mix}(x)$.

CHAPTER VII

Polymer Solubility and Solutions

7.1 INTRODUCTION

The thermodynamics and statistics of polymer solutions is an interesting and important branch of physical chemistry, and is the subject of many good books and large sections of books in itself. It is far beyond the scope of this chapter to attempt to cover the subject in detail. Instead, we will concentrate on topics of practical interest and try to indicate, at least qualitatively, their fundamental bases. Three factors are of general interest:

1. What solvents will dissolve what polymers?
2. How do the interactions between polymer and solvent influence the properties of the solution?
3. To what applications do the interesting properties of polymer solutions lead?

7.2 TYPICAL PHASE BEHAVIOR IN POLYMER–SOLVENT SYSTEMS

Figure 7.1 shows schematically a phase diagram for a typical polymer–solvent system, plotting temperature vs. the fraction polymer in the system. At low temperatures, a two-phase system is formed. The dotted tie lines connect the compositions of phases in equilibrium, a solvent-rich (dilute-solution) phase on the left and a polymer-rich (swollen-polymer or gel) phase on the right. As the temperature is raised, the compositions of the phases become more nearly alike, until at the upper critical solution temperature (UCST) they are identical. Above the UCST, the system forms homogeneous (single-phase) solutions across the entire composition range. The location of the phase boundary depends on the

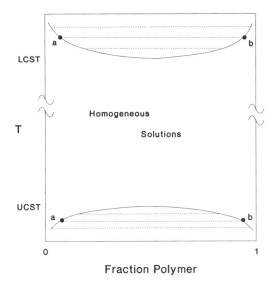

Figure 7.1 Schematic phase diagram for polymer–solvent system: (*a*) dilute solution phase; (*b*) swollen polymer or "gel" phase. UCST, upper critical solution temperature; LCST, lower critical solution temperature.

molecular weight of the polymer and the interaction between the polymer and solvent.

In recent years, a number of systems have been examined that also exhibit a lower critical solution temperature (LCST), as shown in Fig. 7.1. (One might question the nomenclature that puts the LCST *above* the UCST, but that's the way it is). LCSTs are more difficult to observe experimentally because they often lie well above the normal boiling points of the solvents.

When we talk about a polymer being soluble in particular solvent, we generally mean that the system lies between its LCST and UCST; that is, it forms homogeneous solutions over the entire composition range. Keep in mind, however, that homogeneous solutions can still be formed toward the extremes of the composition range below the UCST and above the LCST.

7.3 GENERAL RULES FOR POLYMER SOLUBILITY

Let's begin by listing some general qualitative observations on the dissolution of polymers:

1. Like dissolves like; that is, polar solvents will tend to dissolve polar polymers and nonpolar solvents will tend to dissolve nonpolar polymers. Chemical similarity of polymer and solvent is a fair indication of solubility;

for example, polyvinyl alcohol, $\left[\begin{smallmatrix} H & H \\ | & | \\ -C-C- \\ | & | \\ H & O-H \end{smallmatrix}\right]_x$, will dissolve in water, H–O–H,

and polystyrene, $\left[\begin{smallmatrix} H & H \\ | & | \\ -C-C- \\ | & | \\ H & \phi \end{smallmatrix}\right]_x$, in toluene, ϕ-CH$_3$, but toluene won't dissolve

polyvinyl alcohol and water won't dissolve polystyrene (which is good news for those of us who drink coffee out of foamed polystyrene cups).

2. In a given solvent at a particular temperature, the solubility of a polymer will decrease with increasing molecular weight.

3. **a.** Crosslinking eliminates solubility.

   b. Crystallinity, in general, acts like crosslinking, but it is possible in some cases to find solvents strong enough to overcome the crystalline bonding forces and dissolve the polymer. Heating the polymer toward its crystalline melting point allows its solubility in appropriate solvents. For example, nothing dissolves polyethylene at room temperature. At 100°C, however, it will dissolve in a variety of aliphatic, aromatic, and chlorinated hydrocarbons.

4. The rate of polymer solubility decreases with increasing molecular weight. For reasonably high molecular weight polymers, it can be orders of magnitude slower than that for nonpolymeric solutes.

It is important to note here that items 1, 2, and 3 are *equilibrium* phenomena and are therefore describable thermodynamically (at least in principle), while item 4 is a *rate* phenomenon and is governed by the rates of diffusion of polymer and solvent.

Example 1. The polymers of ω-amino acids are termed "nylon n," where n is the number of consecutive carbon atoms in the chain. Their general formula is

$$\left[\begin{smallmatrix} H & O \\ | & \| \\ N-C \end{smallmatrix}(CH_2)_{n-1}\right]_x$$

The polymers are crystalline, and will not dissolve in either water or hexane at room temperature. They will, however, reach an equilibrium level of absorption when immersed in each liquid. Describe how and why water and hexane absorption will vary with n.

Solution. Water is highly polar liquid; hexane is nonpolar. The polarity of the nylons depends on the relative proportion of polar nylon linkages $\begin{smallmatrix} H & O \\ | & \| \\ -N-C- \end{smallmatrix}$ in the chains. As n increases, the polarity of the chains decreases (they become more hydrocarbon-like), and so hexane absorption increases with n and water absorption decreases.

7.4   THE THERMODYNAMIC BASIS OF POLYMER SOLUBILITY

"To dissolve or not to dissolve. That is the question." (with apologies to W.S.). The answer is determined by the sign of the Gibbs free energy. Consider the process of mixing pure polymer and pure solvent (state 1) at constant pressure and temperature to form a solution (state 2):

$$\Delta G = \Delta H - T \Delta S \tag{7.1}$$

where ΔG = the change in Gibbs free energy
  ΔH = the change in enthalpy
  T = the absolute temperature
  ΔS = the change in entropy

Only if ΔG is negative will the solution process be thermodynamically feasible. The absolute temperature must be positive, and the change in entropy for a solution process is generally positive, because in a solution, the molecules are in a more random state in the solid (this might not always be the case with lyotropic liquid-crystal materials). The positive product is preceded by a negative sign. Thus, the third ($- T \Delta S$) term in (7.1) favors solubility. The change in enthalpy may be either positive or negative. A positive ΔH means the solvent and polymer "prefer their own company," that is, the pure materials are in a lower energy state, while a negative ΔH indicates that the solution is the lower energy state. If the latter obtains, solution is assured. Negative ΔH's usually arise where specific interactions such as hydrogen bonds are formed between the solvent and polymer molecules. But, if ΔH is positive, then $\Delta H < T \Delta S$ if the polymer is to be soluble.

One of the things that makes polymers unusual is that the entropy change in forming a polymer solution is generally much smaller than that which occurs on dissolution of equivalent masses or volumes of low molecular weight solutes. The reasons for this are illustrated qualitatively on a two-dimensional lattice model in Fig. 7.2. With the low molecular weight solute, the solute molecules may be distributed randomly throughout the lattice, the only restriction being that a lattice site cannot be occupied simultaneously by two (or more) molecules. This gives rise to a large number of configurational possibilities, that is, a high entropy. In the polymer solution, however, each chain segment is confined to a lattice site adjacent to the next chain segment, greatly reducing the configurational possibilities. Note also that for a given number of chain segments (equivalent masses or volumes of polymer) the more chains they are split up into, that is, the lower their molecular weight, the higher will be their entropy upon

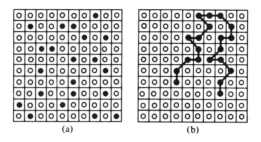

Figure 7.2 Lattice model of solubility: (*a*) low molecular weight solute; (*b*) polymeric solute. ○, Solvent; ●, Solute.

solution. This explains directly observation 2, the decrease in solubility with molecular weight. But in general, for high molecular weight polymers, because the $T\Delta S$ term is so small, if ΔH is positive then it must be even smaller if the polymer is to be soluble. So in the absence of specific interactions, predicting polymer solubility largely boils down to minimizing ΔH.

7.5   THE SOLUBILITY PARAMETER

How can ΔH be estimated? Well, for the formation of *regular* solutions (those in which solute and solvent do not form specific interactions), the change in *internal energy* per unit volume of solution is given by

$$\Delta H \approx \Delta E = \phi_1 \phi_2 (\delta_1 - \delta_2)^2 [\; = \;] \text{cal/cm}^3 \text{ soln} \qquad (7.2)$$

where ΔE = the change in internal energy *per unit volume of solution*
  ϕ_i = volume fractions
  δ_i = *solubility parameters*

The subscripts 1 and 2 usually (but not always!) refer to solvent and solute (polymer), respectively. The *solubility parameter* is defined as follows:

$$\delta = (\text{CED})^{1/2} = (\Delta E_v / v)^{1/2} \qquad (7.3)$$

where CED = *cohesive energy density*, a measure of the strength of the intermolecular forces holding the molecules together in the liquid state
  ΔE_v = molar change in internal energy on vaporization
  v = molar volume of liquid.

Traditionally, solubility parameters have been given in $(\text{cal/cm}^3)^{1/2}$ = hildebrands (in honor of the originator of regular solution theory), but they are now more commonly listed in $(\text{MPa})^{1/2}$ (1 hildebrand = 0.4889 $(\text{MPa})^{1/2}$).

Now, for a process that occurs at constant volume and constant pressure, the changes in internal energy and enthalpy are equal. Since the change in volume on solution is usually quite small, this is a good approximation for the dissolution of polymers under most conditions, so (7.2) provides a means of estimating enthalpies of solution if the solubility parameters of the polymer and solvent are known.

Note that regardless of the magnitudes of δ_1 and δ_2 (they must be positive), the predicted ΔH is always positive, because (7.2) applies only in the absence of the specific interactions that lead to negative ΔH's. Inspection of (7.2) also reveals that ΔH is minimized, and the tendency toward solubility is therefore maximized by matching the solubility parameters as closely as possible. As a very rough rule-of-thumb (or heuristic principle, if you prefer),

$$|\delta_1 - \delta_2| < 1 \; (\text{cal/cm}^3)^{1/2} \quad \text{for solubility} \tag{7.4}$$

Measuring the solubility parameter of a low molecular weight solvent is no problem. Polymers, on the other hand, degrade long before reaching their vaporization temperatures, making it impossible to evaluate ΔE_v directly. Fortunately, there is a way around this impasse. The greatest tendency of a polymer to dissolve occurs when its solubility parameter matches that of the solvent. If the polymer is crosslinked lightly, it cannot dissolve, but only swell. The maximum swelling will be observed when the polymer and solvent solubility parameters are matched. So polymer solubility parameters are determined by soaking lightly crosslinked samples in a series of solvents of known solubility parameters. The value of the solvent are which maximum swelling is observed is taken as the solubility parameter of the polymer (Fig. 7.3).

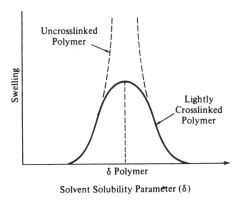

Figure 7.3 Determination of polymer solubility parameter by swelling lightly crosslinked samples in a series of solvents.

Solubility parameters of solvent mixtures can be readily calculated from

$$\delta_{mix} = \frac{\Sigma\, y_i v_i \delta_i}{\Sigma\, y_i v_i} = \Sigma\, \phi_i \delta_i \tag{7.5}$$

where y_i = mole fraction of component i
        v_i = molar volume of component i
        ϕ_i = volume fraction of component i

Equation 7.5 has often been used to prepare a series of mixed solvents for establishing the solubility parameter of a polymer as described above. Care must be exercised in this application, however, because what winds up inside the swollen polymer is not necessarily what you mixed up. In general, the cross-linked polymer will preferentially absorb the better (closer δ) solvent component, a phenomenon known as *coacervation*.

In the absence of specific data on solvents, a group-contribution method is available for estimating both the solubility parameters and molar volumes of liquids.[1]

While the solubility-parameter concept has proved useful, there are unfortunately many exceptions to (7.4). First, regular solution theory which leads to (7.2) has some shortcomings in practice. Second, polymer solubility is too complex a phenomenon to be described quantitatively with a single parameter. Several techniques have been proposed that supplement solubility parameters with quantitative information on hydrogen bonding and dipole moments.[2,3] One of the simplest of these classifies solvents into three categories according to their hydrogen-bonding ability (poor, moderate, strong). Three different δ *ranges* are then listed for each polymer, one for each solvent category. Presumably, a solvent that falls within the δ range for its hydrogen-bonding category will dissolve the polymer. Another technique that has achieved widespread practical application is discussed in the next section.

7.6   HANSEN'S THREE-DIMENSIONAL SOLUBILITY PARAMETER

According to Hansen,[4-7] the total change in internal energy on vaporization, ΔE_v may be considered the sum of three individual contributions: one due to hydrogen bonds ΔE_h, another due to permanent dipole interactions ΔE_p, and a third from dispersion (van der Waals or London) forces ΔE_d:

$$\Delta E_v = \Delta E_d + \Delta E_p + \Delta E_h \tag{7.6}$$

Dividing by the molar volume v gives

$$\frac{\Delta E_v}{v} = \frac{\Delta E_d}{v} + \frac{\Delta E_p}{v} + \frac{\Delta E_h}{v} \tag{7.7}$$

or
$$\delta^2 = \delta_d^2 + \delta_p^2 + \delta_h^2 \qquad (7.8)$$

where
$$\delta_j = (\Delta E_j/v)^{1/2} \qquad j = d, p, h$$

Thus, the solubility parameter δ may be thought of as a vector in a three-dimensional d, p, h space. Equation 7.8 gives the magnitude of the vector in terms of its components. A solvent, therefore, with given values of δ_{p1}, δ_{d1}, and δ_{h1} is represented as a point in space, with δ being the vector from the origin to this point.

A polymer is also characterized by δ_{p2}, δ_{d2}, and δ_{h2}. Furthermore, it has been found on a purely empirical basis that if δ_d is plotted on a scale twice the size as that used for δ_p and δ_h, then all solvents that dissolve that polymer fall within a sphere of radius R surrounding the point (δ_{p2}, δ_{d2}, and δ_{h2}).

Solubility judgments for the determination of R are usually based on visual observation of 0.5 g polymer in 5 cm^3 of solvent at room temperature. Given the concentration and temperature dependence of the phase boundaries in Fig. 7.1, this is somewhat arbitrary, but it seems to work out pretty well in practice, probably because the boundaries are fairly "flat" for polymers of reasonable molecular weight.

The three-dimensional equivalent of (7.4) is obtained by calculating the magnitude of the vector from the center of the polymer sphere (δ_{p2}, δ_{d2}, and δ_{h2}) to the point representing the solvent (δ_{p1}, δ_{d1}, and δ_{h1}). If this is less than R, the polymer is deemed soluble:

$$[(\delta_{p1} - \delta_{p2})^2 + (\delta_{h1} - \delta_{h2})^2 + 4(\delta_{d1} - \delta_{d2})^2]^{1/2} < R \quad \text{for solubility} \qquad (7.9)$$

(The factor of 4 arises from the empirical need to double the δ_d scale to achieve a spherical solubility region).

Figure 7.4 shows the solubility sphere for polystyrene ($\delta_d = 8.6$, $\delta_p = 3.0$, $\delta_h = 2.0$, $R = 3.5$, all in hildebrands).[4] Note that parts of the polystyrene sphere lie outside the first octant. The physical significance of these areas is questionable, at best.

The range of δ_d's spanned by typical polymers and solvents is rather small. In practice, therefore, the three-dimensional scheme is often reduced to two dimensions, with polymers and solvents represented on δ_h–δ_p coordinates with a polymer solubility circle of radius R.

Values of the individual components δ_h, δ_p and δ_d have been developed from measured δ values, theoretical calculations, studies on model compounds, and plenty of computer fitting. They are extensively tabulated for solvents.[4-9] They, along with R, are less readily available for polymers, but have been published.[4,5,9] Mixed solvents are handled by weighting the individual δ_j components according to (7.5).

Despite its semiempirical nature, the three-dimensional solubility parameter has proved of great practical utility, particularly in the paint industry, where the choice of solvents to meet economic, ecological, and safety constraints is of

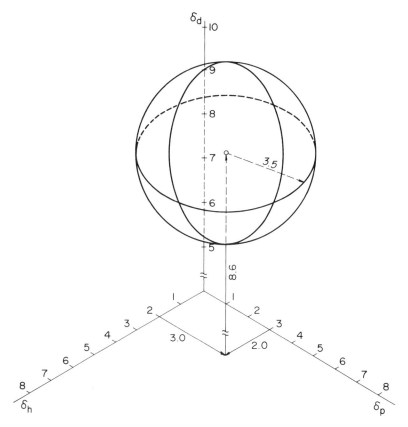

Figure 7.4   The Hansen solubility sphere for polystyrene ($\delta_d = 8.6$, $\delta_p = 3.0$, $\delta_h = 2.0$, $R = 3.5$).[5]

critical importance. It is capable of explaining those cases in which solvent and polymer δ's are almost perfectly matched, yet the polymer won't dissolve (the δ vectors have the same magnitudes, but different directions), or where two nonsolvents can be mixed to form a solvent (the solvent components lie on opposite sides outside the sphere, the mixture within). Inorganic pigments may also be characterized by δ vectors. Pigments whose δ vectors closely match those of a solvent tend to form stable suspensions in that solvent.

Example 2.   A polymer has a solubility parameter $\delta = 9.95$ ($\delta_p = 7.0$, $\delta_d = 5.0$, $\delta_h = 5.0$) and a solubility sphere of radius $R = 3.0$ (all numbers in hildebrands). Will a solvent with $\delta = 10$ ($\delta_p = 8$, $\delta_d = 6$, $\delta_h = 0$) dissolve it?

Solution. No. The solvent point lies in the δ_p–δ_d plane (i.e., $\delta_h = 0$). The closest approach the polymer solubility sphere makes to this plane is $5.0 - 3.0 = 2$. Thus, despite nearly identical δ's, the solvent will not dissolve the polymer.

7.7   THE FLORY–HUGGINS THEORY

Theoretical treatment of polymer solutions was initiated independently and essentially simultaneously by Flory[10] and Huggins[11] in 1942. The *Flory–Huggins theory* is based on the lattice model shown in Fig. 7.2. In the case of the low molecular weight solute, Fig. 7.2a, it is assumed that the solute and solvent molecules have roughly the same volumes; each occupies one lattice site. With the polymeric solute, Fig. 7.2b, a *segment* of the polymer molecule (which corresponds roughly but not necessarily exactly to a repeating unit) has the same volume as a solvent molecule and also occupies one lattice site.

By statistically evaluating the number of arrangements possible on the lattice, Flory and Huggins obtained an expression for the (extensive) configurational entropy change (that due to geometry alone), ΔS^*, in forming a solution from n_1 moles of solvent and n_2 moles of solute:

$$\Delta S^* = -R(n_1 \ln \phi_1 + n_2 \ln \phi_2) \tag{7.10}$$

where the ϕ's are volume fractions,

$$\phi_1 = \frac{x_1 n_1}{x_1 n_1 + x_2 n_2} \tag{7.10a}$$

$$\phi_2 = \frac{x_2 n_2}{x_1 n_1 + x_2 n_2} \tag{7.10b}$$

and the x's are the number of segments in the species. For the usual monomeric solvent, $x_1 = 1$. For a polydisperse polymeric solute, strictly speaking, a term must be included in (7.10) for each individual species in the distribution, but x_2 is usually taken as $\bar{x}_n$, the number-average degree of polymerization, with little error. (Writing the volume fractions in terms of moles implies equal molar segmental volumes.) Note that while ϕ_1, ϕ_2, and n_1 are the same in Figs. 7.2a and b, $n_2 = 20$ molecules for the monomeric solute, but only 1 molecule for the polymeric solute.

Example 3. Estimate the configurational entropy changes that occur when

a. 500 g of toluene (T) are mixed with 500 g of styrene monomer (S)

b. 500 g of toluene are mixed with 500 g of polystyrene (PS), $\bar{M}_n = 100\,000$

c. 500 g of PS, $\bar{M}_n = 100\,000$ are mixed with 500 g of polyphenylene oxide (PPO) (see Chapter II, Example 4K), $\bar{M}_n = 100\,000$. (This is one of the rare examples where two high molecular weight polymers are soluble in one another.)

Solution. $M_T = 92$, $M_S = 104$. Using these values and those given for the polymers, we get $n_i = 500\,g/M_i$. In the absence of other information, we must assume that the number of segments equals the number of repeating units for the polymers. Therefore, $x_i = \bar{M}_{ni}/m_i$, where m_i is the molecular weight of the repeating unit. $m_{PS} = 104$, $m_{PPO} = 120$. These quantities may now be inserted in Eqs. 7.10 and 7.10a and b. The results are summarized below ($R = 1.99$ cal/mol·K):

i	n_i(mol)	x_i	ϕ_i
a. Toluene	5.44	1	0.531
Styrene	4.81	1	0.469
	$\Delta S^* = 14.1$ cal/K		
b. Toluene	5.44	1	0.531
PS	0.005	962	0.469
	$\Delta S^* = 6.85$ cal/K		
c. PS	0.005	962	0.536
PPO	0.005	833	0.464
	$\Delta S^* = 0.0138$ cal/K		

The result for (c) illustrates why polymer–polymer solubility essentially *requires* a negative ΔH.

An expression for the (extensive) enthalpy of mixing, ΔH, was obtained by considering the change in adjacent-neighbor (molecules or segments) interactions on the lattice, specifically the replacement of [1,1] and [2,2] interactions with [1,2] interactions upon mixing:

$$\Delta H = RT\chi\phi_2 n_1 x_1 \quad [=] \text{cal} \tag{7.11}$$

where χ is the *Flory–Huggins interaction parameter*. Initially, χ was interpreted as the enthalpy of interaction per mole of solvent divided by RT. By equating (7.2) and (7.11) (keeping in mind that the enthalpy in (7.2) is based on a unit volume of solution, while that in (7.11) is an extensive quantity) and making use of (7.10a), it may be shown that the Flory–Huggins parameter and solubility parameters are related by

$$\chi = \frac{v(\delta_1 - \delta_2)^2}{RT} \tag{7.12}$$

where v is the molar segmental volume of species 1 and 2 (assumed to be the same). For the dissolution of a polymer in a monomeric solvent, v is taken as the molar volume of the solvent, v_1. From our knowledge of solubility parameters, we see that (7.12) predicts $\chi \geq 0$. Actually, negative values have been observed.

If it is assumed that the entropy of solution is entirely configurational, substitution of (7.10) and (7.11) into (7.1) gives

$$\Delta G = RT(n_1 \ln \phi_1 + n_2 \ln \phi_2 + \chi\phi_2 n_1 x_1) \tag{7.13}$$

Again, for the usual monomeric solvent, $x_1 = 1$. For a polydisperse solute, the middle term in the right side of (7.13) must be replaced by a summation over all the solute species; however, treatment as a single solute with $x_2 = \bar{x}_n$ usually suffices.

In terms of the Flory–Huggins theory, the criterion for complete solubility of a high molecular weight polymer across the composition range is

$$\chi \le 0.5 \quad \text{for solubility} \tag{7.14}$$

It is now recognized that there is an interactive as well as a configurational contribution to the entropy of solution. That is also included in the χ term, so χ is now considered to be a ΔG (rather than strictly a ΔH) of interaction per mole of solvent divided by RT. The first two terms on the right of (7.13) therefore represent the configurational entropy contribution to ΔG, while the third term is the interaction contribution and includes both enthalpy and entropy effects.

The Flory–Huggins theory has been used extensively to describe phase equilibria in polymer systems. It can, for example, qualitatively describe the lower phase boundary (UCST) in Fig. 7.1, though it rarely gives a good quantitative fit of experimental data. Partial differentiation of (7.13) with respect to n_1 (keeping in mind that ϕ_1 and ϕ_2 are functions of n_1) gives the chemical potential of the solvent. This is, of course, a key quantity in phase equilibrium, and also makes χ experimentally accessible. Further development is beyond the scope of this chapter, but the subject is well treated in standard works on polymer solutions.[12–15]

The limitations of the Flory–Huggins theory have been recognized for a long time. It can't predict an LCST (Fig. 7.1). It is perhaps not surprising that χ depends on temperature, but it unfortunately turns out to be a function of concentration and molecular weight as well, limiting practical application of the theory. These deficiencies are thought to arise because the theory assumes no volume change upon mixing and the statistical analysis on which it is based is not valid for very dilute solutions, particularly in poor solvents. There has been considerable subsequent work done to correct these deficiencies and extend lattice-type theories.[16,17]

Experimental values for χ have been tabulated for a number of polymer–solvent systems, both single values[9] and even as a function of composition.[8] They may be used with (7.14) to predict solubility.

7.8 A PROMISING RECENT APPROACH

In 1975, Abrams and Prausnitz[18] published a new equation for the Gibbs free energy of mixtures which they called UNIQUAC (universal quasi-chemical). It contains two interaction parameters per binary pair in the mixture and two parameters characteristic of each component. The utility of UNIQUAC was extended considerably with the development of UNIFAC (UNIQUAC functional-group activity coefficients) by Fredenslund et al.[19] UNIFAC is a group-contribution method based on the UNIQUAC equation. With it, calculations can be made from a knowledge of the molecular structure of the compounds. UNIFAC has been remarkably successful when applied to nonpolymer solutions, and it's beginning to look as if it will be equally successful for treating polymer solutions.[20]

7.9 PROPERTIES OF DILUTE SOLUTIONS

Well, now the polymer is in solution. Let's initially assume it's a fairly dilute solution so we don't have to worry about too many entanglements between the molecules. Entanglements begin to set in when the dimensionless Berry number, the product of intrinsic viscosity and concentration, exceeds unity[21]:

$$\text{Be} = [\eta]c > 1 \quad \text{for entanglements} \tag{7.15}$$

For typical polymer–solvent systems this usually works out to a few percent polymer.

In a "good" solvent (one whose solubility parameter closely matches that of the polymer), the secondary forces between polymer segments and solvent molecules are strong, and the polymer molecules will assume a spread out conformation in solution. In a "poor" solvent, the attractive forces between the segments of the polymer chain will be greater than those between the chain segments and the solvent; in other words, the chain segments "prefer their own company," and the chain will ball up tightly (Fig. 7.5).

Imagine a polymer in a good solvent, for example, polystyrene ($\delta = 9.3$) in chloroform ($\delta = 9.2$). A nonsolvent is now added, say, methanol ($\delta = 14.5$).

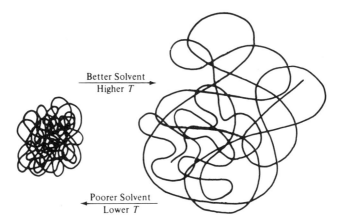

Figure 7.5 The effects of solvent power and temperature on a polymer molecule in solution.

(Despite the large difference in solubility parameters, chloroform and methanol are mutually soluble in all proportions because of the large ΔS of solution for low molecular weight compounds.) Ultimately, a point is reached where the mixed solvent becomes too poor to sustain solution, and the polymer precipitates out, as the attractive forces between polymer segments become much greater than that between polymer and solvent. At some point, the polymer teeters on the brink of solubility, $\Delta G = 0$ and $\Delta H = T\Delta S$. This point obviously depends on the temperature, polymer molecular weight (mainly through its influence on ΔS), and the polymer–solvent system (mainly through its influence on ΔH). Adjusting either the temperature or the polymer–solvent system allows fractionation of the polymer according to molecular weight, as successively smaller molecules precipitate upon lowering the temperature or going to poorer solvent.

In the limit of infinite molecular weight (the minimum possible ΔS), the situation where $\Delta H = T\Delta S$ is known as the θ or Flory condition. For a polymer of infinite molecular weight in a particular solvent, the θ temperature equals the UCST. Under these conditions, the polymer–solvent and polymer–polymer interactions are equal, and the solution behaves in a so-called ideal fashion, with the second virial coefficient equal to 0, etc., and the MHS exponent (Eq. 6.14) $a = 0.5$.

For a given polymer, θ conditions can be reached at a fixed temperature by adjusting the solvent to give a θ solvent, or with a particular solvent by adjusting the temperature to reach the θ or Flory temperature. Any actual polymer will still be soluble under θ conditions, of course, because of its lower-than-infinite molecular weight and consequently larger ΔS.

Getting back to our example, we might ask what happens to the viscosity of the solution upon going from a good solvent to a poor solvent (to make a fair comparison, imagine that as nonsolvent is added, an equivalent amount of good solvent is removed, maintaining constant the polymer concentration). This question can be answered qualitatively by imagining the polymer molecules in solution to be rigid spheres (they really aren't) and applying an equation derived by Einstein[22] for the viscosity of a dilute suspension of rigid spheres:

$$\eta = \eta_s(1 + 2.5\,\phi) \tag{7.16}$$

where η is the viscosity of the suspension (solution), η_s the viscosity of the solvent, and ϕ the volume fraction of spheres. (We further assume that the viscosities of the low molecular weight liquids are comparable, i.e., that η_s doesn't change). Going from good to poor solvent, the polymer molecules ball up, giving a smaller effective ϕ, *lowering* the solution viscosity. Thus, solution viscosity can be controlled by adjusting solvent power.

Example 4. Indicate how solvent "power" (good solvent vs. poor solvent) will influence the following:

a. The intrinsic viscosity of a polymer sample at a particular T.

b. The molecular weight of a polymer sample as determined by membrane osmometry.

Solution. **a.** The measured viscosities of solutions will be less in a "poor" solvent than in a good one, leading to lower intrinsic viscosities.

b. If the membrane were ideal, solvent power would make no difference, since osmometry measures only the number (moles) of solute per unit volume, regardless of geometry. Some of the smallest particles can sneak through any real membrane, however. The poor solvent, causing the molecules to ball up tightly, will allow passage of more molecules through the membrane. This will lower the observed osmotic pressure, causing the calculated $\bar{M}_n$ to err on the high side.

Consider now the effects of temperature on the viscosity of a solution of polymer in a relatively poor solvent. As is the case with all simple liquids, the solvent viscosity η_s decreases with increasing temperature. Increasing temperature, however, imparts more thermal energy to the segments of the polymer molecules, causing the molecules to spread out and assume a larger effective ϕ in solution. Thus, the effects of temperature on η_s and ϕ tend to compensate, giving a solution that has a much smaller change in viscosity with temperature than that of the solvent alone. In fact, the additives that are used to produce the so-called multiviscosity (10W-40) motor oils are polymers with compositions adjusted so that the base oil is a relatively poor solvent at the lowest operating temperatures. As the engine heats up, the polymer molecules uncoil, providing a much greater resistance to thinning than is possible from oil alone.

Viscosities (at low shear rates) of dilute solutions of polymers with known molecular weights may be calculated using Eqs. 6.13 and 6.14, reversing the procedure for obtaining molecular weights from viscosity measurements.

7.10 POLYMER–POLYMER–COMMON SOLVENT SYSTEMS

As was illustrated in Example 3c, ΔS for the dissolution of one polymer in another is extremely small. For this reason, the true solubility of one polymer in another is relatively rare, although many more examples have come to light in recent years. When such solubility does occur, it generally results from strong interactions (e.g., a large, negative ΔH), most often hydrogen bonds, between the polymer pairs. Polymer–polymer miscibility has been extensively reviewed.[23,24]

For the far more common case where two polymers are insoluble in one another, even when a common solvent is added (one that is infinitely soluble with each polymer alone), the two polymers usually cannot coexist in a homogeneous phase beyond a few percent concentration. A schematic phase diagram for such a system is shown in Fig. 7.6. Beyond a few percent polymer (the exact value depends on the chemical nature of the polymers and the solvent and the molecular weights of the polymers), two phases in equilibrium are formed, each phase containing a nearly pure polymer. These results extend to more than two

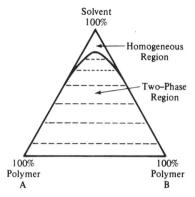

Figure 7.6 Typical ternary phase diagram for a polymer–polymer–common solvent system. The dotted tie lines connect composition of phases in equilibrium.

polymers. In general, each polymer will coexist in a separate phase. This fact was the basis of an early proposal for separating and recycling mixed plastic waste.[25]

7.11 CONCENTRATED SOLUTIONS: PLASTICIZERS

Up to this point, we have considered relatively dilute polymer solutions. Now let's look at the other end of the spectrum, where the polymer is the major constituent of the solution. A pure, amorphous polymer consists of a tangled mass of polymer chains. The ease with which this mass can deform depends on the ability of the polymer chains to untangle and slip past one another. One way of increasing this ability is to raise the temperature, as we have seen. Another is to add a low molecular weight liquid, an *external plasticizer*, to the polymer. By forming secondary bonds to the polymer molecules and spreading them apart, the plasticizer reduces the polymer–polymer chain secondary bonding, and provides more room for the polymer molecules to move around, providing a softer, more easily deformable mass.

Example 5. If you have ever heard baseball announcers comment that "On these humid nights, the ball just isn't hit as hard" or "A pitcher throws a heavier ball on these humid nights," you may have wondered if there was a scientific basis for these statements. They cannot be justified in terms of the properties of the humid air (fluid dynamics). Can you justify them in terms of the baseball properties? A baseball is made largely of tightly wound wool yarn. Wool, a natural polymer, is similar, physically and chemically, to nylons (see Example 1).

Solution. The polar nylon linkages will hydrogen bond to water. The higher the humidity, the greater will be the equilibrium moisture content of the wool, accounting for the heavier ball. The absorbed water will also act as an external plasticizer for the wool, softening it and decreasing its resiliency.

Quantitative confirmation of Example 5 was provided by the noted technical journal, *Sports Illustrated*, which, between swimsuit photos, reported on the effects of moisture on the mass and resilience of baseballs[26]:

balls baked in a 212° F oven for 24 hours lost 12 grams and gained 5.8% in vigor

24 hours in a steam-filled moist room caused them to pick up 11 grams and lose 17.4% of their bounce

A similar reversible plasticization of fibers by heat and moisture is applied in the process of steam ironing fabrics.

A thermodynamically poor plasticizer should be more effective than a good one, giving a lower viscosity at a given level (smaller ϕ, fewer entanglements), but since it is less strongly bound to the polymer, it will have a greater tendency to exude out over a period of time, leaving the polymer mass stiffer. This was a familiar problem with some early imported cars. When parked in the sun, a greasy plasticizer film would condense on the windows, and over a period of time, the upholstery and dashboard would crack from loss of plasticizer. (Much of that new-car smell is due to plasticizers.) Thus, a balance must be struck between plasticizer efficiency and permanence. Also, from the standpoint of permanence, it is necessary that the plasticizer have low volatility. Plasticizers therefore generally have higher molecular weights than solvents, but are still well below the high polymer range in this respect.

A polymer may be *internally plasticized* by random copolymerization with a monomer whose homopolymer is very soft (has a low T_g—see Chapter VIII). There are obviously never any permanence problems when this is done. The composition of the copolymer can be adjusted to give the desired properties at a particular temperature. Above this temperature, the copolymer will be softer than intended, and below this temperature it will be harder. With external plasticizers, the same sort of temperature compensation as in the multi-vis motor oils is obtained, giving materials that maintain the desired flexibility over a broader temperature range.

The most common externally plasticized polymer is polyvinyl chloride (PVC) $\{CH_2-CHCl\}_x$. A typical plasticizer for PVC is dioctyl phthalate (DOP) (actually di(2 ethylhexyl) phthalate, DEHP), the diester of phthalic acid and iso-octyl alcohol:

PVC, because of its high chlorine content, is inherently fire resistance. Dilution of the PVC with a hydrocarbon plasticizer increases flammability. Therefore, chlorinated waxes and phosphate esters, e.g., tricresyl phosphate (TCP), are used as plasticizers which maintain fire resistance. So-called polymeric plasticizers, polymers with molecular weights on the order of 1000, provide low volatility and good permanence. It might also be noted that a low molecular weight fraction in a pure polymer behaves as a plasticizer, often in an undesirable fashion.

Epoxidized plasticizers also perform the important function of stabilizing PVC. When PVC degrades, HCl is given off, and the HCl catalyzes further degradation. Not only that, HCl attacks metallic molding machines, molds, extruders, etc. The epoxy plasticizers are made by epoxidizing polyunsaturated vegetable oils:

$$\underset{\text{C}=\text{C}}{\overset{\text{H}\quad\text{H}}{\sim\!\!\sim}}\!\sim\!\sim \;+\; H_2O_2 \;\xrightarrow{\text{Catalyst}}\; \underset{\underset{\text{O}}{\text{C}-\text{C}}}{\overset{\text{H}\quad\text{H}}{\sim\!\!\sim}}\!\sim\!\sim \;+\; H_2O$$

Oxirane ring

The *oxirane rings* soak up HCl and minimize further degradation (this is an admittedly simplified view of a very complex phenomenon)[27]:

$$\underset{\underset{\text{O}}{\text{C}-\text{C}}}{\overset{\text{H}\quad\text{H}}{\sim\!\!\sim}}\!\sim\!\sim \;+\; HCl \;\longrightarrow\; \underset{\underset{\text{OH}\;\;\text{Cl}}{\text{C}\!-\!-\!\text{C}}}{\overset{\text{H}\quad\quad\text{H}}{\sim\!\!\sim}}\!\sim\!\sim$$

Having their origin in vegetable oils, these epoxidized plasticizers can often obtain FDA clearance for use in food-contact applications.

Unplasticized (or nearly so) PVC is a rigid material used for pipe and fittings, house siding, window frames, and bottles, among other things. The properties of plasticized PVC vary considerably depending on the plasticizer level. Plasticized PVC is familiar as a gasketing material, a leather-like upholstery material (Naugahyde®), wire and cable covering, shower curtains, and packaging film.

Plastisols are an interesting and useful technological application of plasticized PVC. A typical formulation might consist of 100 parts DOP phr (per hundred parts resin) plus some stabilizers, pigments, etc. Although the PVC is thermodynamically soluble in the DOP at room temperature, the rate of dissolution of high molecular weight PVC at room temperature is extremely low. So, initially the plastisol is a milky suspension of finely divided PVC particles in the plasticizer. As a suspension rather than a solution, the viscosity is of the order of magnitude of that of the plasticizer itself (Eq. 7.16), and it can be applied to substrates or molds by brushing, dipping, rolling, etc. When heated to about 350° F (175°C), the increased thermal agitation of the polymer molecules speeds up the solution process greatly. If there is no filler or pigment present, the solution process can be observed as the plastisol becomes transparent. When cooled back to room temperature, the viscosity of the solution is so high that for all practical purposes, it may be considered a flexible solid. Examples of plastisols include doll "skin" and the covering of wire dish racks. Vinyl foam is also made by whipping air into a plastisol prior to fusion.

The viscosity of the initial suspension may be lowered by incorporating a volatile organic diluent for the plasticizer, giving an *organsol*, or by whipping in water to foam a *hydrosol*. In both cases, the diluent vaporizes upon heating.

REFERENCES

1. Fedors, R. F., *Polym. Eng. Sci.* **14**, 147, 472 (1974).

2. Beerbower, A., L. A. Kaye, and D. A. Pattison, *Chem. Eng.* **74**(26), 118 (1967).

3. Seymour, R. B., *Modern Plastics* **48**(10), 150 (1971).

4. Hansen, C. M., *J. Paint. Technol.* **39**, 505, 104 (1967).

5. Hansen, C. M., *The Three Dimensional Solubility Parameter and Solvent Diffusion Coefficient*, Danish Technical Pres, Copenhagen, 1967.

6. Hansen, C. M., and A. Beerbower, Solubility parameters, in *Encyclopedia of Chemical Technology*, Supplement Volume, 2d ed., Wiley, New York, 1971.

7. Hansen, C. M., *Ind. Eng. Chem. Prod. Res. Dev.* **8**, 2 (1969).

8. Brandrup, J., and E. H. Immergut (eds.), *Polymer Handbook*, 3d ed., Wiley-Interscience, New York, 1989.

9. Barton, A. F. M., *Handbook of Polymer–Liquid Interaction Parameters and Solubility Parameters*, CRC Press, Boca Raton, FL, 1990.

10. Flory, P. J., *J. Chem. Phys.* **10**, 51 (1942).

11. Huggins, M. L., *Ann. N.Y. Acad. Sci.* **43**, 1 (1942).

12. Flory, P. J., *Principles of Polymer Chemistry*, Cornell University Press, Ithaca, NY, 1953.

13. Morawetz, H., *Molecules in Solution*, Wiley-Interscience, NewYork, 1975.

14. Tompa, H., *Polymer Solutions*, Butterworths, London, 1956.

15. Kwei, T. K., Macromolecules in solution, Chapter 4 in *Macromolecules: an Introduction to Polymer Science*, F. A. Bovey and F. H. Winslow (eds.), Academic Press, New York, 1979.

16. Sanchez, I. C., and R. H. Lacombe, *Macromolecules* **11**(6), 1145 (1978).

17. Pesci, A.I., and K. F. Freed, *J. Chem. Phys.* **90**(3), 2017 (1989).

18. Abrams, D. S., and J. M. Prausnitz, *AIChE J.* **21**(1), 116 (1975).

19. Fredenslund, A., R. L. Jones, and J. M. Prausnitz, *AIChE J.* **21**, 1086 (1975).

20. Holten-Andersen, J., P. Rasmussen, and A. Fredenslund, *Ind. Eng. Chem. Res.* **26**, 1382 (1987).

21. Hager, B. L., and G. C. Berry, *J. Polym. Sci.: Polym. Phys. Ed.* **20**, 911 (1982).

22. Einstein, A., *Ann. Phys.* **19**, 289 (1906); **34**, 591 (1911).

23. Olabisi, O., L. M. Robeson, and M. T. Shaw, *Polymer–Polymer Miscibility*, Academic, New York, 1979.

24. Coleman, M. M., J. F. Graf, and P. C. Paniter, *Specific Interactions and the Miscibility of Polymer Blends*, Technomic, Lancaster, PA, 1991.

25. Sperber, R. J., and S. L. Rosen, *Polym. Eng. Sci.* **16**(4), 246 (1976).

26. *Sports Illustrated*, Time, Inc., July 20, 1970, p. 22.

27. Anderson, D. F., and D. A. McKenzie, *J. Polymer Sci.: A-1*, **8**, 2905 (1970).

PROBLEMS

1. Crud Chemicals is using SEC for quality-control checks on their polystyrene ($\delta = 9.3$) product. Normally, the column runs at 25°C with THF solvent ($\delta = 9.1$). Unbeknownst to them, a disgruntled former employee contaminated their THF supply with methanol ($\delta = 14.5$) before leaving. Being fairly sophisticated technically, he's loaded it up pretty well without causing the polymer to precipitate out. How and why will this affect their calculated average molecular weights?

2. Crud's analytical lab corrects its solvent problem (see above), but the temperature controller on their SEC column goes haywire, allowing the column temperature to rise 20°C without their knowledge. How and why will this affect their calculated average molecular weights?

3. Crud finally gets its SEC system under control. Now, however, they want to use it to check their new product, a copolymer of styrene (75%) with acrylonitrile (25%). Not having monodisperse samples of this material to recalibrate their system, they go ahead and use the existing polystyrene calibration. How and why will this affect their calculated average molecular weights?

4. Neoprene (polychloroprene) and nitrile (butadiene-acrylonitrile) rubbers are generally favored over SBR (butadiene-styrene) and natural rubbers for automotive underhood applications. Why?

5. According to Hansen's 3-D solubility parameter scheme, which of the following solvents will dissolve polystyrene (Fig. 7.4)? How well do these predictions agree with Eq. 7.4?

Solvent	δ_d	δ_p	δ_h
n-Hexane	7.28	0	0
Styrene	9.09	0.49	2.0
Methyl chloride	7.48	3.0	1.9
Acetone	7.58	5.08	3.4
Ethyl acetate	7.72	2.6	3.5
Isopropanol	7.72	3.0	8.02
Phenol	8.80	2.9	7.28

All δ's are in $(cal/cm^3)^{1/2}$ = hildebrands.

6. Obtain the Flory–Huggins expression for the chemical potential of the solvent.

7. Apply the general thermodynamic criteria for phase equilibria and derive equations that would allow you to describe the lower phase envelope in Fig. 7.1 in terms of the Flory–Huggins theory.

8. Show the derivation of Eq. 7.12.

9. Sketch qualitatively how you would expect the molecular weight of the polymer to affect the phase diagram in Fig. 7.1.

10. Sketch qualitatively how you would expect (**a**) polymer molecular weights and (**b**) temperature to affect the ternary phase diagram in Fig. 7.6.

11. Figure 7.6 illustrates a ternary phase diagram for a polymer–polymer–common solvent system. Sketch a ternary phase diagram for a polymer–solvent–nonsolvent system. Include tie lines and illustrate the effects of temperature and polymer molecular weight. Assume that the solvent and nonsolvent are soluble in all proportions.

12. One of the shortcomings of Hansen's 3-D scheme is that it does not incorporate the effect of polymer molecular weight on solubility. Perhaps it could be extended to do so. What parameter(s) in Hansen's scheme do you think should depend on polymer weight? Sketch qualitatively its (their) expected variation with M.

13. Polymer solubility parameters can also be determined through viscosity measurements. Describe how you would do this.

14. A small amount of finely divided polymer is put into a beaker containing a liquid. The beaker is covered and the contents are magnetically stirred with no change in the milky appearance over several hours. The beaker is then uncovered, with the stirring continued. The contents of the beaker gradually become transparent over a period of several hours. Explain (there are at least two possibilities here).

15. It has been proposed that a mixture of waste thermoplastics could be separated by extraction with a single solvent by using different temperatures. That is, polymer A would be extracted from the mixture at the lowest temperature T_A, polymer B at a higher temperature T_B, and so on, leading to nearly pure recovered polymers. Comment on the thermodynamic feasibility of this proposal.

CHAPTER VIII

Transitions in Polymers

8.1 THE GLASS TRANSITION

It has long been known that amorphous polymers can exhibit two distinctly different types of mechanical behavior. Some, like polymethyl methacrylate (Lucite®, Plexiglas®) and polystyrene, are hard, rigid, *glassy* plastics at room temperature, while others, for example, polybutadiene, polyethyl acrylate, and polyisoprene, are soft, flexible rubbery materials. If, however, polystyrene and polymethyl methacrylate are heated to about 125°C, they exhibit typical rubbery properties, and when a rubber ball is cooled in liquid nitrogen, it becomes rigid and glassy, and shatters when an attempt is made to bounce it. So, there is some temperature, or narrow range of temperatures, below which an amorphous polymer is in a glassy state, and above which it is rubbery. This temperature is known as the *glass transition temperature* T_g. The glass transition temperature is a property of the polymer, and whether the polymer has glassy or rubbery properties depends on whether its application temperature is above or below its glass transition temperature.

8.2 MOLECULAR MOTIONS IN AN AMORPHOUS POLYMER

To understand the molecular basis for the glass transition, the various molecular motions occurring in an amorphous polymer mass may be broken into four categories:

1. Translational motion of entire molecules, which permits flow
2. Cooperative wriggling and jumping of segments of molecules approximately 40 to 50 carbon atoms in length, permitting the flexing and uncoiling which lead to elasticity

3. Motions of a few atoms along the main chain (five or six, or so) or of side groups on the main chains

4. Vibrations of atoms about equilibrium positions, as occurs in crystal lattices, except that the atomic centers are not in a regular arrangement in an amorphous polymer

Motions 1–4 above are arranged in order of decreasing activation energy; that is, smaller amounts of thermal energy kT are required to produce them. The glass transition temperature is thought to be that temperature at which motions 1 and 2 are pretty much "frozen out," and there is only sufficient energy available for motions of types 3 and 4. Of course, not all molecules possess the same energies at a given temperature. The molecular energies follow a Boltzmann distribution, and even below T_g, there will be occasional type 2 and even type 1 motions, which can manifest themselves over extremely long periods of time.

8.3   DETERMINATION OF T_g

How is the glass transition studied? A common method is to observe the variation of some thermodynamic property with T, for example, the specific volume, as shown in Fig. 8.1. Note that the slope of the v vs. T plot increases above the glass transition temperature.

The value of T_g determined in this fashion will vary somewhat with the rate of cooling or heating. This reflects the fact that long, entangled polymer chains cannot respond instantaneously to changes in temperature, and illustrates the

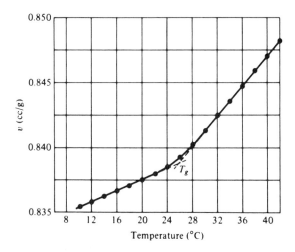

Figure 8.1   Specific volume vs. temperature for polyvinyl acetate.[1]

difficulty in making thermodynamic measurements on polymers. It often takes an extremely long time to reach equilibrium, if indeed it is ever reached, and it is difficult to be sure if and when it is reached. Strictly speaking, the glass transition temperature should be defined in terms of equilibrium properties, or at least those measured with very low rates of temperature change. Also, a sharp "break" in the property is never observed, but T_g can always be established within a couple of degrees by extrapolation of the linear regions. Other properties such as refractive index may also be used to establish T_g.

In contrast to a change in *slope* at the glass transition, a thermodynamic property such as specific volume exhibits a *discontinuity* with temperature at the crystalline melting point in polymers as in other materials (Fig. 8.2). The glass transition is therefore known as a *second-order* thermodynamic transition (v vs. T continuous, dv/dT vs. T discontinuous) in contrast to a first-order transition such as the melting point (v vs. T discontinuous).

T_g characterizes the amorphous phase. Since all polymers have at least some amorphous material (they can't be 100% crystalline), they all have a T_g, but not all polymers have a crystalline melting point—they can't if they don't crystallize.

Transitions in polymers are rapidly and conveniently studied using differential scanning calorimetry (DSC).[3] Small samples of the polymer and an inert reference substance (one that undergoes no transitions in the temperature range of interest) are mounted in a block with a heater for each and thermocouples to monitor temperatures. The thermodynamic property monitored here is the enthalpy. A servo control system adjusts and measures the power to the heaters to maintain the sample and reference temperatures the same as the sample is

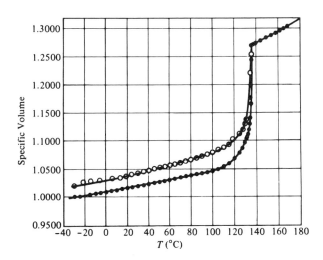

Figure 8.2 Specific volume–temperature relations for linear polyethylene (Marlex-50). ○, Specimen slowly cooled from melt to room temperature prior to fusion; ●, specimen crystallized at 130°C for 40 days, then cooled to room temperature prior to fusion.[2]

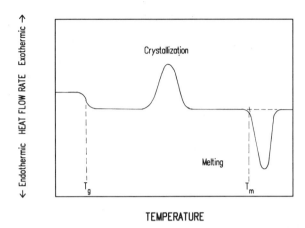

Figure 8.3   Schematic DSC curve. This is what would be observed on heating a material from state 2 (Example 8.8, Fig. 8.5) to state 1.

heated at a pre-programmed rate. At T_g, the heat capacity of the sample suddenly increases, requiring more power (relative to the reference) to maintain the temperatures the same. This differential heat flow to the sample (endothermic) causes a drop in the DSC curve (Fig. 8.3). At T_m, the sample crystals want to melt at constant temperature, so a sudden input of large amounts of heat is required to keep the sample temperature even with the reference temperature. This results in the characteristic endothermic melting peak. Crystallization, in which large amounts of heat are given off at constant temperature, gives rise to a similar but exothermic peak. By measuring the net energy flow to or from the sample, heat capacities and heats of fusion can be determined. Decomposition (endothermic) and oxidation (exothermic) reactions can also be conveniently studied.

A DSC is programmed to heat the sample at a constant rate, such as 5°C/min, 10°C/min, or 20°C/min. The higher the rate, the quicker is the measurement, a practically desirable result. Unfortunately, because polymer chains can't respond instantaneously to the changing temperature, the further from equilibrium the measurement is as well. The dependence of the measured T_g and T_m on heating rate is at least partially responsible for the range of values observed in the literature. To approach the true equilibrium values, very low heating rates should be used, or better yet, several heating rates should be used and the results extrapolated back to zero heating rate. Because of the time and effort involved, this is rarely done. Dynamic mechanical measurements (Chapter XVIII) can also provide useful information on transitions.

8.4   FACTORS THAT INFLUENCE T_g[4]

In general, the glass transition temperature depends on five factors.

1. The *free volume* of the polymer v_f, which is the volume of the polymer mass not actually occupied by the molecules themselves; that is, $v_f = v - v_s$, where v is the specific volume of the polymer mass and v_s is the volume of the solidly packed molecules. The higher v_f, the more room the molecules will have in which to move around, and the lower will be T_g. It has been estimated that for all polymers $v_f/v \approx 0.025$ at T_g.

Example 1. Glass-transition temperatures have been observed to increase at pressures of several thousand psi. Why?

Solution. High pressures compress polymers, reducing v. Since v_s doesn't change appreciably v_f is reduced.

2. The attractive forces between the molecules. The more strongly they are bound together, the more thermal energy will be required to produce motion. Since the solubility parameter δ is a measure of intermolecular forces,

T_g increases with δ. Polyacrylonitrile, $\overset{\displaystyle H\ \ H}{\underset{\displaystyle H\ \ C{\equiv}N}{{+}C{-}C{+}_x}}$,because of the frequent, strong

polar bonding between chains, has a T_g higher than its degradation temperature and therefore, though linear, is not thermoplastic.

3. The internal mobility of the chains, that is, their freedom to rotate about bonds. Figure 8.4 shows potential energy as a function of rotation angle about a bond in a polymer chain. The minimum energy configuration, arbitrarily chosen as $\theta = 0$, is the position where the largest substituents, the rest of the chain, are as far away from each other as possible. As the bond is rotated, the

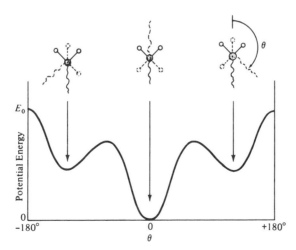

Figure 8.4 Rotation about a bond in a polymer chain backbone, viewed along the bond. The dotted substituents are on the rear carbon atom.

Table 8.1[4]

Polymer	δ	T_g (°C)	E_0 (kcal/mol)
Polydimethyl siloxane	7.3	− 120	~ 0
Polyethylene	7.9	− 85(?)	3.3
Polytetrafluoroethylene	6.2	> 20	4.7

substituent groups are brought into juxtaposition, and energy is required to "push them over the hump." The maximum energy is needed to get the two chain substituents past one another, and this energy must be available if complete rotation is to be obtained. Rotation is necessary for type 1 and 2 motions.

Table 8.1 shows how T_g increases with E_0 for a series of polymers with approximately the same δ. Note how the ether oxygen "swivel" in the silicone chain permits very free rotation.

Example 2. Poly(α-methylstyrene) has a higher T_g than polystyrene. Why?

Polystyrene   Poly(α-methylstyrene)

Solution. The methyl group introduces extra steric hindrance to rotation, giving a higher E_0.

4. The stiffness of the chains. Chains that have difficulty coiling and folding will have higher T_g values. This stiffness usually goes hand-in-hand with high E_0, so it is difficult to separate the effects of 3 and 4.

Chains with parallel bonds in the backbone (ladder-type polymers), e.g., polyimides (Example 4R, Chapter II) and those with highly aromatic backbones have extremely stiff chains, and therefore tend to have high T_g values. This makes these polymers mechanically useful at elevated temperatures, but also very difficult to process.

5. The chain length. As do many mechanical properties of polymers, the glass transition temperature varies according to the empirical relation

$$T_g = T_g^{\infty} - \frac{C}{x} \tag{8.1}$$

where C is a constant for the particular polymer, and T_g^{∞} is the asymptotic value of the glass transition temperature at infinite chain length. This reflects the increased ease of motion for shorter chains. The decrease in T_g with x is only

as ML ↑ T_g ↑

noticeable at relatively low chain lengths. For most commercial polymers, x is high enough so that $T_g \approx T_g^{\infty}$.

Example 3. Measurements such as those described above show that an external plasticizer softens a polymer by reducing its glass transition temperature. Explain.

Solution. The plasticizer molecules pry apart the polymer chains, in essence increasing the free volume available to the chains (although not truly free, the small plasticizer molecules interfere with chain motions much less than would other chains). Also, by forming secondary bonds with the polymer chains, the plasticizer molecules reduce the bonding forces between the chains themselves, easing type 1 and 2 motions.

8.5   THE EFFECT OF COPOLYMERIZATION ON T_g

The glass transition temperatures for random copolymers vary monotonically with composition between those of the homopolymers. They can be approximated fairly well from a knowledge of the T_g values of the homopolymers, T_{g1} and T_{g2}, with the empirical relation

$$\frac{1}{T_g} = \frac{w_1}{T_{g1}} + \frac{w_2}{T_{g2}} \tag{8.2}$$

where the w's are weight fractions of the monomers in the copolymer. This relation forms the basis for a method of estimating the T_g values of highly crystalline polymers, where the properties of the small amount of amorphous material are masked by the majority of crystalline material present. If a series of random copolymers can be produced in which the randomness prevents crystallization over a certain composition range, then Eq. 8.2 can be used to extrapolate to $w_1 = 0$ and $w_1 = 1$, giving the T_g values of the homopolymers. This method is open to question, because it assumes that the presence of major amounts of crystallinity does not restrict the molecular response in the amorphous regions. In fact, the T_g values of highly crystalline polymers (polyethylene, in particular) are still open to debate.

8.6   THE THERMODYNAMICS OF MELTING

The crystalline melting point T_m in polymers is a phase change similar to that observed in low molecular weight organic compounds, metals, and ceramics.

The (Gibbs) free energy of melting is given by

$$\Delta G_m = \Delta H_m - T\Delta S_m \tag{8.3}$$

At the equilibrium crystalline melting point T_m, $\Delta G_m = 0$, so

$$T_m = \frac{\Delta H_m}{\Delta S_m}$$

(8.4)

Now ΔH_m is the energy needed to overcome the crystalline bonding forces at constant T and P, and is essentially independent of chain length for high polymers. For a given mass or volume of polymer, however, the shorter the chains are, the more randomized they become upon melting, giving a higher ΔS_m. (For a more detailed description of this in connection with solutions, see Chapter VII.) Thus, the crystalline melting point decreases with decreasing chain length, and in a polydisperse polymer, the distribution of chain lengths gives a distribution of melting points or rounding off, as noted in Fig. 8.2.

Equation 8.4 also indicates that chains that are strongly bound in the crystal lattice, that is, have a high ΔH_m, will have a high T_m, as expected. Also, the stiffer and less mobile chains, those that can randomize less upon melting and therefore have low ΔS_m, will tend to have higher T_m values.

Example 4. Discuss how the crystalline melting point varies with n in the "nylon n" series (Example 1, Chapter VII).

Solution. Increasing values of n dilute the nylon linkages which are responsible for interchain hydrogen bonding, and thus should lower ΔH_m and the crystalline melting point. As n goes to infinity, the structure approaches that of linear polyethylene. This should represent the asymptotic minimum T_m, with the chains held together only by van der Waal's forces.

Table 8.2 illustrates the variation in T_m and some other properties with n for some commercial members of the series.

Example 5. Consider the following classes of linear, aliphatic polymers:

$$\left[O-\overset{\overset{\displaystyle O}{\|}}{C}-\underset{\underset{\displaystyle H}{|}}{N}-\left(CH_2\right)_n \right]_x \qquad \left[\overset{\overset{\displaystyle O}{\|}}{C}-\underset{\underset{\displaystyle H}{|}}{N}-\left(CH_2\right)_n \right]_x \qquad \left[\underset{\underset{\displaystyle H}{|}}{N}-\overset{\overset{\displaystyle O}{\|}}{C}-\underset{\underset{\displaystyle H}{|}}{N}-\left(CH_2\right)_n \right]_x$$

Polyurethanes		Polyamides		Polyureas
T_m	$<$	T_m	$<$	T_m

For given values of n and x the crystalline melting points increase from left to right, as indicated. Explain.

Solution. The polyurethane chains contain the $-O-$ swivel, thus they are the most flexible, have the largest ΔS_m, and have the lowest T_m. Hydrogen bonding, and thus ΔH_m, is roughly comparable in the polyurethanes and polyamides. The polyureas and polyamides should have chains of comparable flexibility (no

Table 8.2   Variation of Properties with *n* for Nylon *n*'s

n	T_m (°C)	ρ (g/cm^3)	Tensile strength (psi)	Water absorption % in 24 h
6	216	1.14	12 000	1.7
11	185	1.04	8 000	0.3
12	177	1.02	7 500	0.25
∞ (PE)	135	0.97	5 500	nil

swivel), but with the extra –NH–, the polyureas form stronger or more extensive hydrogen bonds, and therefore have a higher ΔH_m than the polyamides.

Example 6. Experiments show that uniaxial orientation (drawing) increases the crystalline melting point. Explain.

Solution. The entropy of melting ΔS_m is the difference between the entropies of the amorphous and crystalline materials:

$$\Delta S_m = S_a - S_c$$

Drawing aligns the amorphous chains in the direction of stretch, increasing order, reducing S_a and thus ΔS_m. According to (8.4), this increases T_m.

Example 7. Liquid–crystalline polyesters (Section 5.7) have significantly higher T_m values than non-LC polyesters. For example, Xydar® has a T_m value in the vicinity of 400°C, whereas polyethylene terephthalate (Example 2.4D) has a T_m of 267°C and polybutylene terephthalate (Example 2.4C) has a T_m of 224°C. Explain.

Solution. Here, one might expect the ordinary polyesters to have higher ΔH_m's, particularly the PET, because of the decreased spacing between the polar ester linkages, so the explanation must lie in the ΔS_m's. Because thermotropic LCPs by definition maintain considerable order in the melt state, they randomize less upon melting and therefore have lower S_m's than non-LC polyesters.

The crystalline melting point also increases a bit with the degree of crystallinity of a polymer. For example, low-density polyethylene (ca. 50% crystalline) has a T_m of about 115°C, whereas high density polyethylene (ca. 80% crystalline) melts at about 135°C. This can be explained by treating the amorphous material as an impurity. It is well known that the introduction of an impurity lowers the melting point of common materials. In a similar fashion, greater amounts of noncrystalline "impurities" lower the crystalline melting point of a polymer.

8.7 THE METASTABLE AMORPHOUS STATE

Since polymer chains are largely immobilized below T_g, if they are cooled rapidly through T_m to below T_g, it is sometimes possible to obtain a metastable amorphous state in polymers which at equilibrium would be crystalline. As long as the material is held below T_g, this metastable amorphous state will persist indefinitely. When annealed above T_g (and below T_m), the polymer will crystallize, as the chains gain the mobility necessary to pack into a lattice.

Polyethylene terephthalate (Chapter II, Example 4D), because of its bulky chain structure, crystallizes sluggishly and is therefore relatively easy to obtain in the metastable amorphous state. When desired, crystallinity can be promoted by slow cooling, annealing, or the addition of nucleating agents. PET has important commercial applications in both the metastable amorphous state (soda bottles) and in the crystalline state (textile fibers, microwaveable food trays, molding resin). On the other hand, no one has yet succeeded in producing an amorphous polyethylene, with its much more flexible chains, although the degree of crystallinity can be reduced substantially by rapid cooling (Fig. 8.2). Interestingly, metallurgists have fairly recently produced amorphous metals by rapidly cooling certain alloys.

Example 8. Polyethylene terephthalate (Mylar®, Dacron®) is cooled rapidly from 300°C (state 1) to room temperature. The resulting material is rigid and perfectly transparent (state 2). The sample is then heated to 100°C (state 3) and maintained at that temperature, during which time it gradually becomes translucent (state 4). It is then cooled down to room temperature and is again found to be rigid, but is now translucent rather than transparent (state 5). For this polymer, $T_m = 267$°C and $T_g = 69$°C. Sketch a general specific-volume vs. temperature curve for a crystallizable polymer, illustrating T_g and T_m, and show the locations of states 1–5 for the sample above.

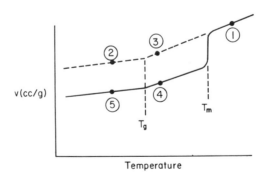

Figure 8.5 Specific volume–temperature relation for crystallizable polymer. Numbers apply to Example 8.

Solution. Figure 8.5 illustrates the general v vs. T curve for a crystallizable polymer. The dotted upper portion represents the metastable amorphous material obtainable by rapid cooling. The history above is shown on the diagram. The metastable amorphous material (transparent, state 2) obtained by rapid cooling to below T_g crystallizes upon annealing between T_g and T_m (state 3 to state 4). The schematic DSC curve in Fig. 8.3 illustrates what would be observed in a DSC measurement starting with a material in the metastable amorphous state (state 2) and heating it to above T_m (state 1).

The greater mobility of chains crystallized just below T_m accounts for their higher degree of crystallinity as observed in Fig. 8.2.

8.8 THE INFLUENCE OF COPOLYMERIZATION ON PROPERTIES

The influence of random copolymerization on T_m and T_g is interesting and technologically important. Occasionally, if two repeating units are similar enough sterically to fit into the same crystal lattice, random copolymerization will result in copolymers whose crystalline melting points vary monotonically with composition between those of the pure homopolymers. Much more common, however, is the case where the homopolymers form different crystal lattices because of steric differences. The random incorporation of minor amounts of repeating unit B with A will disrupt the A lattice, lowering the T_m beneath that of homopolymer A, and vice versa. In an intermediate composition range, the disruption will be so great that no crystallites can form, and the copolymers will be completely amorphous. A phase diagram for such a random copolymer system is shown in Fig. 8.6.

The physical properties of random copolymers are determined by their location on the diagram, that is, by their composition and use temperature. In region 1, the polymer is a homogeneous, amorphous, and, if pure, transparent material. The distinction between melt and rubbery behavior is not sharp: At higher temperatures, the material flows more easily and becomes less elastic in character. It should be kept in mind, though, that the viscosities of polymer melts, even well above T_g or T_m, are far greater than those encountered in nonpolymeric materials.

A copolymer in region 5 is a typical amorphous, glassy polymer: hard, rigid, and usually brittle. Again, if the polymer is pure, it will be perfectly transparent. Polymethyl methacrylate (Lucite®, Plexiglas®) and polystyrene are familiar examples of homopolymers with these properties.

Copolymers in regions 2 and 3 consist of rigid crystallites dispersed in a relatively soft, rubbery, amorphous matrix. Since the refractive indices of the crystalline and amorphous phases are in general different, materials in these regions will be translucent to opaque, depending on the size of the crystallites, the degree of crystallinity, and the thickness of the sample. Since the crystallites

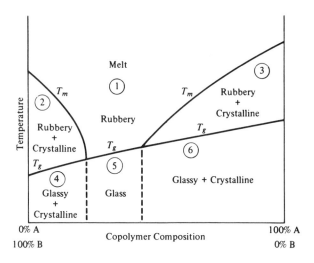

Figure 8.6   Phase diagram for a random copolymer system.

restrict chain mobility, the materials are not elastic, but the rubbery matrix confers flexibility and toughness. The stiffness depends largely on the degree of crystallinity; the more rigid crystalline phase present, the stiffer the polymer. Polyethylene (squeeze bottles, bleach bottles, etc.) is a good example of a homopolymer with these properties, typical of a polymer that is between its T_g and T_m at room temperature.

Copolymers in regions 4 and 6 consist of crystallites in an amorphous, glassy matrix. Since both phases are rigid, the materials are hard, stiff, and rigid. Again, the two phases impart opacity. Nylons 6/6 and 6 are examples of homopolymers in a region below both T_g and T_m at room temperature.

8.9   GENERAL OBSERVATIONS ABOUT T_g and T_m

Some other useful observations regarding T_m and T_g are that for polymers with a symmetrical repeating unit, such as polyethylene $\{CH_2-CH_2\}_x$ and polyvinylidene chloride (Saran®) $\{CH_2-CCl_2\}_x$, $T_g/T_m \approx \frac{1}{2}$ (absolute T's); for unsymmetrical repeating units, such as polypropylene $\{CH_2-CHCH_3\}_x$ and polychlorotrifluoroethylene $\{CF_2-CFCl\}_x$, $T_g/T_m \approx \frac{2}{3}$. In all cases $T_g < T_m$.

8.10   EFFECTS OF CROSSLINKING

To this point, the discussion has centered on non-crosslinked polymers. Light crosslinking, as in rubber bands, won't alter things appreciably. Higher degrees of crosslinking, however, if formed in the amorphous melt state, as is usually the case, will hinder the alignment of chains necessary to form a crystal lattice, and reduce or prevent crystallization. Similarly, crosslinking restricts chain mobility and causes an increase in the apparent T_g. When the crosslinks are more frequent than every 40–50 main-chain atoms, the type 2 motions necessary to reach the rubbery state can never be achieved and the polymer will degrade before reaching T_g.

8.11   OTHER TRANSITIONS

Transitions other than T_g and T_m are sometimes observed in polymers. Some polymers possess more than one crystal form, so there will be an equilibrium temperature of transition from one to another. Similarly, second-order transitions below T_g occur in some materials (T_g is then termed the α transition, the next lower the β, etc.). These are attributed to motions of groups of atoms smaller than those necessary to produce T_g (type 3 motions, Section 8.2). These transitions may strongly influence properties. For example, *tough* amorphous plastics (e.g., polycarbonate) have such a transition well below room temperature, while *brittle* amorphous plastics (e.g., polystyrene and polymethyl methacrylate) do not.

The existence of another transition above T_g has been claimed, but is still the subject of considerable controversy. This T_{ll} (*liquid–liquid* transition) presumably represents the boundary between type 1 and type 2 motions. It has been observed in a number of systems,[5-7] and it has been suggested that $T_{ll} \approx 1.2\, T_g$ for all polymers.[5] For each article that reports T_{ll}, however, it seems that there is another that claims that T_{ll} results from impurities (traces of solvent or unreacted monomer) in the sample or is an artifact of the experimental or data-analysis technique.[8,9]

REFERENCES

1. Meares, P., *Trans. Faraday Soc.* **53**, 31 (1957).

2. Mandelkern, L., *Rubber Chem. Technol.* **32**, 1392 (1959)

3. Wunderlich, B., *Thermal Analysis*, Academic, San Diego, 1990.

4. Tobolsky, A. V., *Properties and Structure of Polymers*, Wiley, New York, 1960, Chapter 2.

5. Kumar, P. L., et al., *Org. Coat. Plast. Chem.* **44**, 396 (1981).

6. Ibar, J. P., *Polym. Prepr.* **22**(2), 405 (1981).

7. Boyer, R. F., *Macromolecules* **15**(6), 1498 (1982).

8. Plazek, D. J., et al., *J. Polym. Sci., Polym. Phys. Ed.* **20**(9), 1533, 1551, 1565, 1575 (1982).

9. Loomis, L. D., and P. Zollar, *J. Polym. Sci., Polym. Phys. Ed.* **21**(2), 241 (1983).

PROBLEMS

1. Pure nylon 6/6 (Example 4E, Chapter II) is a white, opaque plastic at room temperature. If isophthalic acid (two acid groups *meta* on a benzene ring) is

substituted for the adipic acid, a rigid, transparent plastic is obtained. By varying the ratio of the two acids, one can get a series of polyamide copolymers. Sketch a phase diagram like Fig. 8.6 for these copolymers. Be sure to show which repeating unit goes with each composition extreme, and show room temperature.

2. A series of DSC runs is made on an amorphous polymer starting at room temperature but using different heating rates ($1°C/min$, $2°C/min$, $5°C/min$, etc.). Sketch qualitatively how you think the observed T_g will vary with heating rate.

3. Injection molding consists of squirting a molten polymer into a cold mold. When thick parts are molded from a crystalline polymer, e.g., polypropylene, they sometimes exhibit "sink marks," where the surface of the part has actually sunk away from the mold wall. Explain why. *Hint*: Polymers have very low thermal diffusivities.

4. Where it is necessary to maintain precise dimensional control over injection-molded (see above) parts, amorphous polymers are generally preferred to crystalline polymers. Why?

5. Two diols, ethylene glycol (Example 4D, Chapter II) and bisphenol-A (Example 4N, Chapter II) are commercially available at low cost. Which would you choose for polyesterification with a diacid if your object was to

   a. Produce a transparent polyester?

   b. Obtain the higher T_g?

6. The rate of crystallization can be conveniently studied by observing spherulite growth on a temperature-controlled microscope stage. The sample is heated well above T_m and then cooled rapidly to a constant temperature at which the isothermal rate of crystallization is observed. Sketch qualitatively the rate of crystallization vs. crystallization temperature. Be sure to show T_g and T_m on your sketch.

7. Most plastic soda bottles are made from polyethylene terephthalate (PET) (Example 4D, Chapter II). One manufacturer is pushing a copolymer in which some of the ethylene glycol is replaced by cyclohexanedimethanol (CHDM) for that application. What advantage(s) do you think it might have?

8. Sketch a phase diagram like Fig. 8.6 for the ethylene–propylene copolymer system. Label the composition axis. Show the location of ethylene–propylene rubber (EPR), a 65/35 copolymer, on your sketch.

9. An amorphous emulsion copolymer consisting of 60% (by weight) methyl methacrylate (homopolymer $T_g = 105°C$) and 40% ethyl acrylate (homopolymer $T_g = -23°C$) has been proposed as the basis of a latex paint formulation. A latex consists of tiny ($\approx 1–10\,\mu m$) polymer particles suspended in water. After application of the paint, these particles must coalesce

to form a film upon evaporation of the water.

a. Would this copolymer be suitable?

b. An actual paint formulation based on this copolymer contains some medium-volatility solvent dissolved in the latex particles. This solvent is designed to evaporate over a period of a few hours after application of the paint. What's it doing there?

10. A patent claims a new polymer which forms strong, highly crystalline parts when injection molded. Furthermore, it is claimed that $T_g > T_m$ for this material. Comment.

11. High molecular weight linear polyesters from 1,4-butanediol $(HO{(CH_2)}_4OH)$ and terephthalic acid $(para\text{-}\phi(COOH)_2)$ are successful engineering plastics. They are not used as a blister (see-through) packaging material, however. A polymer for the blister-packaging market is made from the two monomers above plus isophthalic acid $(meta\text{-}\phi(COOH)_2)$. Explain the difference in applications.

12. Addition copolymers of styrene with minor amounts of maleic anhydride (Example 3, Chapter II) are used in automotive applications. What advantages do you think these materials would have over homopolystyrene? (*Note*: This problem can be solved in the context of Chapter VII as well as Chapter VIII.)

13. Polyethylene $(T_m = 135°C, \; T_g < T_{room})$ may be lightly crosslinked by a chemical reaction with an organic peroxide at $175°C$. Heat-shrink tubing is made from such a crosslinked polyethylene. When heated from room temperature to about $150°C$, its diameter shrinks by a factor of three or four. How is it made and why does it behave as it does? *Hints*: Consider what keeps it from shrinking at room temperature and what the driving force for shrinkage might be.

14. A diagram like Fig. 8.1 is prepared by heating a polymer well above its T_g and then rapidly cooling it to the desired temperature and holding it there as v is measured. Sketch v vs. T curves for v measurements taken 1 min after cooling and 100 h after cooling. Illustrate how this would affect the value of T_g obtained.

15. Professor Irving Inept of MIT (Monongahela Institute of Technology) figures that he can pad his publication list by publishing a series of polymer tables. Like the steam tables, they will contain the thermodynamic properties of various polymers as functions of temperature and pressure. Discuss the difficulties associated with

a. Obtaining the necessary data

b. Applying the published numbers in practice.

16. Sketch on a copy of Fig. 8.5 the path of a DSC test on a material starting in state 2 and heated to above T_m.

PART **II**

POLYMER SYNTHESIS

CHAPTER IX

Step-Growth (Condensation) Polymerization

9.1 INTRODUCTION

Polymerization reactions may be divided into two categories according to the mechanism by which the chains grow. In *step-growth polymerization*, also known as condensation polymerization, chains of *any* lengths x and y combine to form longer chains:

$$x\text{-mer} + y\text{-mer} \rightarrow (x + y)\text{-mer} \qquad \text{(step growth)} \qquad (9.1a)$$

In *chain-growth* (addition) *polymerization*, a chain of any length x can only add a monomer molecule and continue its growth:

$$x\text{-mer} + \text{monomer} \rightarrow (x + 1)\text{-mer} \qquad \text{(chain growth)} \qquad (9.1b)$$

In Chapter II, we saw some examples of reactions that lead to what are normally considered condensation polymers, but that do not split out a molecule of condensation (Example 4F and Q). By classifying the reactions according to either of the above mechanisms, we avoid any confusion. Step-growth polymerization is treated in this chapter and various types of chain growth are discussed in Chapters X and XI.

Regardless of the type of polymerization reaction, quantitative treatments are usually based on the assumption that the reactivity of the functional group at a chain end is independent of the length of the chain, x and y in (9.1a) and (9.1b) above. Experimentally, this is an excellent assumption for x's greater than about five or six. Since most polymers must develop x's on the order of a hundred or so to be of practical value, this assumption introduces little error while enormously simplifying the mathematical treatment. With these basic concepts in mind, we proceed to a more detailed and quantitative treatment of polymerization reactions.

9.2   STATISTICS OF LINEAR STEP-GROWTH POLYMERIZATION

Consider the two equivalent linear step-growth reactions, assuming difunctional monomers and stoichiometric equivalence,

$$\frac{x}{2}\text{ARA} + \frac{x}{2}\text{BR}'\text{B} \rightarrow \text{ARA}\text{[BR}'\text{B–ARA]}_{\left(\frac{x}{2}-1\right)}\text{BR}'\text{B*} \tag{9.2a}$$

$$x\text{ARB} \rightarrow \text{ARB[ARB]}_{x-2}\text{ARB} \tag{9.2b}$$

where A and B represent the complementary functional groups. It is important to note that x is used here to denote the number of *monomer residues* or *structural units* in the chains, rather than repeating units.

Each of the polymer molecules above contains a total of x A groups:

1 *unreacted* A group (on the end)

$x - 1$ *reacted* A groups

Let p = probability of finding a reacted A group

= number or mole fraction of reacted A groups present

= *conversion* or extent of reaction of A groups

$(1 - p)$ = the probability of finding an unreacted A group.

N = the total number of molecules present in the reaction mass (of all sizes)

n_x = the number of molecules containing x A groups, both reacted and unreacted $(N = \sum\limits_{x=1}^{\infty} n_x)$.

The total probability of finding a molecule with x A groups (reacted and unreacted) is equal to the mole or number fraction of those molecules present in the reaction mass n_x/N. This in turn is equal to the probability of finding a molecule with $x - 1$ reacted A groups and one unreacted A group. Since the overall probability is the product of the individual probabilities

$$\frac{n_x}{N} = p^{(x-1)}(1 - p) = \text{mole (number) fraction } x\text{-mer} \tag{9.3}$$

This is the so-called most probable distribution. It results from the random

---

* Some of the molecules formed in this type of reaction will be capped with two A groups. For each of these, however, stoichiometric equivalence requires that there be one capped by two B groups, so (9.2a) represents the *average* reaction.

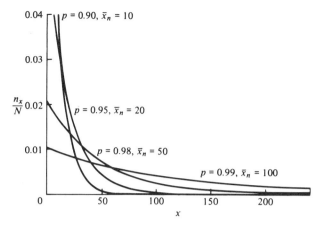

Figure 9.1 Number- (mole-) fraction distributions for linear step-growth polymers (Eq. 9.3).

nature of the reaction between chains of different length. The distribution is plotted in Fig. 9.1 for several conversions. Note that at any conversion, the shorter chains are always more numerous, that is, the longer the chain, the fewer there are. Also, the fraction of short chains decreases with conversion, while the fraction of long chains increases, because increasing conversion hooks together shorter chains to form longer chains.

9.3 NUMBER-AVERAGE CHAIN LENGTHS

Since a distribution of chain lengths always arises in these reactions, the average chain length is often of interest. Each molecule in the reaction mass has, on the average, one unreacted A group. If N_0 is the original number of molecules present in the reaction mass, at the start of the reaction there are N_0 unreacted A groups present. At some later time during the reaction there are N molecules, and therefore N unreacted A groups present, so $N_0 - N$ of the A groups have reacted. Thus,

$$p = \frac{N_0 - N}{N_0} = \text{fraction of reacted A groups present} \qquad (9.4)$$

Because the N_0 original monomer molecules are distributed among the N molecules present in the reaction mass, the average number of monomer residues per molecule $\bar{x}_n$ is given by

$$\bar{x}_n = \frac{N_0}{N} = \text{average number of monomer residues per molecule} \qquad (9.5)$$

Combining (9.4) and (9.5) gives

$$\bar{x}_n = \frac{1}{1-p} \qquad \text{(Carothers' equation)} \qquad (9.6)$$

This rather simple conclusion was reached by W. H. Carothers, the discoverer of nylon and one of the founders of polymer science, in the 1930s, but it is often ignored to this day. (The same result may be obtained in a more laborious fashion by inserting (9.3) into (6.20) and turning the crank.) Its importance becomes obvious when it is realized that typical linear polymers must have $\bar{x}_n$ values on the order of 100 to achieve useful mechanical properties. This requires a conversion of at least 99%, assuming difunctional monomers in perfect stoichiometric equivalence. Such high conversions are almost unheard of in most organic reactions, but are necessary to achieve high molecular weight condensation polymers.

Since all reactions are reversible, many polycondensations would reach equilibrium at low conversions if the molecule of condensation were not efficiently removed (e.g., with heat and vacuum or by a second reaction) to drive the reaction to high conversions.

Example 1. Crud Chemicals is producing a linear polyester in a batch reactor by the condensation of a hydroxy acid HORCOOH. It has been proposed to follow the progress of the reaction by measuring the amount of water removed from the reaction mass. Assuming pure monomer, and that all the water of condensation can be removed and measured, derive for them an equation relating $\bar{x}_n$ of the polymer to N_0, the moles of monomer charged, and M, the total moles of water evolved since the start of the reaction.

Solution. In this polycondensation reaction, each reacted –OH (or –COOH) group produces one molecule of water. Therefore, the moles of reacted –OH (or –COOH) groups is M. The probability of finding a reacted group (of either kind) is

$$p = \frac{M}{N_0} = \frac{\text{moles reacted –OH groups}}{\text{total moles –OH groups}}$$

Plugging into Carothers' equation (9.6) yields

$$\bar{x}_n = \frac{1}{1 - (M/N_0)}$$

If the restriction on stoichiometric equivalence is removed, similar (but more involved) reasoning leads to

$$\bar{x}_n = \frac{1+r}{2r(1-p) + (1-r)} = \frac{1+r}{1+r-2rp} \qquad (9.7)$$

where $r = N_{A0}/N_{B0}$ is the stoichiometric ratio of functional groups present (A is taken to be the limiting reactant, so r is less than 1, and p represents the fraction of A groups reacted).

To examine the effect of stoichiometric imbalance, consider the limiting case of complete conversion, $p = 1$, for which Eq. 9.7 reduces to

$$(\bar{x}_n)_{max} = \frac{1 + r}{1 - r} \qquad \text{(for } p = 1) \tag{9.8}$$

At $r = 1$, all the reactant molecules are combined in a single molecule of essentially infinite molecular weight. Reducing r to 0.99 cuts $(\bar{x}_n)_{max}$ to 199, and to 0.95 all the way down to 39. So to reach high chain lengths, a close approach to stoichiometric equivalence is required in addition to high conversions. Considering the usual industrial purity levels and the precision of weighing techniques, this is not always easy to achieve. Fortunately, in the production of nylon 6/6 from adipic acid and hexamethylene diamine (Chapter II, Example 4E), the monomers form an ionic salt of perfect 1:1 ratio ("nylon salt"), which is carefully purified by crystallization before polymerization. Also, chain length may sometimes be pushed up despite an initial stoichiometric imbalance by eliminating an excess reactant. For example, if a polyester is made from a diacid plus an excess of glycol, the reaction must stop at a point when all the chains are capped by OH groups. If the glycol is volatile enough to be driven off with the application of heat and vacuum, however, the chains may combine further with the elimination of excess glycol:

$$\text{HO–R–O} \left[\overset{\overset{\text{O}}{\|}}{\text{C}} \text{–R}' \text{–} \overset{\overset{\text{O}}{\|}}{\text{C}} \text{–O–R–O} \right]_x \text{H} + \text{H} \left[\text{O–R–O–} \overset{\overset{\text{O}}{\|}}{\text{C}} \text{–R}' \text{–} \overset{\overset{\text{O}}{\|}}{\text{C}} \right]_y \text{O–R–OH}$$

$$\longrightarrow \text{HO–R–O} \left[\overset{\overset{\text{O}}{\|}}{\text{C}} \text{–R}' \text{–} \overset{\overset{\text{O}}{\|}}{\text{C}} \text{–O–R–O} \right]_{(x+y)} \text{H} + \text{HO–R–OH}$$

On the other hand, reducing r by the deliberate introduction of an excess of one of the monomers or of some monofunctional material provides a convenient means of limiting chain length when desired (e.g., see Example 3, Chapter II).

9.4   CHAIN LENGTHS ON A WEIGHT BASIS

The previous development of the distribution of chain lengths in a linear step-growth polymer, although perfectly legitimate, is in some ways misleading, because it describes the *number* of molecules of a given chain length present, and counts equally both monomer units and chains containing many hundreds of monomer units; that is, each is one molecule. For example, in a mixture consisting of one monomer molecule and one 100-mer, the number or mole fraction is $\frac{1}{2}$. Another way of looking at it is to inquire about the relative *weights* of the various chain lengths present. On this basis, the weight fraction monomer in the mixture is only $1/101$. (This assumes that the monomer and repeating unit have the same molecular weight.) By neglecting the weight of the small molecule split out in condensation reactions, we can most simply obtain the weight-fraction distribution of chain lengths by combining (6.29), (9.3), and (9.6) to get

$$\frac{w_x}{W} = x p^{(x-1)} (1-p)^2 \tag{9.9}$$

This "most-probable" weight-fraction distribution is shown in Fig. 9.2. Although the monomer molecules are the most numerous, their combined weight is an insignificant portion of the total weight. The peak is at $x = -1/(\ln p) \approx 1/(1-p)$; that is, the x-mer present in greatest weight is approximately $\bar{x}_n$. The neglect of a molecule of condensation (reacted and unreacted A groups don't have the same weight, but each is part of one monomer residue) in this derivation can lead to significant errors at low x's. The exact solution is quite complex.[1]

The weight-average chain length, $\bar{x}_w$, can be obtained by inserting the number distribution, (9.3), into (6.23) and integrating at constant p:

$$\bar{x}_w = \frac{\int_0^\infty (n_x/N) x^2 \, dx}{\int_0^\infty (n_x/N) x \, dx} = \frac{\int_0^\infty x^2 \, p^{(x-1)} \, dx}{\int_0^\infty x p^{(x-1)} \, dx} = \frac{1+p}{1-p} \tag{9.10}$$

(Evaluation of the above integrals is a good exercise for mathematical masochists.) Combining (9.10) and (9.6) gives an expression for the polydispersity

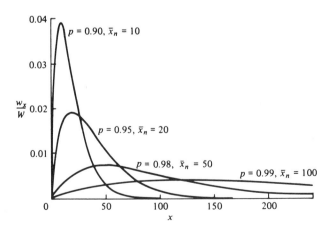

Figure 9.2 Weight-fraction distributions for linear step-growth polymers (Eq. 9.9).

index in terms of conversion:

$$\frac{\bar{x}_w}{\bar{x}_n} = 1 + p \qquad (0 \le p \le 1) \tag{9.11}$$

9.5   GEL FORMATION[2]

If a step-growth polymerization batch contains some monomer with a functionality $f > 2$, and if the reaction is carried to a high enough conversion, a cross-linked network or *gel* may be formed. In this context, a gel is defined as a molecule of essentially infinite molecular weight, extending throughout the reaction mass. In the production of thermosetting polymers, the reaction must be terminated short of the conversion at which a gel is formed, or the product could not be molded or processed further (cross-linking is later completed in the mold). Hence, the prediction of this *gel point* conversion is of great practical importance.

The multifunctional monomer is represented by A_f; for example,

$$
\begin{array}{cc}
\text{A} & \text{A} \\
\overset{|}{\underset{\text{A}\quad\text{A}}{\diagup\hspace{-0.3em}\diagdown}} & \text{A}\!-\!\!\!\overset{|}{\underset{|}{+}}\!\!\!-\!\text{A} \\
& \text{A} \\
f = 3 & f = 4
\end{array}
$$

In a gel network, the multifunctional monomer acts as a *branch unit*. These branch units are connected to *chains*, which are linear segments of difunctional units that lead either to another branch unit or to an unreacted end:

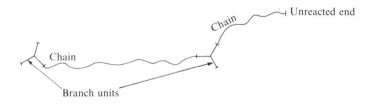

The *branching coefficient* α is defined as the probability that a given functional group on a branch unit is connected to another branch unit by a chain (rather than to an unreacted end).

Further analysis is based on two assumptions: (1) All functional groups are equally reactive. This might not always be true. For example, the secondary (middle) hydroxyl group in glycerine is probably not as reactive as the primary (end) hydroxyls. Furthermore (2), intramolecular condensations (i.e., reactions between A and B groups *on the same molecule*) do not occur. In other words,

$$\bar{f} = \frac{\sum N_i \cdot f_i}{\sum N_i}$$

each reaction between an A and a B reduces the number of molecules in the reaction mass by one.

Consider the reaction

$$AA + BB + A_f \rightarrow A_{f-1}A-[BB-AA-]_i BB-AA_{f-1}$$

Branch unit	Linear chain	Branch unit
↑ 1	↑ 2 ↑ 3	↑ 4

Define ρ = fraction of the *total* A groups in the branch units A_f

$(1 - \rho)$ = fraction of the total A groups in the difunctional AA molecules

p_A = probability of finding a reacted A group

p_B = probability of finding a reacted B group

The probabilities of finding each of the numbered types of bonds in the molecule shown on the right above are tabulated below:

Bond	Probability of formation	
1	p_A,	probability of A reacting with B
2	$p_B(1 - \rho)$,	probability of B reacting with A on AA
3	p_A,	probability of A reacting with B
4	$p_B\rho$	probability of B reacting with A on A_f

Therefore, the overall probability of finding the chain shown, which, counting from left to right, contains 1 type 1 bond, *i* type 2 bonds, *i* type 3 bonds, and one type 4 bond is

$$p_A[p_B(1 - \rho)p_A]^i p_B\rho$$

From the standpoint of connecting branch units, *i* may have any value between zero and infinity. Thus, to obtain the probability of finding a chain of *any i* that connects the branch units, we must sum over all *i*:

$$\alpha = \sum_{i=0}^{\infty} p_A[p_B(1 - \rho)p_A]^i p_B\rho = \frac{p_A p_B\rho}{1 - p_A p_B(1 - \rho)} \tag{9.12}$$

Now,

$$r = \frac{N_{A0}}{N_{B0}} \tag{9.13}$$

$$p_A = \frac{N_{A0} - N_A}{N_{A0}} \tag{9.14}$$

$$p_B = \frac{N_{B0} - N_B}{N_{B0}} \qquad (9.15)$$

From the stoichiometry of the reaction, the number of reacted A groups must equal the number of reacted B groups:

$$N_{A0} - N_A = N_{B0} - N_B \qquad (9.16)$$

Therefore,

$$p_B = \frac{N_{A0}}{N_{B0}} p_A = rp_A \qquad (9.17)$$

and

$$\alpha = \frac{rp_A^2 \rho}{1 - rp_A^2(1 - \rho)} = \frac{p_B^2 \rho}{r - p_B^2(1 - \rho)} \qquad (9.18)$$

Each of the terminal branch units in the molecule on the right above has $(f - 1)$ unreacted functional groups. If at least one of these unreacted A groups is then connected to another branch unit, a gel is formed. Since all the unreacted A groups on branch units are statistically identical, all the branch units in the reaction mass will be connected together by chains, which is our definition of a gel. (In general, there will still be some monomer and linear chains floating throughout the gel network, i.e., the reaction mass is not a single molecule at the gel point.) This occurs when $\alpha(f - 1) \geq 1$, that is, when the product of the probability of a chain connecting a given functional group to another branch unit and the number of remaining unreacted functional groups becomes a certainty. Thus, the critical value of α for gelation is

$$\alpha_c = \frac{1}{f - 1} \qquad (9.19)$$

Combining (9.18) and (9.19) gives the gel-point conversion.

Example 2. Calculate the gel-point conversion of a reaction mass consisting of 2 mol of glycerine plus 3 mol of phthalic anhydride

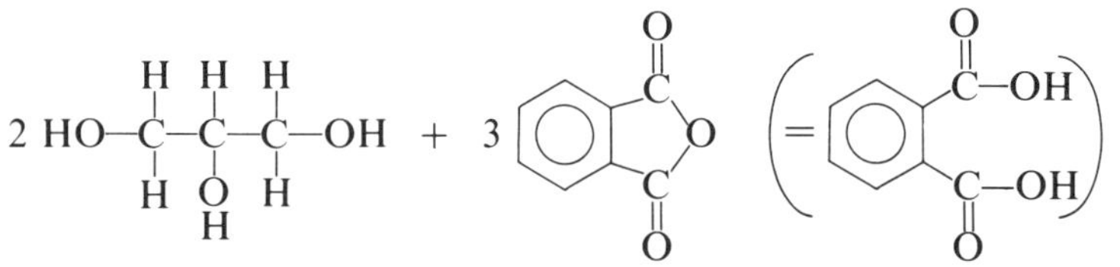

Solution. By convention, A should be chosen as the functional group on the multifunctional monomer. Here, A = OH, $f = 3$, and $\alpha_c = 0.5$ (9.19). Also,

$N_{A0} = N_{B0}$, so $r = 1$. Since all the A (–OH) groups are on the multifunctional (glycerine) molecules, $\rho = 1$. Plugging these values into (9.18) and solving gives $p_A = 0.707$.

Experimentally, the gel-point conversion for this system was found to be $p_A = 0.765$. This was done by noting the conversion at which bubbles ceased to rise in the reaction mass. The difference between the experimental and theoretical results can be accounted for by some intramolecular condensation and the lower reactivity of the secondary hydroxyls. Also, Bobalek et al.[3] showed that as large molecules grew within the reaction mass, their molecular weights became high enough to cause them to precipitate from solution before a true, infinite gel network was formed (a particle just 1 μm in diameter has a molecular weight on the order of 10^{11} g/mol). Higher conversions were needed to link these precipitated particles together to prevent the rise of bubbles.

Regardless of the reasons, however, the theoretical calculation usually gives the *lower limit* or most conservative value for the gel-point conversion, providing a margin of safety in practice. An apparent exception sometimes occurs in the preparation of unsaturated polyester resins (see Example 3, Chapter II). In these calculations, the double bond would be assumed inert in the condensation reaction. Under certain conditions, some addition crosslinking through the double bonds occurs inadvertently during the condensation reaction. This, in effect, makes the value of f much greater than that for condensation alone, which results in premature gelling. Addition crosslinking is normally counteracted by incorporating an inhibitor (see Chapter X) for addition reactions during the polycondensation reaction.

Example 3. Calculate the minimum number of moles of ethylene glycol (I) needed to produce a gel when reacted with one mole of phthalic anhydride (II) and one mole of BTDA (III)

$$HO-CH_2-CH_2-OH$$

(I)   (II)   (III)

Solution. First, recognize that BTDA (III) is a dianhydride, and therefore can be considered a tetrafunctional acid. Again, A is by convention the functional group on the multifunctional monomer. Here A = –COOH, and (I) = BB, (II) = AA, and (III) = A_4.

The tendency toward gelation increases with conversion, and with the minimum moles of (I) present, A groups will be in stoichiometric excess so that B is the limiting reactant. Therefore,

$$(p_B)_{max} = 1.0 \text{ for minimum (I)}$$

$$\alpha_c = \frac{1}{f-1} = \frac{1}{4-1} = \frac{1}{3} \tag{9.19}$$

Here,

$$\rho = \frac{\text{moles A on (III)}}{\text{moles A on [(II) + (III)]}} = \frac{4}{6} = \frac{2}{3}$$

At the gel point, with $p_B = 1$ and $\rho = \frac{2}{3}$,

$$\frac{1}{3} = \frac{rp_A^2\rho}{1 - rp_A^2(1-\rho)} = \frac{p_B^2\rho}{r - p_B^2(1-\rho)} \tag{9.18}$$

$$r = \frac{7}{3} = \frac{N_{A0}}{N_{B0}}$$

$$N_{B0} = \frac{N_{A0}}{r} = 6\left(\frac{3}{7}\right) = \frac{18}{7}$$

But, $N_{B0} = 2N_I$; that is, there are two moles of hydroxyl per mole of glycol. Therefore,

$$N_I = \frac{N_{B0}}{2} = \frac{18}{14} = \frac{9}{7} \text{ mol}$$

9.6 KINETICS OF POLYCONDENSATION

The kinetics of polycondensation reactions is similar to that of ordinary condensation reactions. Since the average chain length is related to conversion in linear polycondensation by (9.7), and conversion is given as a function of time by the kinetic expression, $\bar{x}_n$ is directly related to the reaction time, and can thus be controlled by limiting the reaction time. Similarly, the time to reach a gel point is related by the rate expression, (9.18) and (9.19).

Example 4. For a second-order, irreversible polycondensation reaction with rate proportional to the concentrations of reactive A and B groups, obtain expressions for conversion and number-average chain length as a function of time for a stoichiometrically equivalent batch.

Solution. For a second-order reaction, assuming constant volume,

$$\frac{-d[A]}{dt} = k[A][B] \tag{9.20}$$

where k is the reaction-rate constant and the brackets indicate concentrations. For $[A]_0 = [B]_0$, $[A] = [B]$ at all times, and

$$\frac{-d[A]}{dt} = k[A]^2 \tag{9.21}$$

Separating variables and integrating between the limits when $t = 0$, $[A] = [A]_0$ and when $t = t$, $[A] = [A]$ gives

$$\frac{1}{[A]} - \frac{1}{[A]_0} = kt \tag{9.22}$$

Since

$$p = ([A]_0 - [A])/[A]_0 = ([A]_0 kt)/(1 + [A]_0 kt). \tag{9.23}$$

combining (9.6) and (9.23) gives

$$\bar{x}_n = 1 + [A]_0 kt \tag{9.24}$$

so the number-average chain length increases linearly with time.

Two cautions are in order about the preceding example. First, by writing an *irreversible* rate expression, we have assumed that any molecule of condensation is being continuously and efficiently removed from the reaction mass. Second, not all step-growth reactions are second order. Some polyesterifications, for example, are catalyzed by their own acid groups and are therefore first order in hydroxyl concentration, second order in acid, and third order overall. The rate may also be proportional to the concentration of an added catalyst (usually acids or bases), if used.

Reference 4 provides a more comprehensive and detailed look at step-growth polymerizations, extending and applying many of the concepts introduced here.

REFERENCES

1. Grethlein, H. E., *Ind. Eng. Chem. Fund.*, **8**, 206 (1969).
2. Flory, P. J., *Principles of Polymer Chemistry*, Cornell UP, Ithaca, NY, 1953.
3. Bobalek, E. G., et al., *J. Appl. Polym. Sci.*, **8**, 625 (1964).
4. Gupta, S. K., and A. Kumar, *Reaction Engineering of Step-Growth Polymerizations*, Plenum, New York, 1987.

$$\frac{1}{[A]^2} - \frac{1}{[A]_0^2} = 2kt$$

$$[A] = [A]_0(1-p)$$

$$\bar{x}_n^2 = 1 + 2[A]_0^2 kt$$

PROBLEMS

1. Crud Chemicals wishes to limit chain length in their linear polycondensation reaction by adding monofunctional B to the equimolar AA, BB reactant mix. Obtain an expression for the maximum number-average chain length possible, $(\bar{x}_n)_{max}$, when N_B moles of B are added per mole of AA or BB.

2. One mole of a hydroxy acid, HO–R–COOH, is reacted to 50% conversion. How many moles of unreacted monomer remain in the reaction mass?

3.*a*. See Example 3. Here, calculate the maximum number of moles of ethylene glycol that will form a gel when reacted with one mole of phthalic anhydride and one mole of BTDA.

b. Sketch qualitatively a ternary diagram for the A_4, AA, BB system at maximum conversion which shows the gel and no-gel regions. Locate this solution and the solution to Example 3 on your sketch.

c. Describe what happens to prevent gel formation "just across the border" in each case.

4.*a*. Modify Eq. 9.18 to handle a condensation reaction batch that contains monofunctional B in addition to the A_f, A–A, and B–B present in the original analysis. You will need to include one additional stoichiometric parameter:

$$\beta = \text{fraction of total B's on the B–B molecules}$$

b. Oil-base paints are made by condensing glycerine, phthalic anhydride, and a drying oil. The drying oil is an unsaturated monoorganic acid R–COOH (e.g., linseed oil). The paint film is crosslinked after application by an addition reaction involving the double bonds in R. Since the polymer must be soluble for application, it cannot be allowed to gel during the condensation reaction. With a 2/3 ratio of glycerine/phthalic anhydride, what is the minimum amount of drying oil that must be incorporated in the reaction mass to prevent gel formation during the condensation reaction?

c. For the type of A_f, B–B, B system in (*b*), sketch qualitatively a ternary diagram showing the gel/no-gel composition regions at maximum conversion. Locate the solution to (*b*) on the diagram.

5.*a*. How many moles of monofunctional A must be added to a mixture of one mole of BB and one mole of A_3 to prevent gelation at maximum conversion?

b. Sketch a ternary diagram for the A_3, BB, A system showing the gel/no-gel regions at maximum conversion. Locate the solution to (*a*) on your sketch.

6. Modify the bond–probability of formation table in Section 9.5 to handle a reaction mass that contains monofunctional A in addition to the AA, BB, and A_f present in the original analysis. You will need one additional stoichiometric parameter:

$$a = \text{fraction of A groups on the monofunctional A's.}$$

7. See Example 2. Given 1 mol of glycerine, over what range may the moles of phthalic anhydride be varied with the possibility of forming a gel?

8. Calculate the gel/no gel boundaries and plot the ternary diagram showing the gel/no gel regions for the A_3, BB, B system at maximum conversion.

9. A step-growth polymerization is started with the following initial composition:

A_3	1 mol
AA	1 mol
B	1 mol
BB	2 mol

At $p_A = 0.5$, what is the numerical probability of finding one branch unit connected to another by a single BB?

10. Crud Chemicals is producing a casting resin that consists of an AB-type monomer to which a catalyst is added to promote polymerization in the mold. The material has many useful properties, but is deficient in heat and solvent resistance. A proposal is made to improve heat and solvent resistance by adding a tetrafunctional monomer A_4 to the mix to provide crosslinks. Unfortunately, no one in-house knows if this will work, or how to calculate the minimum amount of A_4 needed to form a gel, etc., so they call you in as a consultant. Evaluate the proposal for them.

11. Calculate the minimum number of moles of phthalic anhydride needed to form a gel when reacted with one mole each of glycerine, ethylene glycol, and propylene glycol.

12. Equal masses of two samples of a linear step-growth polymer are mixed without reaction. One sample has $\bar{x}_n = 100$ and the other $\bar{x}_n = 50$. Calculate the mole fraction monomer in the mixture.

13. Obtain the expression that relates $\bar{x}_n$ to time for a linear polyesterification that is catalyzed by its own acid groups, that is, is second order in COOH and first order in OH. Assume an irreverible reaction at constant volume and $[A]_0 = [B]_0$.

14. Assume that the reaction in Example 2 follows the kinetic scheme outlined in Section 9.6. Obtain an expression (in terms of the rate constant k) for the time to reach the gel point t_{gel}.

CHAPTER X

Free-Radical Addition (Chain-Growth) Polymerization

10.1 INTRODUCTION

One of the most important types of addition polymerization is initiated by the action of *free radicals*, electrically neutral species with an *unshared electron*. In the developments to follow, a dot (·) will represent a single electron. The single bond, a pair of shared electrons, will be denoted by a double dot (:) or, where it is not necessary to indicate electronic configurations, by the usual sign (–). A double bond, *two* shared electron pairs, is indicated by : : or =.

Free radicals for the initiation of addition polymerization are usually generated by the thermal decomposition of organic peroxides or azo compounds. Two common examples are

Benzoyl peroxide

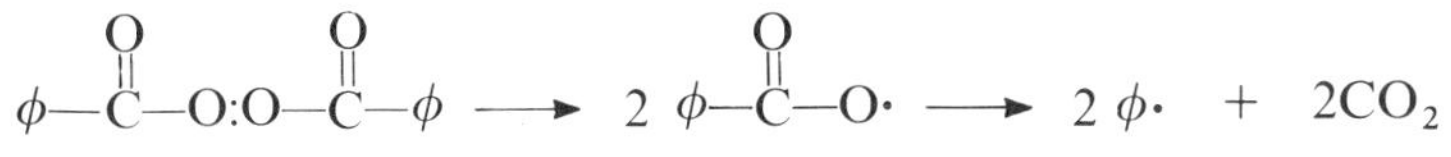

$$\phi—\overset{\overset{O}{\|}}{C}—O:O—\overset{\overset{O}{\|}}{C}—\phi \longrightarrow 2\ \phi—\overset{\overset{O}{\|}}{C}—O· \longrightarrow 2\ \phi· \ + \ 2CO_2$$

Azobisisobutyronitrile

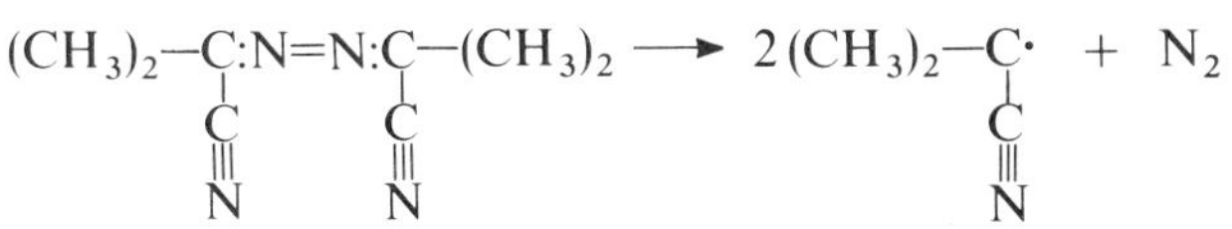

$$(CH_3)_2—\underset{\underset{N}{\overset{\|}{\underset{\|}{C}}}}{C}:N=N:\underset{\underset{N}{\overset{\|}{\underset{\|}{C}}}}{C}—(CH_3)_2 \longrightarrow 2(CH_3)_2—\underset{\underset{N}{\overset{\|}{\underset{\|}{C}}}}{C}· \ + \ N_2$$

10.2 MECHANISM OF POLYMERIZATION

The initiator molecule, represented by I, undergoes a first-order decomposition with a rate constant k_d to give two free radicals, $R·$:

$$I \xrightarrow{k_d} 2\,R\cdot \qquad \text{Decomposition} \qquad (10.1)$$

The radical then adds a monomer by grabbing an electron from the electron-rich double bond, forming a single bond with the monomer, but leaving an unshared electron at the other end:

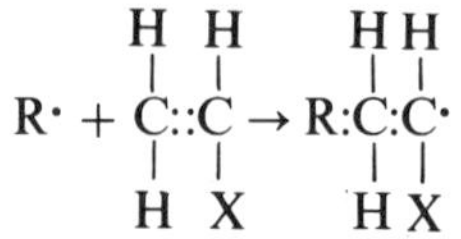

$$
\begin{array}{cc}
\text{H H} & \text{H H} \\
| \ | & | \ | \\
R\cdot + \text{C::C} & \rightarrow \text{R:C:C}\cdot \\
| \ | & | \ | \\
\text{H X} & \text{H X}
\end{array}
$$

This may be abbreviated by

$$R\cdot + M \xrightarrow{k_a} P_1\cdot \qquad \text{Addition} \qquad (10.2)$$

where $P_1\cdot$ represents a growing polymer chain with 1 repeating unit. Note that the product of the addition reaction is still a free radical; it proceeds to propagate the chain by adding another monomer unit:

$$P_1\cdot + M \xrightarrow{k_p} P_2\cdot$$

again maintaining the unshared electron at the chain end, which adds another monomer unit:

$$P_2\cdot + M \xrightarrow{k_p} P_3\cdot$$

and so on. In general, the propagation reaction is written as

$$P_x\cdot + M \xrightarrow{k_p} P_{(x+1)}\cdot \qquad \text{Propagation} \qquad (10.3)$$

We have again assumed that reactivity is independent of chain length by using the same k_p for each propagation step.

Growing chains can be terminated in one of two ways. Two can bump together and stick, their unshared electrons combining to form a single bond between them (*combination*):

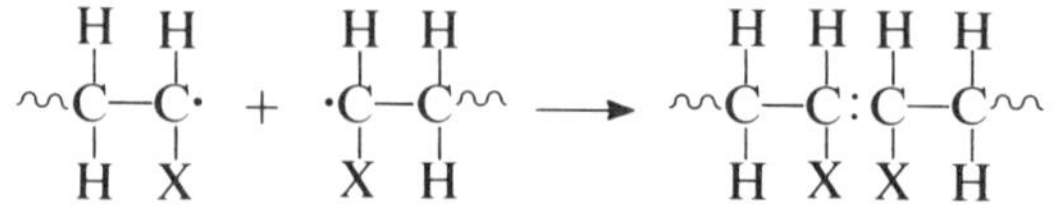

$$
\begin{array}{ccccc}
\text{H H} & & \text{H H} & & \text{H H H H} \\
| \ | & & | \ | & & | \ | \ | \ | \\
\sim\!\!\text{C}-\text{C}\cdot & + & \cdot\text{C}-\text{C}\!\!\sim & \longrightarrow & \sim\!\!\text{C}-\text{C:C}-\text{C}\!\!\sim \\
| \ | & & | \ | & & | \ | \ | \ | \\
\text{H X} & & \text{X H} & & \text{H X X H}
\end{array}
$$

$$P_x\cdot + P_y\cdot \xrightarrow{k_{tc}} P_{(x+y)} \qquad \text{Combination} \qquad (10.4a)$$

where $P_{(x+y)}$ is a *dead* polymer chain of $(x + y)$ repeating units. Or, one can abstract a proton from the penultimate carbon of the other (*disproportionation*):

$$
\underset{\substack{| \; | \\ H \; X}}{\overset{\substack{H \; H \\ | \; |}}{\sim\!\!C\!-\!C\cdot}} \;+\; \underset{\substack{| \; | \\ X \; H}}{\overset{\substack{H \; H \\ | \; |}}{\cdot C\!-\!C\!\sim}} \;\longrightarrow\; \underset{\substack{| \\ X}}{\overset{\substack{H \; H \\ | \; |}}{\sim\!\!C\!=\!C}} \;+\; \underset{\substack{| \; | \\ X \; H}}{\overset{\substack{H \; H \\ | \; |}}{H\!:\!C\!-\!C\!\sim}}
$$

$$P_x\cdot + P_y\cdot \xrightarrow{k_{td}} P_x + P_y \qquad \text{Disproportionation} \qquad (10.4b)$$

The relative proportion of each termination mode depends on the particular polymer and the reaction temperature, but in most cases, one or the other predominates.

Example 1. A free radical initiator $R:R$ $(R:R \to 2R\cdot)$ and a vinyl monomer $H_2C=CHX$ are polymerized in a homogeneous reaction mass. The resulting molecules all have the structures

$$
A\!\!\underset{\substack{| \; | \\ H \; X}}{\overset{\substack{H \; H \\ | \; |}}{\left[C\!-\!C\right]_x}}\!\!B \quad \text{or} \quad D\!\!\underset{\substack{| \; | \\ H \; X}}{\overset{\substack{H \; H \\ | \; |}}{\left[C\!-\!C\right]_x}}\!\!\underset{\substack{| \; | \\ X \; H}}{\overset{\substack{H \; H \\ | \; |}}{\left[C\!-\!C\right]_y}}\!\!E
$$

$$\text{(I)} \qquad\qquad\qquad \text{(II)}$$

Tabulate all possible combinations of (**a**) groups A and B, and (**b**) of groups D and E. Interchanging A and B or D and E are not considered separate combinations.

Solution. Molecules of type I are terminated by disproportionation; those of type II by combination. Therefore,

a.	A	B
	R	H
	R	$-\overset{\substack{H \; H \\ | \; |}}{\underset{\substack{| \\ X}}{C\!=\!C}}$

b.	D	E
	R	R

10.3   KINETICS OF HOMOGENEOUS POLYMERIZATION

In practice, not all the radicals generated in reaction 10.1 actually initiate chain growth as in reaction 10.2. Some recombine or are used up by side reactions.

Letting f = fraction of radicals generated that actually do initiate chain growth, the rate of formation of growing chain radicals through reactions 10.1 and 10.2 is

$$r_i = \left(\frac{1}{V}\right)\left(\frac{dP_1 \cdot}{dt}\right) = 2fk_a[I] \qquad \text{Initiation rate} \qquad (10.5)$$

where V is the volume of the reaction mass, $P_1\cdot$ is the moles of chain radicals with $x = 1$, and $[I]$ is the initiator concentration (moles/volume). In the notation used here, the quantity Q of a species without brackets represents the moles of the species and when enclosed in square brackets, $[Q]$, the *molar concentration* of the species. They are related by

$$[Q] = \frac{Q}{V} \qquad (10.6)$$

Although it is rarely mentioned explicitly, (10.5) is based on the generally valid assumption that the rate of reaction 10.2 is much greater than that of (10.1), that is, that initiator decomposition is *rate controlling*. Thus, as soon as an initiator radical is formed, it grabs a monomer molecule, starting chain growth, so k_a does not appear in the expression.

According to reaction 10.3, the rate of monomer removal in the propagation step is

$$r_p = -\left(\frac{1}{V}\right)\left(\frac{dM}{dt}\right)_p = k_p[M][P\cdot] \qquad \text{Propagation rate} \qquad (10.7)$$

where M is the moles of monomer and $[P\cdot]$ is the total concentration of growing chain radicals,

$$[P\cdot] = \sum_{x=1}^{\infty} [P_x\cdot] \qquad (10.8)$$

The rate of removal of chain radicals is the sum of the rates of the two termination reactions. Since both are second order

$$r_t = -\left(\frac{1}{V}\right)\left(\frac{dP\cdot}{dt}\right) = 2k_t[P\cdot]^2 \qquad \text{Termination rate} \qquad (10.9)$$

where

$$k_t = k_{tc} + k_{td} \qquad (10.10)$$

Unfortunately, the unknown quantity $[P\cdot]$ is present in the equations. It can be removed by making the standard kinetic assumption of a steady-state

QSSA $r_i = r_t$

concentration of a transient species, in this case, the chain radicals. For $[P\cdot]$ to remain constant, chain radicals must be generated at the same rate at which they are removed, $r_i = r_t$, or

$$2fk_d[I] = 2k_t[P\cdot]^2 \tag{10.11}$$

This gives the chain-radical concentration

$$[P\cdot] = \left(\frac{fk_d[I]}{k_t}\right)^{1/2} \tag{10.12}$$

which, when inserted in (10.7) provides an expression for the rate of monomer removal in the propagation reaction

$$r_p = -\left(\frac{1}{V}\right)\left(\frac{dM}{dt}\right)_p = k_p\left(\frac{fk_d}{k_t}\right)^{1/2}[I]^{1/2}\,[M] \tag{10.13}$$

One molecule of monomer is consumed in the addition step (10.2) also, but for long chains this is insignificant, so that (10.13) may be taken as the overall rate of polymerization, the rate at which monomer is converted to polymer.

Equation 10.13 is the classical rate expression for a *homogeneous*, free-radical polymerization. It has been an extremely useful approximation over the years, particularly in fairly dilute solutions, but deviations from it are sometimes observed. It is now evident that many deviations are due to the fact that the termination reaction is diffusion controlled, so k_t is not really a constant. In certain cases of practical interest, these deviations can be of major significance. They are discussed in Chapter XIII on polymerization practice.

It is often more convenient to work in terms of *monomer conversion X*. By making use of the definition of conversion

$$X \equiv \frac{M_0 - M}{M_0} \tag{10.14}$$

and (10.6) for M, we can rewrite (10.13) as

$$\frac{dX}{dt} = k_p\left(\frac{fk_d}{k_t}\right)^{1/2}[I]^{1/2}\,(1 - X) \tag{10.13a}$$

Note that the conversion X used here is a very different quantity than the p used in Chapter IX. Here, X is the fraction of *monomer molecules* that have reacted to form polymer. The quantity p represents the fraction of *functional groups* that have reacted. In step-growth polymerization, each monomer molecule has several functional groups that react independently, so p is not the fraction of reacted monomer.

Integration of (10.13a) gives conversion as a function of time. A form occasionally used in studying the initial stages of an isothermal, batch reaction is obtained by assuming a constant initiator concentration $[I]_0$ with the initial condition $X = 0$ at $t = 0$

$$X = 1 - \exp\left\{ -k_p \left(\frac{fk_d}{k_t}\right)^{1/2} [I]_0^{1/2} t \right\} \qquad \text{(constant } [I]) \qquad (10.15)$$

The assumption of constant $[I]$ may not be realistic at high conversions for two reasons. First, the initiator concentration can decrease significantly with time. Second, the volume of the reaction mass will, in general, change with conversion. Most liquid-phase addition polymerizations undergo a density increase of 10–20% from pure monomer ($X = 0$) to polymer ($X = 1$) (carrying out the reaction in an inert solvent reduces this, of course).

We can include the first-order initiator decay

$$\left(\frac{1}{V}\right)\left(\frac{dI}{dt}\right) = -k_d[I] = -k_d\left(\frac{I}{V}\right) \qquad (10.16a)$$

and integrate from $t = 0$ with $I = I_0$ to get

$$I = I_0 \exp(-k_d t) \qquad (10.16b)$$

$$M = M_0(1-X)$$

We then assume that the volume of the reaction mass is linear in conversion, as outlined by Levenspiel[1]:

$$V = V_0(1 + \varepsilon X) \qquad (10.17)$$

where ε is the fractional change in volume from $X = 0$ to $X = 1$ and V_0 is the volume of the reaction mass at $X = 0$. Combining (10.6) (for I), (10.13), (10.16b), and (10.17) gives

$$\frac{dX}{dt} = k_p \left(\frac{fk_d}{k_t}\right)^{1/2} [I]_0^{1/2} \frac{1 - X}{(1 + \varepsilon X)^{1/2}} \exp\left(\frac{-k_d t}{2}\right) \qquad (10.18)$$

The variables in (10.18) may be separated and it may be integrated analytically with respect to time, at least, to give

$$\int_0^X \frac{(1 + \varepsilon X)^{1/2}}{1 - X} dX = \left(\frac{2k_p}{k_d}\right)\left(\frac{fk_d}{k_t}\right)^{1/2} [I]_0^{1/2} \left[1 - \exp\left(\frac{-k_d t}{2}\right)\right] \qquad (10.19)$$

Unfortunately, the left side of (10.19) can't be integrated analytically by normal human beings. Even for abnormal humans or symbolic math routines it would be difficult to extract an explicit expression for X as a function of t, so calculation of $X(t)$ would have to be done numerically in any case.

If we neglect volume change ($\varepsilon = 0$), we get

$$X = 1 - \exp\left\{\left(\frac{2k_p}{k_d}\right)\left(\frac{fk_d}{k_t}\right)^{1/2}[I]_0^{1/2}\left[\exp\left(\frac{-k_d t}{2}\right) - 1\right]\right\} \quad (\varepsilon = 0)$$

(10.20)

This expression has some interesting and important implications when compared with (10.15). Setting $t = \infty$ reveals that there is a maximum attainable conversion that depends on $[I]_0$:

$$X_{\text{max}} = 1 - \exp\left\{-\left(\frac{2k_p}{k_d}\right)\left(\frac{fk_d}{k_t}\right)^{1/2}[I]_0^{1/2}\right\} \quad (\varepsilon = 0) \qquad (10.21)$$

This "dead-stop" situation is basically a matter of the initiator being used up before the monomer, but it is not revealed by (10.15), which always predicts complete conversion in the limit of infinite time. Thus, the use of (10.15) instead of (10.19) or (10.20) can result in considerable error at high conversions, although the results approach one another at low conversions.

Example 2. For the homogeneous polymerization of pure styrene with azobisisobutyronitrile at 60°C in an isothermal, batch reactor,

   a. Compare equations (10.15), (10.19), and (10.20) by plotting conversion vs. time, using the data given below.

   b. Calculate and compare the times needed to reach 90% conversion according to (10.15), (10.19), and (10.20).

   c. Determine the minimum $[I]_0$ needed to achieve 90% conversion. Note that according to (10.15), there is *no* minimum.

Data: $k_p^2/k_t = 1.18 \times 10^{-3}$ L/mol,

$k_d = 0.96 \times 10^{-5}$ s^{-1}

$[I]_0 = 0.05$ mol/L

$f = 1$

$\varepsilon = -0.136$ (13.6% shrinkage from monomer to polymer)

Solution. **a.** Equations 10.15 ($[I]$ constant) and 10.20 (V constant) may be solved analytically, but (10.19) (both $[I]$ and V changing) requires a numerical solution. The results are shown in Fig. 10.1. Graphically, (10.19) and (10.20) are almost indistinguishable. Note that volume shrinkage increases concentrations and speeds up the reaction slightly.

   b. Numerical solution of (10.19) gives a time of 34.0 h to reach 90% conversion. Neglecting initiator decay and solving (10.15) for t gives 26.9 h (21% low)

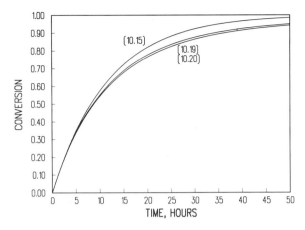

Figure 10.1 Conversion vs. time for an isothermal, free-radical batch polymerization (data of Example 2): Eq. 10.15, $[I]$ = constant; Eq. 10.19, $\varepsilon = -0.136$; Eq. 10.20, $\varepsilon = 0$.

and neglecting volume change (10.20) gives 36.1 h (6.2% high). In many cases, the error introduced by neglecting volume change is probably less than the precision with which the rate constants are known. The error will increase with higher conversions, however.

c. Numerical solution of (10.19) with $t = \infty$ gives a minimum $[I]_0$ of 0.009 89 mol/L for 90% conversion. Solving (10.21) with $X_{max} = 0.90$ gives $[I]_0 = 0.0108$, a 9.07% error from neglecting volume change.

Side reactions sometimes cause deviations from the classical kinetics. Agents that cause these reactions are generally categorized as *inhibitors* or *retarders*. An

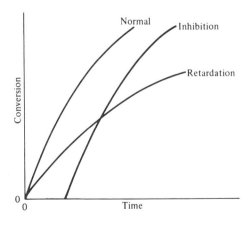

Figure 10.2   Inhibition and retardation.

inhibitor delays the start of the reaction, but once begun, the reaction proceeds at the normal rate. Liquid vinyl monomers are usually shipped with a few parts per million inhibitor to prevent polymerization in transit. A retarder slows down the reaction rate. Some chemicals combine both effects. These are illustrated in Fig. 10.2. Oxygen can be an effective inhibitor for free-radical polymerization, so the reactions are normally carried out under a blanket of nitrogen.

10.4   INSTANTANEOUS AVERAGE CHAIN LENGTHS

As in step-growth polymerization, a distribution of chain lengths is always obtained in a free-radical addition polymerization because of the inherently random nature of the termination reaction with regard to chain length. Expressions for the number-average chain length are usually couched in terms of the *kinetic chain length* v, which is the rate of monomer addition to growing chains over the rate at which chains are started by initiator radicals, that is, the average number of monomer units per growing chain radical at a particular instant. It thus expresses the efficiency of the initiator radicals in polymerizing monomer:

$$v = \frac{r_p}{r_i} = \frac{k_p[M]}{2(fk_dk_t)^{1/2}[I]^{1/2}} \tag{10.22}$$

If the growing chains terminate exclusively by disproportionation, they undergo no change in length in the process; but if combination is the exclusive mode of termination, the growing chains, on the average, double in length upon termination. Therefore,

$$\bar{x}_n = v \qquad \text{(termination by disproportionation)} \tag{10.23a}$$

$$\bar{x}_n = 2v \qquad \text{(termination by combination)} \tag{10.23b}$$

The average chain length may be expressed more generally in terms of a quantity ξ, the average number of dead chains produced per termination reaction, which equals the rate of dead chain formation/rate of termination reactions. Since each disproportionation reaction produces *two* dead chains and each combination reaction *one*,

$$\text{Rate of dead chain formation} = (2k_{td} + k_{tc})[P\cdot]^2 \tag{10.24}$$

$$\text{Rate of termination reactions} = (k_{tc} + k_{td})[P\cdot]^2 \tag{10.25}$$

$$\xi = \frac{k_{tc} + 2k_{td}}{k_{tc} + k_{td}} = \frac{k_{tc} + 2k_{td}}{k_t} \tag{10.26}$$

The instantaneous number-average chain length is the rate of addition of monomer units to all chains (r_p) over the rate of dead chain formation

$$\bar{x}_n = \frac{k_p[P\cdot][M]}{(2k_{td} + k_{tc})[P\cdot]^2} = \frac{k_p[M]}{\xi(fk_dk_t)^{1/2}[I]^{1/2}} \tag{10.27}$$

When $k_{td} \gg k_{tc}$, $\xi = 2$ and when $k_{tc} \gg k_{td}$, $\xi = 1$, duplicating the previous result, but (10.27) can also handle various degrees of mixed termination.

Keeping in mind that $\bar{x}_n$ is one of the most important factors in determining certain mechanical properties of polymers, what is the significance of (10.27)? A growing chain may react with another growing chain and terminate, or it may add another monomer unit and continue its growth. The more monomer molecules in the vicinity of the chain radical, the higher the probability of another monomer addition, hence the proportionality to $[M]$. On the other hand, the more initiator radicals there are present competing for the available monomer, the shorter the chains will be, on the average, causing the inverse proportionality to the square root of $[I]$. Equation 10.13 shows that the rate of polymerization can be increased (always an economically desirable goal) by increasing both $[M]$ and $[I]$, the former being more efficient than the latter. However, there is an upper limit to $[M]$ set by the density of the pure monomer at the reaction conditions. So it is often tempting to increase $[I]$ to attain higher rates. But according to (10.27), this unavoidably lowers $\bar{x}_n$. You can't have your cake and eat it, too.

10.5 TEMPERATURE DEPENDENCE OF RATE AND CHAIN LENGTH

We may assume that the temperature dependence of the individual rate constants in (10.13) is given by the Arrhenius expression:

$$k_i = A_i\exp\left(\frac{-E_i}{RT}\right) \tag{10.28}$$

where k_i is the rate constant for a particular elementary reaction, A_i its frequency factor, and E_i its activation energy. If we neglect the temperature dependence of f, (10.13) becomes

$$r_p = C\exp\left[\frac{-(E_p + E_d/2 - E_t/2)}{RT}\right] \tag{10.29}$$

where C combines temperature-independent quantities into a single constant and $E_p + E_d/2 - E_t/2$ is the *effective* activation energy for polymerization. Therefore, between two (absolute) temperatures T_1 and T_2

$$\ln\frac{r_p(T_2)}{r_p(T_1)} = \frac{-(E_p + E_d/2 - E_t/2)}{R}\left(\frac{1}{T_2} - \frac{1}{T_1}\right) \tag{10.30}$$

Similar treatment of (10.27), with the assumption that ξ is independent of temperature, results in

$$\ln \frac{\bar{x}_n(T_2)}{\bar{x}_n(T_1)} = \frac{-(E_p - E_d/2 - E_t/2)}{R} \left(\frac{1}{T_2} - \frac{1}{T_1} \right) \tag{10.31}$$

where $E_p - E_d/2 - E_t/2$ is the *effective* activation energy for number-average chain length.

Example 3. For a typical free-radical addition reaction, $E_p - E_t/2 \approx 5$ kcal/mol and $E_d \approx 30$ kcal/mol. Estimate the changes in r_p and $\bar{x}_n$ for a typical homogeneous, free-radical addition polymerization on going from 60 to 70°C.

Solution. The effective activation energy for polymerization is $+20$ kcal/mol. Inserting this into (10.30) along with $T_1 = 273 + 60 = 333$ K, $T_2 = 273 + 70 = 343$ K, and $R = 1.99$ cal/mol·K gives

$$\ln \frac{r_p(70°C)}{r_p(60°C)} = 0.879 \quad \text{and} \quad \frac{r_p(70°C)}{r_p(60°C)} = 2.41 \ (!)$$

that is, the rate increases by 141% over the stated 10° temperature interval. This rather spectacular (and sometimes dangerous) sensitivity is due mainly to the large activation energy for the rate-controlling initiator decomposition.

The effective activation energy for $\bar{x}_n$ is -10 kcal/mol. With (10.31) we get

$$\ln \frac{\bar{x}_n(70°C)}{\bar{x}_n(60°C)} = -0.439 \quad \text{and} \quad \frac{\bar{x}_n(70°C)}{\bar{x}_n(60°C)} = 0.644$$

so $\bar{x}_n$ is subject to a 35.6% *decrease* over the same temperature range.

10.6   CHAIN TRANSFER

In practice, another type of reaction sometimes occurs in free-radical addition polymerizations. These *chain-transfer* reactions kill a growing chain radical and start a new one in its place

$$R' \colon\! \ddot{H} + P_x \cdot \xrightarrow{k_{tr}} P_x + R' \cdot \tag{10.32a}$$

$$R' \cdot + M \xrightarrow{k'_a} P_1 \cdot \quad \text{etc.} \tag{10.32b}$$

Thus, chain transfer results in shorter chains, and if reactions 10.32 are not too frequent compared to the propagation reaction and don't have very low rate constants, chain transfer will not change the overall rate of polymerization appreciably.

The compound R′·H is known as a *chain-transfer agent*. Under appropriate conditions, almost anything in the reaction mass may act as a chain-transfer agent, including initiator, monomer, solvent and dead polymer.

Example 4. Show, using dots to represent the electrons involved, how chain transfer to dead polymer leads to long-chain branching in polyethylene.

Solution.

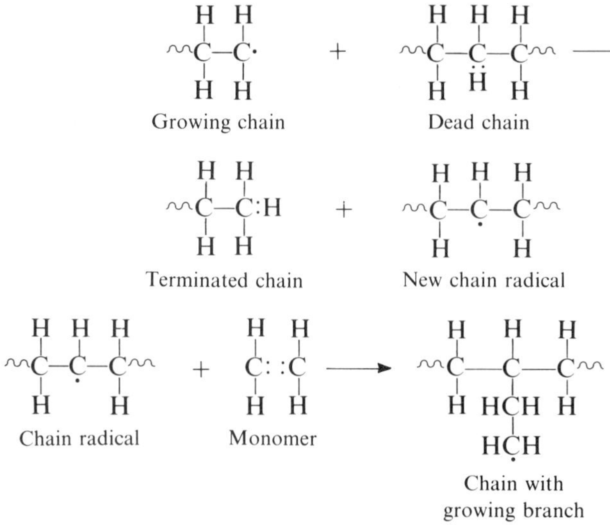

Example 5. Show, as above, how short branches arise in polyethylene when a growing chain "bites its own back," that is, transfers to an atom a few (5–8) carbon atoms down the chain.

Solution.

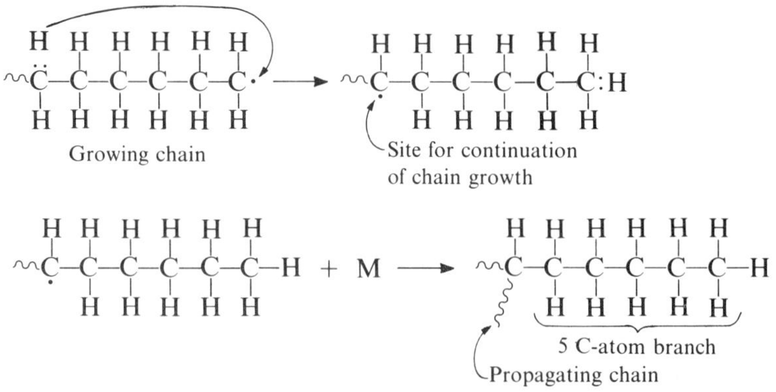

Most frequently, mercaptans, the sulfur analogs of alcohols (R′S:H) are added to the reaction mass as effective chain-transfer agents to lower the average chain length.

Example 6. A mercaptan, R′S:H, is added to the reaction in Example 1. Tabulate the additional possibilities for (***a***) groups A and B and (***b***) groups D and E.

Solution. Chains may now be terminated by the chain-transfer agent and new ones started by it, giving rise to the additional possibilities

a.

A	B	
R′S	H	(started by chain transfer, terminated by transfer or disproportionation)

R′S	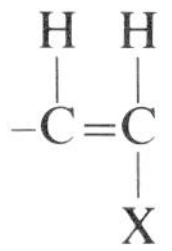 $-\overset{\overset{\displaystyle H}{\mid}}{C}=\overset{\overset{\displaystyle H}{\mid}}{\underset{\underset{\displaystyle X}{\mid}}{C}}$	(started by transfer, terminated by disproportionation)

b.

D	E	
R′S	R′S	(both ends started by transfer)
R	R′S	(one end started by initiator, other by transfer)

The rate of dead chain formation by chain transfer

$$r_{\mathrm{tr}} = -\left(\frac{1}{V}\right)\left(\frac{dP\cdot}{dt}\right)_{\mathrm{tr}} = k_{\mathrm{tr}}[\mathrm{R}':\mathrm{H}][P\cdot] \tag{10.33}$$

must be added to the denominator of (10.27) to give the total rate of dead chain formation:

$$\bar{x}_n = \frac{k_p[P\cdot][M]}{(2k_{td}+k_{tc})[P\cdot]^2 + k_{\mathrm{tr}}[\mathrm{R}':\mathrm{H}][P\cdot]} \tag{10.34}$$

Using (10.12) and (10.26) gives

$$\bar{x}_n = \frac{k_p[M]}{\xi(fk_dk_t[I])^{1/2} + k_{\mathrm{tr}}[\mathrm{R}':\mathrm{H}]} \tag{10.35}$$

Taking the reciprocal of (10.35) gives

$$\frac{1}{\bar{x}_n} = \frac{1}{(\bar{x}_n)_0} + C\frac{[\mathrm{R}':\mathrm{H}]}{[M]} \tag{10.36}$$

where C is the *chain-transfer constant* $= k_{tr}/k_p$ and $(\bar{x}_n)_0$ is simply the average chain length in the absence of transfer, (10.27). Thus, a plot of $1/\bar{x}_n$ vs. $[R':H]/[M]$ is linear with a slope of C and an intercept of $1/(\bar{x}_n)_0$.

Example 7. Prove that the use of a chain-transfer agent with $C = 1$ will maintain the ratio $[R':H]/[M]$ constant in a batch reaction.

Solution

$$\frac{r_{tr}}{r_p} = \frac{-(1/V)(dR':H/dt)}{-(1/V)(dM/dt)} = \frac{d[R':H]}{d[M]} = \frac{k_{tr}[R':H][P\cdot]}{k_p[M][P\cdot]} = C\frac{[R':H]}{[M]}$$

Separating variables and integrating gives

$$\int_{[R':H]_0}^{[R':H]} \frac{d[R':H]}{[R':H]} = C\int_{[M]_0}^{[M]} \frac{d[M]}{[M]}$$

$[M] = [M]_0 (1-x)$ $\dfrac{[R':H]}{[R':H]_0} = \left(\dfrac{[M]}{[M]_0}\right)^C = (1-x)^C$ (10.37)

Therefore, only if $C = 1$ will $([R':H]/[M]) = ([R':H]_0/[M]_0) = $ constant.

Values of C greater than one use up chain-transfer agent too quickly, leaving nothing to modify the polymer formed at high conversions, while low values leave a lot of unreacted chain-transfer agent around toward the end of the reaction when enough is used to be effective at the beginning. If the agent happens to be a mercaptan, and if you have ever smelled a mercaptan, you can appreciate what a problem that can be. These problems are circumvented by (1) adding an active $(C > 1)$ agent over the course of the reaction, and (2) using mixtures of active $(C > 1)$ and sluggish $(C < 1)$ agents in the initial charge.

Example 8. Generalize Eq. 10.36 to incorporate transfer to a variety of different possible transfer agents, S_i:

$$P_x\cdot + S_i \xrightarrow{k_{tr\cdot i}} P_x + S_i\cdot$$

Solution. The rate of dead chain formation from transfer to each agent S_i must be added to the denominator of (10.27):

$$\bar{x}_n = \frac{k_p[P\cdot][M]}{(2k_{td} + k_{tc})[P\cdot]^2 + \sum_i k_{tr,i}[S_i][P\cdot]}$$

Invoking (10.12) and (10.26) gives

$$\bar{x}_n = \frac{k_p[M]}{\xi(fk_dk_t[I])^{1/2} + \sum_i k_{tr,i}[S_i]}$$

or

$$\frac{1}{\bar{x}_n} = \frac{1}{(\bar{x}_n)_0} + \sum_i C_i \frac{[S_i]}{[M]}$$

where

$$C_i = \frac{k_{tr,i}}{k_p}$$

(10.7) INSTANTANEOUS DISTRIBUTIONS IN FREE-RADICAL ADDITION POLYMERIZATION

A growing chain $P_x\cdot$ can either propagate, terminate, or transfer. The *propagation probability* q is the probability that it will propagate rather than transfer or terminate and is given by

$$q = \frac{\text{Rate of propagation}}{\text{Rate of propagation} + \text{Rate of transfer} + \text{Rate of termination}}$$

$$q = \frac{k_p[M][P\cdot]}{k_p[M][P\cdot] + k_{tr}[R':H][P\cdot] + 2k_t[P\cdot]^2} \qquad (10.38)$$

which becomes, with the aid of (10.12),

$$q = \frac{k_p[M]}{k_p[M] + k_{tr}[R':H] + 2(fk_dk_t[I])^{1/2}} \qquad (10.39)$$

To get chain-length distributions from q we must separately consider termination by disproportionation and combination because the two mechanisms give rise to different distributions. First, consider only those chains whose growth is terminated by either disproportionation ($k_t = k_{td}$, $\xi = 2$) or chain transfer. The resulting distributions are the same, because a growing chain doesn't know if it has been killed by an encounter with another growing chain or with a chain-transfer agent. Either way, its growth stops with no change in length (that's not the case for combination). By taking the reciprocal of (10.39) we get

$$\frac{1}{q} = 1 + \frac{k_{tr}[R':H]}{k_p[M]} + \frac{2(fk_d k_{td}[I])^{1/2}}{k_p[M]} \tag{10.40}$$

$$\underbrace{\qquad\qquad\qquad\qquad\qquad\qquad\qquad\qquad}_{1/\bar{x}_n \text{ for } \xi = 2 \text{ (10.35)}}$$

Therefore,

$$\bar{x}_n = \frac{q}{(1-q)} \tag{10.41}$$

Now, a polymer molecule containing x units, P_x, has been formed by $x - 1$ propagation steps (the remaining unit was incorporated in the addition step), each with a probability q, and one disproportionation or transfer step with a probability $1 - q$. The probability of finding such a molecule is equal to its number (mole) fraction; therefore,

$$\frac{n_x}{N} = (1 - q)\, q^{(x-1)} \tag{10.42}$$

which, lo and behold, is the "most probable" distribution (9.3) again. But, although (10.42) and (9.3) appear the same, p and q are totally different quantities. In batch step-growth polymerization, p increases monotonically from 0 to 1 as the reaction proceeds. In free-radical addition, however, q depends only indirectly on conversion, since $[I]$, $[M]$, and $[R':H]$ will, in general, vary with conversion. In the usual case, q is always close to 1, as indeed it must be to form high molecular weight polymer.

Example 9. Show the structures represented by $x = 1$ in (10.42) for the polymerization of a vinyl monomer $H_2C{=}CHX$. Assume no chain transfer and initiation according to

$$R{:}R \rightarrow 2R\cdot$$

Solution. Since chains here are killed only by disproportionation, dead chains of length $x = 1$ would be

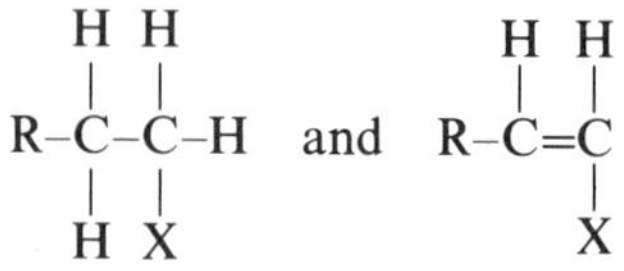

$$
\begin{array}{cc}
\text{H} \ \text{H} & \text{H} \ \text{H} \\
| \ \ | & | \ \ | \\
\text{R--C--C--H} & \text{R--C=C} \\
| \ \ | & | \\
\text{H} \ \text{X} & \text{X}
\end{array}
\quad \text{and}
$$

(With its double bond, the molecule on the right above could conceivably participate as a comonomer in subsequent polymerization. The mechanism discussed here ignores that possibility, however.)

This points out another important difference between the seemingly similar equations (10.42) and (9.3). The distributions for free-radical addition include only terminated chains, not unreacted monomer. In a step-growth reaction, there are no terminated chains short of complete conversion, and $x = 1$ in (9.3) represents unreacted monomer.

Note that by analogy to (9.6), $\bar{x}_n$ for the most-probable distribution should be

$$\bar{x}_n = \frac{1}{1 - q} \tag{10.43}$$

which is greater by one than the value given by (10.41). This discrepancy arises from the fact that the kinetic derivation of (10.41) neglects the monomer molecule added to the chain in the addition step, while that molecule is specifically considered in the derivation of (10.42). From a practical standpoint, this makes little difference, because in the usual free-radical addition polymerization, $q \rightarrow 1$ and $\bar{x}_n \gg 1$.

By analogy to (9.9), (9.10), and (9.11) for step-growth polymerization

$$\frac{w_x}{W} = xq^{(x-1)}(1-q)^2 \tag{10.44}$$

$$\bar{x}_w = \frac{1+q}{1-q} \tag{10.45}$$

$$\frac{\bar{x}_w}{\bar{x}_n} = 1 + q \tag{10.46}$$

Here, however, unlike the case in step growth, the approximations $q \rightarrow 1$, $\bar{x}_n \gg 1$, and $\ln q \approx q - 1$ are almost always applicable, and (10.42), (10.44), and (10.46) simplify to (recall Example 5, Chapter 6)

$$\frac{n_x}{N} \approx \frac{1}{\bar{x}_n} \exp\left(\frac{-x}{\bar{x}_n}\right) \tag{10.47}$$

$$\frac{w_x}{W} \approx \frac{x}{\bar{x}_n^2} \exp\left(\frac{-x}{\bar{x}_n}\right) \tag{10.48}$$

$$\frac{\bar{x}_w}{\bar{x}_n} \approx 2 \tag{10.49}$$

Now let's consider chains which terminate *exclusively* by combination ($k_t = k_{tc}$, $\xi = 1$)

$$q = \frac{k_p[M]}{k_p[M] + 2(fk_dk_{tc}[I])^{1/2}} \tag{10.50}$$

$$\frac{1}{q} = 1 + \frac{2(f k_a k_{tc}[I])^{1/2}}{k_p[M]} \tag{10.51}$$

$$\underbrace{\phantom{\frac{2(f k_a k_{tc}[I])^{1/2}}{k_p[M]}}}$$

$$(2/\bar{x}_n) \quad \text{for } \xi = 1 \quad (10.27)$$

from which

$$\bar{x}_n = \frac{2q}{1 - q} \tag{10.52}$$

A dead chain consisting of x units is formed by the combination of two growing chains, one containing y units and the other containing $x - y$ units:

$$P_y{}^\cdot + P_{(x-y)}{}^\cdot \to P_x \tag{10.53}$$

One growing chains has undergone $y - 1$ propagation steps, each of probability q. The other has undergone $x - y - 1$ propagation steps, each of probability q. Two growing chains are terminated, each with a probability $1 - q$. But, each dead chain consisting of x total units could have been formed in $x - 1$ different ways; for example,

x	Possible combinations of $y + (x - y)$
2	$1 + 1$
3	$2 + 1, \quad 1 + 2$
4	$3 + 1, \quad 2 + 2, \quad 1 + 3$
5	$4 + 1, \quad 3 + 2, \quad 2 + 3, \quad 1 + 4$
	etc.

and all possible ways to form a chain of x units total must be counted. Therefore,

$$\frac{n_x}{N} = (x - 1)q^{(y-1)}q^{(x-y-1)}(1 - q)^2 = (x - 1)(1 - q)^2 q^{(x-2)} \tag{10.54}$$

For this distribution,

$$\bar{x}_n = \frac{2}{1 - q} \tag{10.55}$$

which is greater by two than the value given by (10.52), for the reasons noted above, and

$$\bar{x}_w = \frac{2 + q}{1 - q} \tag{10.56}$$

$$\frac{\bar{x}_w}{\bar{x}_n} = \frac{2 + q}{2} \tag{10.57}$$

For the usual free-radical case, where $q \to 1$, $\bar{x}_n \gg 1$, and $\ln q \approx q - 1$

$$\frac{n_x}{N} \approx \frac{4x}{\bar{x}_n^2} \exp\left(-\frac{2x}{\bar{x}_n}\right) \tag{10.58}$$

$$\frac{w_x}{W} \approx \frac{4x^2}{\bar{x}_n^3} \exp\left(-\frac{2x}{\bar{x}_n}\right) \tag{10.59}$$

$$\frac{\bar{x}_w}{\bar{x}_n} \approx 1.5 \tag{10.60}$$

These distributions are plotted in reduced form in Fig. 10.3. Note that termination by combination sharpens the distributions because of the low probability of two very short or two very long macroradicals combining.

As shown above, chains terminated by combination follow a different distribution than those terminated by disproportionation or chain transfer. If a material contains chains formed according to both distributions (as would happen, for example, if a chain-transfer agent were added to a system that terminates inherently by combination), the distributions must be summed according to the proportion (mole or mass fraction) of each, as detailed in Kenat et al.[2] Let ψ = mole (number) fraction of the chains that have been terminated by combination. The weight fraction of chains terminated by combination is $2\psi/(\psi + 1)$. In terms of kinetic parameters, ψ is given by

$$\psi = \frac{2 - \xi}{[\xi + \{k_{tr}[R':H]/(k_t f k_d [I])^{1/2}\}]} \tag{10.61}$$

where ξ is defined by (10.26). In terms of ψ, the general distributions become

$$\bar{x}_n\left(\frac{n_x}{N}\right) = \left[\psi(\psi + 1)^2 \left(\frac{x}{\bar{x}_n}\right) + (1 - \psi^2)\right] \exp\left[-(\psi + 1)\left(\frac{x}{\bar{x}_n}\right)\right] \tag{10.62}$$

$$\bar{x}_n\left(\frac{w_x}{W}\right) = \left[\psi(\psi + 1)^2 \left(\frac{x}{\bar{x}_n}\right) + (1 - \psi^2)\right]\left(\frac{x}{\bar{x}_n}\right) \exp\left[-(\psi + 1)\left(\frac{x}{\bar{x}_n}\right)\right]$$

$$\tag{10.63}$$

$$\bar{x}_w/\bar{x}_n = \frac{4\psi + 2}{(\psi + 1)^2} \tag{10.64}$$

For termination by disproportionation and/or chain transfer, $\psi = 0$ and (10.62)–(10.64) reduce to (10.47)–(10.49); for termination exclusively by combination, $\psi = 1$, they reduce to (10.58)–(10.60), but they are also able to describe various degrees of mixed termination.

Example 10. Calculate q and $\bar{x}_n$ for the styrene polymerization in Example 10.2 *for conditions at the start of the reaction.* Neglect chain transfer. Styrene terminates by combination at 60°C so that $k_t = k_{tc}$ ($\xi = 1$). $[M]_0 = 8.72$ mol/L.

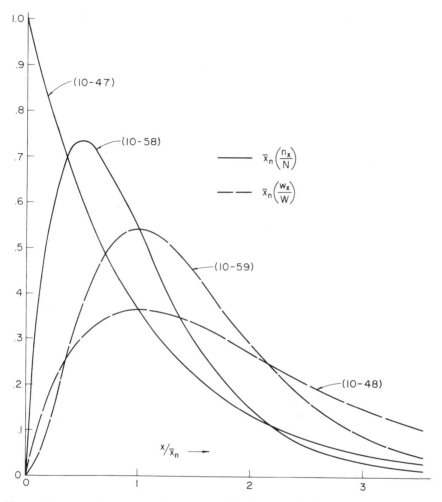

Figure 10.3 Instantaneous number- and weight-fraction distributions of chain lengths in free-radical addition polymerization. Equations 10.47 and 10.48, termination by disproportionation and/or chain transfer; Eqs. 10.58 and 10.59, termination by combination.

Solution. Using the rate constants from Example 10.2 along with $[M] = [M]_0$ and $[I] = [I]_0$ from above in (10.39) gives $q = 0.9954$. This illustrates the point made above that $q \to 1$ right from the beginning of a typical free-radical addition polymerization.

Because termination is by combination, we use (10.55) to get $\bar{x}_n = 434$ ($\bar{M}_n = 45\,200$), so we're getting high molecular weight polymer right off the bat, too.

10.8 / INSTANTANEOUS QUANTITIES

It is obvious from (10.13) that the rate of polymerization is an instantaneous quantity; it depends on the particular values of $[M]$, $[I]$, and T (through the temperature dependence of the rate constants) that exist *at a particular instant* (and location, for that matter) in a reactor. In a uniform, isothermal batch reaction (Example 2), the rate of polymerization decreases monotonically because of the decreases in both $[M]$ and $[I]$ with time. In a similar fashion, $\bar{x}_n$ according to (10.35) is a function of $[M]$, $[I]$, T, and $[R':H]$, all of which may vary with time (and/or location) in a reactor. But is the concept of an instantaneous $\bar{x}_n$ valid? How much do these quantities change during the lifetimes of individual chains? This important point is clarified in the following example.

Example 11. Consider the uniform, isothermal batch polymerization of Example 2. For conditions *at the start of the reaction,* calculate

a. The average lifetime of a growing chain, $\bar{t}$, and compare it with the half lives of monomer and initiator.

b. The percent decrease in initiator concentration during the interval $\bar{t}$.

c. The conversion during the time interval $\bar{t}$.

d. The number of growing chains per liter of reaction mass.
Additional necessary data: $k_p = 176$ L/mol·s

Solution.

a. This is always conceptually difficult. Think of it as calculating the time required for a car traveling at a constant velocity to cover a specified distance. Here, the distance corresponds to the kinetic chain length v (10.22), the average number of monomer units added to a chain during its lifetime (note that the mode of termination is immaterial). The velocity corresponds to the rate at which monomer is added to a single chain. Monomer is added to all chains (in a unit volume of reaction mass) at a rate r_p. In that unit volume of reaction mass, there are $[P\cdot]$ growing chains. Therefore, the rate of monomer addition to a single chain $= r_p/[P\cdot]$ and

$$\bar{t} = \frac{v[P\cdot]}{r_p}$$

With (10.12), (10.13), and (10.22), this becomes

$$\bar{t} = \frac{1}{2(fk_dk_t[I])^{1/2}}$$

Inserting the given rate constants and $[I] = [I]_0 = 0.05$ mol/L gives $\bar{t} = 0.141$ *seconds* (!). From Figure 10.1, the time to reach 50% conversion (monomer half life) is about 8 *hours*. From (10.16*b*), the time to reach $(I/I_0) = 0.5$ is $-(\ln 0.5)/k_d$ or 20 h.

b. We can confidently neglect any volume change over the extremely short average chain lifetime so that

$(I/I_0) = ([I]/[I]_0) = \exp(-k_d \bar{t})$ (10.16*b*). With $\bar{t} = 0.141$ s,

$$\% \text{ decrease in } [I] = 100\left(1 - \frac{[I]}{[I]_0}\right) = 1.35 \times 10^{-4}$$

c. The result of (**b**) justifies the use of (10.15). With $\bar{t} = 0.141$ s, $X = 3.36 \times 10^{-6}$ or $3.36 \times 10^{-4}\%$.

d. From (10.12), we get $[P\cdot] = 1.35 \times 10^{-7}$ mol growing chains/L. Or, multiplying by Avogadro's number, 6.02×10^{23} growing chains/mol growing chains, it's 8.14×10^{16} growing chains per liter.

Some important conclusions can be drawn from the preceding example. In a typical, homogeneous, free-radical addition polymerization, there are lots of chains growing at any instant. The average lifetime of a growing chain, however, is extremely short, many orders of magnitude smaller than the half-lives of either monomer or initiator. Once a chain has been initiated by the decomposition of an initiator molecule (recall that this is the slow step), it grows and dies in a flash, and once terminated, it plays no further role in the reaction (unless it happens to act as a chain-transfer agent, Example 4) and merely sits around inertly as more chains are formed. This is to be contrasted with step-growth polymerization, in which the chains always maintain their terminal reactivity and continue to grow throughout the reaction.

The lifetime of free-radical chains is, in fact, so short that changes in concentrations are entirely negligible during a chain lifetime. Hence, it is perfectly proper to characterize chains formed at any instant when $[M]$, $[I]$, T, and $[R':H]$ have a particular set of values. All the quantities defined to this point $(\bar{x}_n, \bar{x}_w, q)$, and therefore the distributions, are just such instantaneous quantities. For this reason, it was necessary to specify conditions "at the start of the reaction" in Examples 10 and 11 to permit their calculation.

10.9 CUMULATIVE QUANTITIES

Unfortunately (for the sake of simplicity), as conditions vary within a polymerization reactor, so do the instantaneous quantities. The polymer in the reactor is a mixture of material formed under varying conditions of temperature and concentrations, and therefore must be characterized by *cumulative* quantities, which are integrated averages of the instantaneous quantities of the material formed up until the reactor is sampled. The cumulative number-average chain length, $\langle \bar{x}_n \rangle$, is simply the total moles of monomer polymerized over the total moles of dead chains formed. In a batch reactor, for example,

$$\text{Moles of monomer in dead chains} = M_0 - M(t) = M_0 X(t)$$

$$\text{Moles of dead chains formed} = f\xi\{I_0 - I(t)\} + \{R':H_0 - R':H(t)\}$$

(For termination by disproportionation, each initiator molecule results in two dead chains, and for combination, one; hence, the ξ. Each reacted molecule of chain-transfer agent results in one dead chain.) Therefore,

$$\langle \bar{x}_n \rangle(t) = \frac{M_0 X(t)}{f\xi\{I_0 - I(t)\} + \{R':H_0 - R':H(t)\}} \tag{10.65}$$

(If you wish to neglect volume change, all the molar quantities can be enclosed in square brackets to give concentrations.) Note that the preceding expression is indeterminate at $t = 0$. At this point, however, $\bar{x}_n = \langle \bar{x}_n \rangle$; both are given by (10.35).

A good analogy for visualizing the relation between $\bar{x}_n$ and $\langle \bar{x}_n \rangle$ is the "well-stirred, adiabatic bathtub." $\bar{x}_n$ corresponds to the temperature of the water leaving the tap and $\langle \bar{x}_n \rangle$ to the temperature of the water in the tub. The amount of water that has left the tap corresponds to the amount of polymer formed. As long as the tub is not full, the analogy applies to a batch reactor. If the tub is allowed to fill and overflow, it applies to a continuous stirred-tank reactor as well.

Example 12. For the reaction in Examples 2 and 11, calculate and plot $\bar{x}_n$ and $\langle \bar{x}_n \rangle$ vs. X.

Solution. For this homogeneous, isothermal, batch reaction, X is given as a function of time by (10.19) (incorporating volume change) or (10.20) (neglecting volume change) as illustrated in Example 2. $I(t)$ is given by (10.16b). By combining (10.27), (10.14), (10.16b), (10.6), and (10.17) we get

$$\bar{x}_n(t) = \frac{k_p[M]_0\{1 - X(t)\}}{\xi(fk_dk_t)^{1/2}[I]_0^{1/2}\{1 + \varepsilon X(t)\}^{1/2}\exp(-k_dt/2)} \tag{10.27'}$$

Had it been necessary to include chain transfer, we would have used (10.35) with [R':H] related to [M] with (10.37). Similar treatment of (10.65) gives

$$\langle \bar{x}_n \rangle(t) = \frac{[M]_0 X(t)}{f\xi[I]_0\{1 - \exp(-k_dt)\}} \tag{10.65'}$$

Calculations are facilitated by taking time as the independent variable and relating X, $\bar{x}_n$, and $\langle \bar{x}_n \rangle$ to t through (10.19) or (10.20), (10.27'), and (10.65').

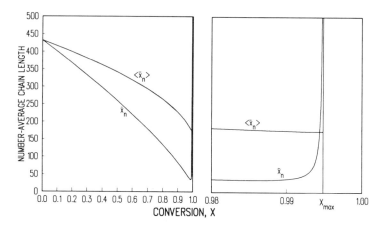

Figure 10.4   Instantaneous and cumulative number-average chain lengths vs. conversion for an isothermal, free-radical, batch polymerization (data of Example 12).

Alternatively, X may be chosen as the independent variable and (10.19) or (10.20) solved for the corresponding times, which are then used in (10.27′) and (10.65′). The results are shown in Fig. 10.4. Note that in this case, $\bar{x}_n$ goes through a minimum at high conversions. In fact, it ultimately goes to infinity, because in this example of "dead-stop" polymerization $[I]$ goes to zero while $[M]$ remains finite. Not so obvious on this scale is the fact that $\langle \bar{x}_n \rangle$ must also go through a minimum. As long as $\bar{x}_n$ is less than $\langle \bar{x}_n \rangle$, the former will continue to drag the latter down (adding colder water from the tap cools off the water in the tub). When $\bar{x}_n$ is greater, it must pull $\langle \bar{x}_n \rangle$ up (you warm up the water in the tub by adding hotter water from the tap). Therefore, there must be a minimum in $\langle \bar{x}_n \rangle$ where the two curves cross, although $\langle \bar{x}_n \rangle$ does reach a finite value of 173.5 at the maximum conversion of 99.48%. (The numbers quoted apply for $\varepsilon = -0.136$. For $\varepsilon = 0$ they are 173.2 and 99.30%.) Even though the last tiny drop from the tap may be very hot, it's not going to raise significantly the temperature of the 50 gallons already in the tub.

It is instructive to compare the results of Example 12 with Eq. 9.6 for step-growth polymerization. There, $\bar{x}_n$ increases monotonically with conversion (p), and high conversions are necessary for high chain lengths. This is not true for free-radical addition, where chains formed early in the reaction have fully developed chain lengths.

Keep in mind also that if samples are removed from the reactor and analyzed for number-average chain length, the quantity determined is $\langle \bar{x}_n \rangle$. The cumulative quantity is what characterizes the reactor contents. The only way to measure $\bar{x}_n$ would be to sample at a very low conversion, where $\bar{x}_n \approx \langle \bar{x}_n \rangle$.

10.10 RELATIONS BETWEEN INSTANTANEOUS AND CUMULATIVE AVERAGE CHAIN LENGTHS FOR A BATCH REACTOR

Given any instantaneous or cumulative average chain length as a function of conversion in a batch reactor, the others may be calculated as follows: Recall that conversion is defined as

$$X \equiv \frac{M_0 - M}{M_0} \tag{10.14}$$

The moles of monomer polymerized *up to* conversion X is

$$M_0 - M = M_0 X$$

while the moles of monomer polymerized in a conversion *increment* dX is

$$dM = M_0 \, dX$$

The moles of *polymer chains* formed in the conversion increment dX is

$$dN = \frac{dM}{\bar{x}_n} = \frac{M_0 \, dX}{\bar{x}_n}$$

The total moles of polymer chains formed *up to* conversion X is obtained by integrating dN:

$$N = \int_0^X dN = \int_0^X \frac{M_0 \, dX}{\bar{x}_n}$$

But, by definition of $\langle \bar{x}_n \rangle$, N is also given by

$$N = \frac{\text{moles monomer polymerized to conversion } X}{\langle \bar{x}_n \rangle} = \frac{M_0 X}{\langle \bar{x}_n \rangle}$$

Equating the two expressions for N gives

$$\langle \bar{x}_n \rangle = \frac{X}{\displaystyle\int_0^X (dX/\bar{x}_n)} \tag{10.66}$$

Differentiating (10.66) reverses the result

$$\bar{x}_n = \frac{dX}{d(X/\langle \bar{x}_n \rangle)} \tag{10.67}$$

Equation 6.6 gives the weight average of a mixture in terms of a summation of the weight averages of the finite components of the mixture. For a mixture of *differential* components of weight-average chain length $\bar{x}_w$ this may be generalized to

$$\langle \bar{x}_w \rangle = \frac{\int_0^X \bar{x}_w \, dW_p}{\int_0^X dW_p} = \frac{1}{W_p} \int_0^X \bar{x}_w \, dW_p$$

where dW_p = weight of polymer formed in conversion increment dX
W_p = total weight of polymer formed up to conversion X

Now $W_p = W_0 X$ and $dW_p = W_0 dX$, where W_0 is the weight of monomer charged; so

$$\langle \bar{x}_w \rangle = \frac{1}{X} \int_0^X \bar{x}_w \, dX \tag{10.68}$$

and by differentiating (10.68), we get

$$\bar{x}_w = \langle \bar{x}_w \rangle + X \left(\frac{d \langle \bar{x}_w \rangle}{dX} \right) \tag{10.69}$$

The relation between $\bar{x}_n$ and $\bar{x}_w$ is, of course, determined on a microscopic scale by the nature of the instantaneous distribution, which fixes the instantaneous polydispersity index, $\bar{x}_w/\bar{x}_n$, given by (10.64). However, it is not the instantaneous polydispersity index that characterizes the reactor product, but rather the cumulative polydispersity index $\langle \bar{x}_w \rangle / \langle \bar{x}_n \rangle$. Even though the instantaneous polydispersity index may remain constant, if the instantaneous averages vary during the reaction (as in Example 12), the cumulative distribution must be broader than the instantaneous, increasing the cumulative polydispersity index (Chapter VI, Example 2 provides a simple quantitative illustration of this). This is one of the reasons why commercial polymers often have polydispersity indices much greater than the minimum value given by (10.64).

The "road map" below summarizes how to get between the various averages when one is known as a function of conversion:

$$\bar{x}_n(X) \xrightarrow[\text{(10.67)}]{\text{(10.66)}} \langle \bar{x}_n \rangle (X)$$

$$\Big\uparrow \text{(10.64)} \Big\downarrow$$

$$\bar{x}_w(X) \xrightarrow[\text{(10.69)}]{\text{(10.68)}} \langle \bar{x}_w \rangle (X)$$

Remember that $\langle \bar{x}_n \rangle$ and $\langle \bar{x}_w \rangle$ can be experimentally determined by sampling that reactor, but $\bar{x}_n$ and $\bar{x}_w$ can be obtained only by calculation at finite conversions.

Example 13. Which type of isothermal reactor would produce the narrowest possible distribution of chain lengths in a free-radical addition polymerization: continuous stirred tank (backmix), batch (assume perfect stirring in each of the previous), plug-flow tubular, or laminar-flow tubular?

Solution. Only in an ideal continuous stirred tank reactor are $[M]$ and $[I]$, and therefore $\bar{x}_n$, constant, giving the narrowest possible distribution, due only to the microscopically random nature of the reaction. With an ideal CSTR, $\langle \bar{x}_w \rangle / \langle \bar{x}_n \rangle = \bar{x}_w / \bar{x}_n$. In a batch reactor, $[M]$ and $[I]$ vary with time, and in tubular reactors, with position, causing a shift in $\bar{x}_n$ with conversion, broadening the distribution so that $\langle \bar{x}_w \rangle / \langle \bar{x}_n \rangle > \bar{x}_w / \bar{x}_n$. Note that changes in temperature, which cause the rate constants to vary, will also contribute to broadening of the distribution.

Example 14. Demonstrate that even though the individual averages $\bar{x}_w$ and $\bar{x}_n$ may vary with conversion, as long as their ratio, the instantaneous polydispersity index, remains constant (as will often be the case), the cumulative polydispersity index will always be proportional to the instantaneous polydispersity index.

Solution. If we divide (10.68) by (10.66) we get for the cumulative polydispersity index

$$\frac{\langle \bar{x}_w \rangle}{\langle \bar{x}_n \rangle} = \frac{1}{X^2} \int_0^X \bar{x}_w \, dX \int_0^X \frac{dX}{\bar{x}_n} \tag{10.70}$$

With the ratio $\bar{x}_w / \bar{x}_n$ constant, this becomes

$$\frac{\langle \bar{x}_w \rangle}{\langle \bar{x}_n \rangle} = \frac{\bar{x}_w}{\bar{x}_n} \frac{1}{X^2} \int_0^X \bar{x}_n \, dX \int_0^X \frac{dX}{\bar{x}_n} \qquad \left(\frac{\bar{x}_w}{\bar{x}_n} \text{ constant} \right) \tag{10.71}$$

Example 15. Consider the isothermal, free-radical batch polymerization of a monomer with $[M]_0 = 1.6901$ gmol/L and $[I]_0 = 1.6901 \times 10^{-3}$ gmol/L. At the reaction temperature, $k_p^2 / k_t = 26.1$ L/mol·s and $k_d = 4.369 \times 10^{-7}$ s^{-1}. This particular polymer terminates by disproportionation ($\xi = 2$). Neglect chain transfer. Assume perfect reactor stirring and $f = 1$. Because this reaction is carried out in a rather dilute solution, volume change is negligible. Calculate and plot $\bar{x}_n$, $\langle \bar{x}_n \rangle$, $\bar{x}_w$, $\langle \bar{x}_w \rangle$, and $\langle \bar{x}_w \rangle / \langle \bar{x}_n \rangle$ vs. conversion for this system.

Hint: For this particular system (k_p^2 / k_t) is very large and k_d is small (compare with Example 2). This has some important consequences. First, the chain lengths will be tremendous. (Although this example is based on a real system, in real life it is likely that chain lengths would be transfer limited; that is, the chains would transfer to *something* in the reaction mass before reaching the lengths calculated.) Second, $[I]$ remains essentially constant throughout the course of the reaction. As a result $\bar{x}_n$ is linear in X (combine (10.27) and (10.14) to see why). This in turn allows easy analytical evaluation of (10.66)–(10.69), but also renders (10.65) useless.

Solution. Following the suggestion in the hint yields

$$\bar{x}_n = \frac{k_p[M]_0}{\xi(fk_dk_t)^{1/2}[I]^{1/2}}(1 - X) = 158\,900\,(1 - X)$$

Plugging this relation into (10.66) gives

$$\langle \bar{x}_n \rangle = \frac{158\,900\,X}{\displaystyle\int_0^X [dX/(1 - X)]} = \frac{158\,900\,X}{-\ln(1 - X)}$$

Now for $\xi = 2$, $\psi = 0$, (10.64) or (10.49) gives $\bar{x}_w = 2\bar{x}_n$, so that

$$\bar{x}_w = 317\,800\,(1 - X)$$

When this relation is plugged into (10.68), we get

$$\langle \bar{x}_w \rangle = \frac{1}{X}\int_0^X \bar{x}_w\,dX = \frac{317\,800}{X}\int_0^X (1 - X)\,dX = 317\,800\left(1 - \frac{X}{2}\right)$$

Plots based on these equations are shown in Figure 10.5. Note that at high conversions, the polydispersity index begins to shoot up well above the minimum (instantaneous) value of 2.0.

If it is necessary to maintain minimum polydispersity (albeit with shorter chains), the drift in instantaneous chain length can often be compensated for by varying the rate of addition of chain-transfer agent to the reactor. In the example above, a high initial rate of a very active agent could be reduced as conversion increased, thereby counteracting the decrease in $\bar{x}_n$ (and increase in polydispersity) that would otherwise occur. Adjusting the rate of addition of chain-transfer agent to maintain $\bar{x}_n$ constant has been discussed for batch[3] and continuous[2] reactors.

Example 15 constitutes one of the simplest possible illustrations of what is sometimes termed "polymer reaction engineering." Even a minor complication, such as substitution of the parameters of Example 2 or the inclusion of chain transfer would necessitate a numerical solution. While the basic principles are there, additional detail is beyond the scope of this chapter, but you might wish to consider the application of these principles to a nonisothermal reactor, etc.

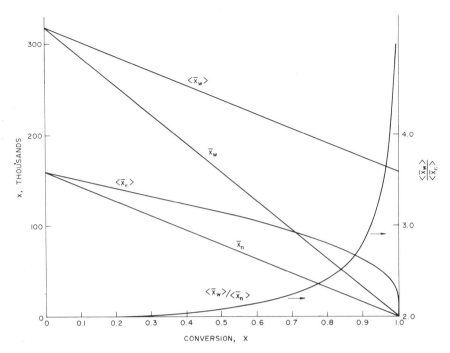

Figure 10.5 Instantaneous and cumulative number and weight average chain lengths and the cumulative polydispersity index for an isothermal, free-radical, batch polymerization (data of Example 15).

10.11 EMULSION POLYMERIZATION

The preceding discussion of free-radical addition polymerization has considered only homogeneous reactions. Considerable polymer is produced commercially by a complex *heterogeneous* free-radical addition process known as emulsion polymerization. This process was developed in the United States during World War II to manufacture synthetic rubber. A rational explanation of the mechanism of emulsion polymerization was proposed by Harkins[4] and quantified by Smith and Ewart[5] after the war, when information gathered at various locations could be freely exchanged. Perhaps the best way to introduce the subject is to list a typical reactor charge:

Typical Emulsion Polymerization Charge

100	parts (by weight) monomer (water insoluble)
180	parts water
2–5	parts fatty acid soap
0.1–0.5	part *water-soluble* initiator
0–1	part chain-transfer agent (monomer soluble)

The first question is, What's the soap for? Soaps are the sodium or potassium salts of organic acids or sulfates:

$$[R\text{--}\overset{\displaystyle O}{\overset{\|}{C}}\text{--}O]^- \; Na^+$$

When they are added to water in low concentrations, they ionize and float around freely much as sodium chloride ions would. The anions, however, consist of a highly polar hydrophilic (water-seeking) "head" ($COO]^-$) and an organic, hydrophobic (water-fearing) "tail" (R). As the soap concentration is increased. a value is suddenly reached where the anions begin to agglomerate in *micelles* rather than float around individually. These micelles have dimensions on the order of 5–6 nm, far too small to be seen with a light microscope. They consist of a tangle of the hydrophobic tails in the interior (getting as far away from the water as possible) with the hydrophilic heads on the outside. This process is easily observed by following the variation of a number of solution properties with soap concentration, for example, electrical conductivity or surface tension (Fig. 10.6). The break occurs when micelles start to form, and is known as the *critical micelle concentration* or CMC.

When an organic monomer is added to an aqueous micelle solution, it naturally prefers the organic environment within the micelles. Some of it congregates there, swelling the micelles until an equilibrium is reached with the contraction force of surface tension. Most of the monomer, however, is distributed in the form of much larger (1 μm or 1000 nm) droplets stabilized by soap. This complex mixture is an *emulsion*. The cleaning action of soaps depends on their ability to emulsify oils and greases.

Despite the fact that most of the monomer is present in the droplets, the swollen micelles, because of their much smaller size, present a much larger surface area than the droplets. This is easily seen by assuming a micelle volume to drop volume ratio of 1/10 and using the ballpark figures given above. Since

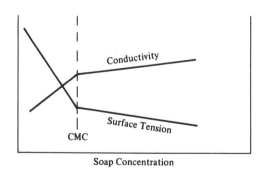

Figure 10.6 Variation of solution properties at the critical micelle concentration.

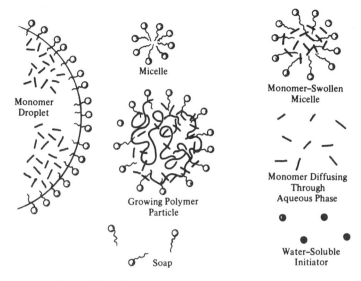

Figure 10.7   Structures in emulsion polymerization.

the surface/volume ratio of a sphere is $3/R$,

$$\frac{S_{\text{micelle}}}{S_{\text{droplet}}} = \left(\frac{V_{\text{mic.}}}{V_{\text{drop.}}}\right)\left(\frac{R_{\text{drop.}}}{R_{\text{mic.}}}\right) \approx \left(\frac{1}{10}\right)\left(\frac{1000}{5}\right) = 20$$

Figure 10.7 illustrates the structures present during emulsion polymerization.

Free radicals for classical emulsion polymerizations are generated *in the aqueous phase* by the decomposition of water-soluble initiators, usually potassium or ammonium persulfate:

$$S_2O_8^{2-} \qquad \rightarrow \qquad 2\,^-SO_4$$

Persulfate   Sulfate ion radical

Redox systems, so called because they involve the alternate oxidation and reduction of a trace catalyst, are a newer and more efficient means of generating radicals. For example,

1. $S_2O_8^{2-} + HSO_3^- \rightarrow SO_4^{2-} + {}^-SO_4\cdot + HSO_3\cdot$
   Persulfate Bisulfite

2. $S_2O_8^{2-} + Fe^{2+} \rightarrow SO_4^{2-} + Fe^{3+} + {}^-SO_4\cdot$

3. $HSO_3^- + Fe^{3+} \rightarrow HSO_3\cdot + Fe^{2+}$

Net $S_2O_8^{2-} + HSO_3^- \rightarrow SO_4^{2-} + {}^-SO_4\cdot + HSO_3\cdot$

The original wartime GR-S (polybutadiene-*co*-styrene) polymerization was carried out at 50°C using potassium persulfate initiator ("hot" rubber.) The use of the more efficient redox systems allowed a reduction in polymerization temperature to 5°C ("cold" rubber). The latter has superior properties because the lower polymerization temperature promotes *cis* 1–4 addition of the butadiene.

The radicals thus generated in the aqueous phase bounce around until they encounter some monomer. Since the surface area presented by the monomer-swollen micelles is so much greater than that of the droplets, the probability of a radical entering a monomer-swollen micelle rather than a droplet is large. As soon as the radical encounters the monomer within the micelle, it initiates polymerization. The conversion of monomer to polymer within the growing micelle lowers the monomer concentration therein, and monomer begins to diffuse from uninitiated micelles and monomer droplets to the growing, polymer-containing micelles. Those monomer-swollen micelles not struck by a radical during the early stages of conversion thus disappear, losing their monomer and soap to those that have been initiated. This first phase of the reaction was termed *Stage I* by Smith and Ewart. The reaction mass now consists of a stable number of growing polymer particles (originally micelles) and the monomer droplets, *Stage II*.

In Stage II, the monomer droplets simply act as reservoirs supplying monomer to the growing polymer particles by diffusion through the water. The monomer concentration in the growing particles maintains a nearly constant dynamic equilibrium value dictated by the tendency toward further dilution (increasing the entropy) and the opposing effect of surface tension attempting to minimize the surface area. (Although most organic monomers are normally thought of as being water "insoluble," their concentrations in the aqueous phase, though small, are sufficient to permit a high enough diffusion flux to maintain the monomer concentration in the polymerizing particles.[6]) Smith and Ewart then subdivided Stage II into three subcases. The description that follows applies to their case 2, which often happens in practice.

A monomer-swollen micelle that has been struck by a radical contains *one* growing chain. With only one radical per particle, there is no way in which the chain can terminate, and it continues to grow until a second initiator radical enters the particle. Under conditions prevailing within the particle in case 2 (tiny particles and highly reactive radicals), the rate of termination is much greater than the rate of propagation, so the chain growth is terminated essentially immediately after the entrance of the second initiator radical.[6] The particle then remains dormant until a third initiator radical enters, initiating the growth of a second chain. This second chain grows until it is terminated by the entry of the fourth radical, and so on.

10.12   KINETICS OF EMULSION POLYMERIZATION IN STAGE II, CASE 2

Thus, in Stage II, the reaction mass consists of a stable number of monomer-swollen polymer particles which are the loci of all polymerization. At any given time (for case 2), a particle contains either one growing chain or no growing chains. Statistically, then, if there are N particles per liter of reaction mass, there are $N/2$ growing chains per liter of reaction mass.

The polymerization rate is given as before by

$$r_p = k_p [M] [P\cdot] \tag{10.7}$$

where k_p is the usual homogeneous propagation rate constant for polymerization within the particles and $[M]$ is the equilibrium monomer concentration within a particle. Now,

$$[P\cdot] = \frac{N}{2A} \text{ moles radicals}/\textit{liter of reaction mass} \tag{10.72}$$

where A is Avogadro's number $(6.02 \times 10^{23}$ radicals/mol radicals). The rate of polymerization is then

$$r_p = k_p \frac{N}{2A} [M] \tag{10.73}$$

where typical units for the various terms are as follows:

$$k_p [=] \frac{\dfrac{\text{moles monomer}}{\text{liter of particles s}}}{\dfrac{\text{moles monomer}}{\text{liter of particles}} \times \dfrac{\text{moles radicals}}{\text{liter of particles}}}$$

$$\frac{N}{2A} [=] \frac{\text{moles radicals}}{\text{liter of reaction mass}}$$

$$[M] [=] \frac{\text{moles monomer}}{\text{liter of particles}}$$

$$r_p [=] \frac{\text{moles of monomer}}{\text{liter of reaction mass s}}$$

Equation 10.73 thus gives the rate of polymerization *per total volume of reaction mass.* Surprisingly, it predicts the rate to be independent of initiator concentration. Moreover, since both N and $[M]$ are constant in Stage II, a constant rate is predicted. This is borne out experimentally, as shown in Fig. 10.8.[4] Deviations from linearity are observed at low conversions as N is being stabilized in Stage I, and at high conversions in Stage III, when the monomer droplets are used up and are no longer able to supply the monomer necessary to maintain $[M]$ constant in the growing particles. Thus, in Stage III, the rate drops off as the monomer is exhausted within the particles.

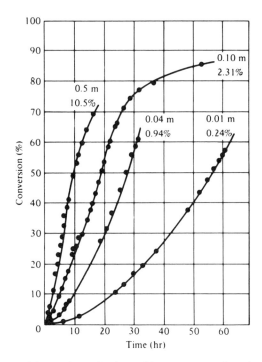

Figure 10.8   The emulsion polymerization of isoprene as a function soap (potassium laurate) concentration. Reprinted from Harkins.[4] Copyright 1947 by the American Chemical Society. Reprinted by permission of the copyright owner.

Note the increase in rate with soap concentration. The more soap used, the more micelles are established in Stage I, and the higher N will be in Stage II. The predicted independence of rate on initiator concentration must be viewed with caution. It's valid as long as N is held constant, but in practice, N increases with $[I]_0$. The more initiator added at the start of a reaction, the greater is the number of monomer-swollen micelles that start growing before N is stabilized. Methods are available for estimating N.[6-8]

Example 16. The Putrid Paint Division of Crud Chemicals, Inc., has available a 10% (by weight) of polyvinyl acetate latex containing 1×10^{14} particles/cm³. To obtain optimum characteristics as an interior wall paint, a larger particle size and higher concentration of polymer are needed. It is proposed to obtain these by adding an additional 4 parts (by weight) of monomer per part of polymer to the latex and polymerizing without further addition of soap. The reaction will be carried to 85% conversion, and the unreacted monomer will be stem-stripped and recovered. (The conditions of this "seeded" polymerization are set up so that the entire reaction proceeds in Stage II.) Estimate the time required for the reaction, and the rate of heat removal (in Btu/gal of original latex, per hour) necessary to maintain an essentially isothermal reaction at 60°C. At what conversion would the monomer droplets disappear and the rate cease to be constant?

$$\text{Data:} \quad k_p = 3700 \text{ L/mol} \cdot \text{s at } 60°\text{C}$$

$$\Delta H_p = -21 \text{ kcal/mol monomer unit}$$

$$\rho \text{ (polymer)} = 1.2 \text{ g/cm}^3$$

$$\rho \text{ (monomer)} = 0.8 \text{ g/cm}^3$$

$$M \text{ (monomer)} = 86 \text{ g/mol}$$

Concentration of monomer in growing polymer particles = 10% (by weight).

Solution. It must first be assumed that in the absence of additional soap, no new micelles will be established, so the original latex particles act as the exclusive loci for further polymerization. Then, $[M]$ and N must be obtained for use in (10.73).

By assuming additivity of monomer and polymer volumes in the latex particles, we get

$$[M] = 1.33 \text{ mol monomer/L of particles}$$

Since $N = 1 \times 10^{17}$ particles/L of original latex,

$$\frac{N}{2A} = 8.31 \times 10^{-8} \text{ mol free radicals/L of original latex}$$

and (10.73) gives

$$r_p = 4.09 \times 10^{-4} \text{ mol monomer/L of original latex, s}$$

$$= 1.47 \text{ mol monomer/L of original latex, h}$$

By again assuming additivity of water and polymer volumes, the density of the original latex is found to be 1.02 g/cm^3. One liter of the original latex therefore has a mass of 1020 g and contains 102 g of polymer, to which are added $102 \times 4 = 408$ g or 4.74 mol of monomer. The reaction converts $4.74 \times 0.85 = 4.02$ mol of monomer, which takes

$$\frac{4.02 \text{ mol monomer}}{\text{liter of original latex}} \times \frac{\text{liter of original latex, h}}{1.48 \text{ mol monomer}} = 2.72 \text{ h}$$

For an isothermal reaction, rate of heat removal = rate of heat generation and is the product of ΔH_p and r_p, or 31 000 cal/L of original latex, h or 467 Btu/gal original latex, h.

When the monomer droplets just disappear, the reaction enters Stage III, and all the monomer and polymer will be in the swollen polymer particles, a total of 102 g (original polymer) + 408 g (added monomer) = 510 g/L of original latex. At this point, the particles still contain 10% monomer, so there are (0.1) (510) = 51 g unconverted monomer. Therefore,

$$X = 1 - \frac{M}{M_0} = 1 - \left(\frac{51}{408}\right) = 0.875 \text{ or } 87.5\%$$

Beyond this point, the rate will drop off as the monomer concentration in the particles falls.

Despite the secondary effect of $[I]_0$ on the rate, it has a strong influence on the average chain length. The greater the rate of radical generation, the greater will be the frequency of alternation between growth and death in a particle, resulting in a lower chain length. If r_c represents the rate of radical capture/liter of reaction mass (half of which produces dead chains), in the absence of transfer

$$\bar{x}_n = \frac{k_p(N/2A)[M]}{(r_c/2)} = \frac{k_p N[M]}{A r_c} \tag{10.74}$$

The rate of generation of radicals is based on the presence of initiator in the aqueous phase alone:

$$r_{gen}\left(\frac{\text{mol radicals}}{\text{liter of aq. phase, s}}\right) = 2k_d(s^{-1})[I]\left(\frac{\text{mol initiator}}{\text{liter of aq. phase}}\right) \tag{10.75}$$

If a steady-state in radicals is assumed,

$$r_c\left(\frac{\text{mol radicals}}{\text{liter of reac. mass, s}}\right) = r_{gen}\left(\frac{\text{mol radicals}}{\text{liter of aq. phase, s}}\right)$$
$$\times \phi_a\left(\frac{\text{liter of aq. phase}}{\text{liter of reaction mass}}\right) \tag{10.76}$$

where ϕ_a is the volume fraction aqueous phase. Thus,

$$r_c = 2k_d[I]\phi_a = 2k_d[I'] \tag{10.77}$$

where $[I'] = [I]\phi_a$, the moles of initiator per total volume of the reaction mass, and

$$\bar{x}_n = \frac{k_p N[M]}{2Ak_d[I']} \tag{10.78}$$

The chain length is, therefore, inversely proportional to the *first* power of initiator concentration (compare with (10.27)).

It must be emphasized that the preceding is a simplified view of an extremely complex process. Even within Stage II, Smith and Ewart pointed out that their case 2 was merely the middle of a spectrum. At one end of the spectrum, case 1, the average number of radicals per particle is much less than 1/2. This is believed to occur sometimes because of radical escape from the particles. At the other extreme, the average number of radicals per particle is much greater than 1/2. This comes about from large particles and/or small k_t, both of which slow termination. Under these conditions, each particle acts as a tiny homogeneous reactor, according to the kinetics and mechanism previously developed for homogeneous free-radical addition. The transition from case 2 to case 3 kinetics is often observed in seeded emulsion polymerizations, where the particles are fairly large to begin with, and grow from there as the reaction proceeds.

The Smith–Ewart theory was developed for monomers such as styrene, with very low water solubility. Monomers such as acrylonitrile, with appreciable water solubility (on the order of 10%), may undergo significant homogeneous initiation in the aqueous phase. In some emulsion systems, the particles flocculate (coalesce) during polymerization, not only making a kinetic description difficult, but sometimes badly fouling reactors. Good reviews of the subject are available,[8-11] as well as complete books.[7, 12-14]

REFERENCES

1. Levenspiel, O., *Chemical Reaction Engineering*, 2d ed., Wiley, New York, 1972.

2. Kenat, T., R. I. Kermode, and S. L. Rosen, *J. Appl. Polym. Sci.* **13**, 1353 (1969).

3. Hoffman, B. F., S. Schreiber, and G. Rosen, *Ind. Eng. Chem.* **56**(5), 51 (1964).

4. Harkins, W. D., *J. Am. Chem. Soc.* **69**, 1428 (1947).

5. Smith, W. V., and R. H. Ewart, *J. Chem. Phys.* **16**(6), 592 (1948).

6. Flory, P. J., *Principles of Polymer Chemistry*, Cornell UP, Ithaca, NY, 1953, Chapter V-3.

7. Bovey, F. A., et al., *Emulsion Polymerization*, Interscience, New York, 1955.

8. Gardon, J. L., Emulsion polymerization, Chapter 6 in *Polymerization Processes*, C. E. Schildknecht and I. Skeist (eds.), Wiley, New York, 1977.

9. Ugelstad, J., and F. K. Hansen, *Rubber Chem. Technol.* **49**, 536 (1976).

10. Blackley, D. C., *Macromol. Chem.* (*London*) **2**, 31 (1982).

11. Poehlein, G. W., Chapter 6 in *Applied Polymer Science*, 2d ed., R. W. Tess and G. W. Poehlein (eds.), ACS Symposium Series 285, American Chemical Society, Washington, DC, 1985.

12. Blackley, D. C., *Emulsion Polymerization*, Wiley, New York, 1975.

13. Piirma, I., and J. L. Gardon (eds.), *Emulsion Polymerization*, ACS Symposium Series 24, American Chemical Society, Washington, DC, 1976.

14. Bassett, D. R., and A. E. Hamielec (eds.), *Emulsion Polymers and Emulsion Polymerization*, ACS Symposium Series 165, American Chemical Society, Washington, DC, 1981.

15. Piirma, I. (ed.), *Emulsion Polymerization*, Academic, New York, 1982.

PROBLEMS

1. Equation 10.14 rigorously defines monomer conversion in terms of M and M_0. Strictly speaking, only when the volume is constant can the moles be bracketed to give concentrations. Obtain the rigorous expression for X in terms of $[M]$, $[M]_0$, and ε.

2. Pure styrene undergoes significant thermal polymerization without any added initiator above about 100°C. A mechanism has been postulated to explain thermal polymerization which involves a reaction between two monomer molecules to form diradicals that initiate chain growth from each end.

   a. Assume that the above reaction is rate controlling and derive an expression for the rate of polymerization according to this mechanism.

   b. Styrene terminates by disproportionation at these temperatures. Obtain an expression for $\bar{x}_n$ according to this mechanism. Neglect chain transfer.

   c. Interpret your answer to (*b*) in terms of the relation between $\bar{x}_n$, $\langle \bar{x}_n \rangle$, and their variation with conversion in an isothermal batch reactor.

   d. What do you think the distribution of chain lengths would be for this mechanism?

3. We now know that the termination reaction in free-radical addition polymerization is diffusion controlled right down to 0% conversion. This is not really surprising given the extremely high chemical reactivity of free-radical pairs and the size of the reacting species.

   a. Below the entanglement region ($[\eta]c < 1$) it's logical to assume that k_t depends on the length of the terminating chains. Let's assume empirically that

$$k_t = k_{t0}(v)^n$$

where k_{t0} depends only on temperature. We would expect n to be negative. Neglect chain transfer (it really screws up the algebra) and combine the above equation with (10.13) and (10.22) to get an expression for the polymerization rate that reflects the chain-length dependence of termination.

   b. Qualitatively, how would you expect the solvent power of the medium to affect polymerization rate, other things being equal?

   c. Qualitatively, how would you expect the viscosity of the medium to affect polymerization rate, other things being equal?

4. Consider a free-radical addition reaction in which there is negligible conventional termination. Instead, growing chains are killed by a degradative chain-transfer reaction:

$$P_x\cdot + D \xrightarrow{k_t} P_x + D'$$

where D' is a stable radical that does not reinitiate polymerization.

a. Obtain the rate expression for this mechanism.

b. Obtain the expression for number-average chain length according to this mechanism.

5. Crud Chemicals has been carrying out an isothermal batch polymerization of styrene exactly like the one treated in Examples 2 and 12 of this chapter. One day, however, they get a little impatient and decide to "kick" the reaction by adding a second charge of initiator, identical to the first, at 50% conversion ($t = 8.36$ h) (this is actually a fairly common practice). To help them understand the effects of this,

a. Calculate the ratio of $\bar{x}_n$ just after the kick to $\bar{x}_n$ just before the kick.

b. Resketch Fig. 10.4 qualitatively for this process.

6. Consider the isothermal batch polymerization of styrene treated in Examples 2 and 12 of this chapter. Suppose now that a chain-transfer agent with $C = 1.0$ is added to the initial charge at a concentration of $[R:H]_0 = 0.01$ mol/L.

Recalculate $\bar{x}_n(t \rightarrow \infty)$ and $\langle \bar{x}_n \rangle (t \rightarrow \infty)$ for this batch. You may neglect volume change if you wish.

7. Chain-transfer agent is included in a homogeneous, batch free-radical polymerization. What fraction of the initial amount of chain-transfer agent remains at 99% conversion for $C = 0.1, 1.0, 10$?

8. The classical kinetic scheme for free-radical addition polymerization neglects the possibility of primary-radical termination (PRT), in which a growing chain is killed by reaction with a free-radical $R\cdot$ from initiator decomposition

$$P_x\cdot + R\cdot \xrightarrow{k_{prt}} P_x$$

For the polymerization of styrene at 60°C in Examples 2, 11, and 12 estimate at $X = 0$

a. The concentration of primary radicals $[R\cdot]$

b. The ratio of rates of regular termination (r_t) to primary-radical termination (r_{prt})

We don't know the rate constants k_a and k_{prt}. However, because of the similarity of the species involved, let's assume, at least as an order-of-magnitude approximation, that

$$k_a \approx k_p \quad \text{and} \quad k_{prt} \approx k_t$$

9. Equations 10.30 and 10.31 give the temperature dependence of rate- and number-average chain length for a homogeneous free-radical polymerization. Obtain the analogous expressions for classical Stage II, case 2 emulsion polymerization. Use the following ballpark figures to calculate $r_p(70°C)/r_p(60°C)$ and $\bar{x}_n(70°C)/\bar{x}_n(60°C)$ for the emulsion polymerization:

$$E_p \approx 6 \text{ kcal/mol} \quad E_d \approx 30 \text{ kcal/mol} \quad E_t \approx 2 \text{ kcal/mol}$$

10. Calculate and plot as a function of conversion $\bar{x}_w$, $\langle \bar{x}_w \rangle$, and the polydispersity index $\langle \bar{x}_w \rangle / \langle \bar{x}_n \rangle$ for the homogeneous, isothermal styrene polymerization of Examples 2 and 12.

11. Crud Chemicals is producing a polymer with minimum polydispersity by homogeneous, free-radical polymerization in a CSTR. The polymer terminates inherently by disproportionation. To maintain and control the desired average chain length, they continuously feed chain-transfer agent to the reactor through a control valve. Things are going along nicely, with the reactor in steady-state operation, when suddenly, the valve in the chain-transfer feed line sticks open, adding chain-transfer agent at a much higher rate than previously. They fiddle with the valve for a while but are unable to free it, so before the reactor has had a chance to reach a new steady state, they simply close a gate valve in the line, shutting off the flow of chain-transfer agent entirely. The reactor then proceeds to a new steady state in the absence of chain-transfer agent.

Sketch qualitatively $\bar{x}_n$ and $\langle \bar{x}_n \rangle$ on the same coordinates vs. time for the period described.

12. Modify Eq. 10.78 to include chain transfer. Define any terms introduced.

13. Obtain an expression for conversion vs. time in Stage III emulsion polymerization. Define any terms introduced.

14. A classical, Stage II, case 2 emulsion polymerization of styrene has the following parameters:

$$N = 3.2 \times 10^{17} \text{ particles/L of reaction mass}$$
$$[M] = 1.5 \text{ mol/L} \quad [I] = 1.0 \times 10^{-3} \text{ mol/L}$$
$$k_p = 176 \text{ L/mol s}$$
$$k_d = 4.96 \times 10^{-6} \text{ s}^{-1}$$
$$\phi_a = 2/3$$

Calculate

a. The rate of polymerization.

b. $\bar{x}_n$.

c. The average lifetime of a growing chain.

15. Calculate and compare the average chain lifetime at the start of the reaction with the monomer half-life for the system in Example 15. Is the concept of an instantaneous quantity valid for this system? $k_t = 14.5 \times 10^6$ L/mol s.

16. A batch, free-radical polymerization is carried out with $[M]_0 = 1.0$ mol/L and $[I]_0 = 2 \times 10^{-4}$ mol/L. The reaction reaches a limiting conversion of 90%. Assume $f = 1$.

    a. Calculate $\langle \bar{x}_n \rangle (t \to \infty)$ assuming termination by disproportionation.

    b. Repeat (a) with the addition of 6×10^{-4} mol/L of a mercaptan to the initial charge.

17. A batch reactor producing a free-radical addition polymer is sampled at various conversions. Analysis of these samples reveals that the weight-average chain length varies linearly with conversion from 10 000 at $X = 0$ to 5000 at $X = 1$. The polymer is known to terminate by combination and there is no significant chain transfer occurring. Obtain an expression for the polydispersity index of the product as a function of conversion.

18. Certain double-bond containing monomers can be made to undergo homogeneous, free-radical addition polymerization in the absence of initiators by activation with ultraviolet radiation:

$$M + hv \to M\cdot + H\cdot$$

Both radical species initiate chain growth. The rate of generation of radicals is proportional to the amount of light absorbed by the monomer, which is in turn proportional to the monomer concentration and the intensity of the incident UV radiation I_0 (assume constant).

    a. Develop an expression for the rate of UV photopolymerization.

    b. Develop an expression for $\bar{x}_n$ according to this mechanism.

    c. Assume that the rate of the photolytic process above is essentially independent of temperature. Compare the temperature sensitivity of polymerization rate for this process with that for chemically initiated free-radical addition.

$$E_p \approx 6 \text{ kcal/mol} \qquad E_t \approx 2 \text{ kcal/mol} \qquad E_d \approx 30 \text{ kcal/mol}$$

CHAPTER XI

Nonradical Addition Polymerization

11.1 CATIONIC POLYMERIZATION[1-4]

Strong Lewis acids, that is, electron acceptors, are often capable of initiating addition polymerization of monomers with electron-rich substituents adjacent to the double bond. Cationic catlysts are most commonly metal trihalides such as $AlCl_3$ or BF_3. These compounds, although electrically neutral, are two electrons short of having a complete valence shell of eight electrons. They were found to require traces of a cocatalyst, usually water, to initiate polymerization, first by grabbing a pair of electrons from the cocatalyst:

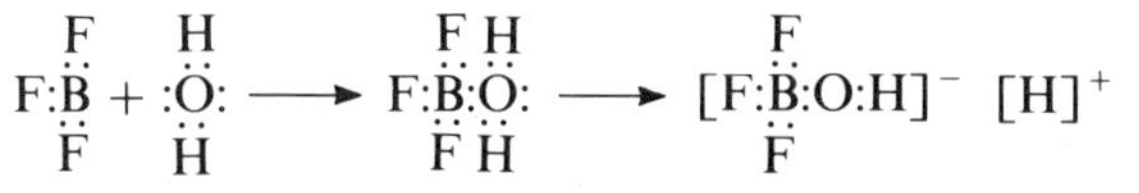

$$
\begin{array}{ccc}
\text{F} \quad \text{H} & \text{F H} & \text{F} \\
\text{F:B} + \text{:O:} \longrightarrow & \text{F:B:O:} \longrightarrow & \text{[F:B:O:H]}^- \ \text{[H]}^+ \\
\text{F} \quad \text{H} & \text{F H} & \text{F}
\end{array}
$$

The leftover proton is thought to be the actual initiating species, abstracting a pair of electrons from the monomer and leaving a cationic chain end which reacts with additional monomer molecules.

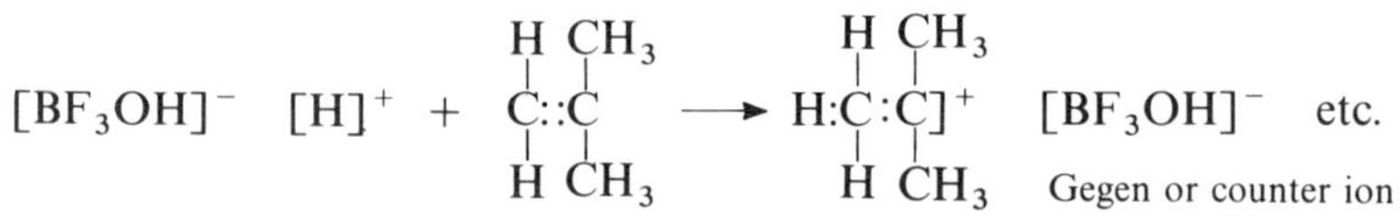

$$
\begin{array}{cccc}
& & \text{H CH}_3 & \text{H CH}_3 \\
\text{[BF}_3\text{OH]}^- \ \text{[H]}^+ + & \text{C::C} & \longrightarrow \text{H:C:C]}^+ & \text{[BF}_3\text{OH]}^- \quad \text{etc.} \\
& & \text{H CH}_3 & \text{H CH}_3 \quad \text{Gegen or counter ion}
\end{array}
$$

An important point here is that the *gegen* or *counter* ion is electrostatically held near the growing chain end and so can exert a steric influence on the addition of monomer units. Termination is thought to occur by a disproportionation-like reaction which regenerates the catalyst complex. The complex, therefore, is a true catalyst, unlike free-radical initiators:

$$\overset{\overset{\displaystyle H}{|}\ \overset{\displaystyle CH_3}{|}}{\underset{\overset{\displaystyle |}{H}\ \overset{\displaystyle |}{CH_3}}{\sim\sim C-C}]^+}\ [BF_3OH]^- \longrightarrow \overset{\overset{\displaystyle H}{|}\ \overset{\displaystyle CH_3}{|}}{\underset{\overset{\displaystyle |}{H}}{\sim\sim C-C}=CH_2}\ +\ BF_3\cdot H_2O$$

The kinetics of these reactions is not well understood, but they proceed very rapidly at low temperatures. For example, the polymerization of isobutylene illustrated above is carried out commercially at $-150°$ F. The average chain length increases as the temperature is lowered.

Cationic initiation is successful only with monomers like isobutylene having electron-rich substitutents adjacent to the double bond, such as

$$\overset{\overset{\displaystyle H}{|}\ \overset{\displaystyle H}{|}}{\underset{\overset{\displaystyle |}{H}\ \overset{\displaystyle |}{O-R}}{C=C}}$$

Alkyl vinyl ethers

$$\overset{\overset{\displaystyle H}{|}\ \overset{\displaystyle CH_3}{|}}{\underset{\overset{\displaystyle |}{H}}{C=C}}$$

α-Methyl styrene

None of these monomers can be polymerized to high molecular weight with free-radical initiators.

11.2   ANIONIC POLYMERIZATION[5-7]

Addition polymerization may also be initiated by anions. Anionic polymerization has achieved tremendous commercial importance in the past two decades because of its ability to control molecular structure during polymerization, allowing the synthesis of materials that were previously difficult or impossible to obtain. A variety of anionic initiators has been investigated, but the organic alkali-metal salts are perhaps most common, as illustrated below for the polymerization of styrene with n-butyllithium:

$$\overset{\overset{\displaystyle H}{|}\ \overset{\displaystyle H}{|}\ \overset{\displaystyle H}{|}\ \overset{\displaystyle H}{|}}{\underset{\overset{\displaystyle |}{H}\ \overset{\displaystyle |}{H}\ \overset{\displaystyle |}{H}\ \overset{\displaystyle |}{H}}{H-C-C-C-C:}]^-\ [Li]^+}\ +\ \overset{\overset{\displaystyle H}{|}\ \overset{\displaystyle H}{|}}{\underset{\overset{\displaystyle |}{H}\ \overset{\displaystyle |}{\phi}}{C=C}} \longrightarrow \overset{\overset{\displaystyle H}{|}\ \overset{\displaystyle H}{|}\ \overset{\displaystyle H}{|}\ \overset{\displaystyle H}{|}\overset{\displaystyle H}{|}\overset{\displaystyle H}{|}}{\underset{\overset{\displaystyle |}{H}\ \overset{\displaystyle |}{H}\ \overset{\displaystyle |}{H}\ \overset{\displaystyle |}{H}\overset{\displaystyle |}{H}\overset{\displaystyle |}{\phi}}{H-C-C-C-C:C:C:}]^-\ [Li]^+$$

n-Butyllithium (n-BuLi)   Styrene

The anionic chain end then propagates the chain by adding another monomer molecule. Again, the gegen ion can sterically influence the reaction.

Sodium and lithium *metals* were used to polymerize butadiene in Germany during World War II. After the war, in the United States, it was discovered that under appropriate conditions, dispersions of lithium could lead to largely *cis*-1,4 addition of butadiene and isoprene (the latter being the synthetic counterpart of natural rubber). In these processes, a metal atom first reacts with the monomer to form an anion radical:

$$
\text{Li·} + \overset{\underset{\displaystyle |}{\text{H H H H}}}{\text{C::C:C::C}} \longrightarrow \overset{\underset{\displaystyle |}{\text{H H H H}}}{\text{·C:C:C:C:}}]^{-} \ [\text{Li}]^{+}
$$

Anion radical

These anion radicals then react in either of two ways. One may react with another atom of lithium,

$$
\text{Li·} + \overset{\text{H H H H}}{\text{·C:C:C:C:}}]^{-} \ [\text{Li}]^{+} \longrightarrow {}^{+}[\text{Li}] \ ^{-}[\text{:C:C:C:C:}]^{-} \ [\text{Li}]^{+}
$$

and/or two may rapidly undergo radical recombination,

$$
2 \ \overset{\text{H H H H}}{\text{·C:C:C:C:}}]^{-} \ [\text{Li}]^{+} \longrightarrow {}^{+}[\text{Li}] \ ^{-}[\text{:C—C=C—C:C—C=C—C:}]^{-} \ [\text{Li}]^{+}
$$

Either way, the result is a dianion that propagates a chain from each end. Other dianionic initiators have been developed.[5-8]

There are a couple of very interesting aspects to these reactions. First, the rates of initiation and propagation vary with the monomer, gegen ion and solvent. In general, the reactions proceed more rapidly in polar solvents as the species are more highly ionized. (The polarity of the solvent also strongly influences the stereospecific nature of the polymer.) In many important cases, the rate of initiation is comparable to the rate of propagation, unlike free-radical addition, in which $r_i \ll r_p$. This means that the initiator (in this case it is an initiator rather than a catalyst) starts chains growing promptly. Second, in the absence of impurities, *there is no termination step*. The chains continue to grow until the monomer supply is exhausted. The ionic chain end is perfectly stable, and the growth of the chains will resume upon addition of more monomer. For this reason, these materials were aptly termed "*living*" polymers by Professor Swarc. Proton-donating impurities such as water or acids quickly kill (terminate) them, however:

$$
\sim\sim\overset{\text{H H}}{\text{C—C:}}]^{-} \ [\text{Li}]^{+} + H_2O \longrightarrow \sim\sim\overset{\text{H H}}{\text{C—C:H}} + \text{LiOH}
$$

This unique mechanism has a number of practically important consequences.

A   Block Copolymerization

If a second monomer is introduced after the initial monomer charge is exhausted, the living chains resume propagation with the second monomer, neatly giving a block copolymer. Monomers can be alternated as desired to give AB, ABA, or other more complicated block structures, conceivably even including three or more monomers.

B   Synthetic Flexibility

Anionic polymerization allows the synthesis of all sorts of interesting and useful molecules. For example, bubbling carbon dioxide through a batch of living chains followed by exposure to water produces a carboxyl-terminated polymer:

$$
\sim\!\!\overset{\overset{\displaystyle H}{|}}{\underset{\underset{\displaystyle H}{|}}{C}}\!\!:]^-\ [Li]^+\ +\ CO_2\ \longrightarrow\ \sim\!\!\overset{\overset{\displaystyle H}{|}}{\underset{\underset{\displaystyle H}{|}}{C}}\!\!-\!\!\overset{\overset{\displaystyle O}{\|}}{C}\!\!-\!\!O\!:]^-\ [Li]^+
$$

$$
\sim\!\!\overset{\overset{\displaystyle H}{|}}{\underset{\underset{\displaystyle H}{|}}{C}}\!\!-\!\!\overset{\overset{\displaystyle O}{\|}}{C}\!\!-\!\!O\!:]^-\ [Li]^+\ +\ H_2O\ \longrightarrow\ \sim\!\!\overset{\overset{\displaystyle H}{|}}{\underset{\underset{\displaystyle H}{|}}{C}}\!\!-\!\!\overset{\overset{\displaystyle O}{\|}}{C}\!\!-\!\!OH\ +\ LiOH
$$

Similarly, ethylene oxide gives hydroxyl-terminated chains:

$$
\sim\!\!\overset{\overset{\displaystyle H}{|}}{\underset{\underset{\displaystyle H}{|}}{C}}\!\!:]^-\ [Li]^+\ +\ H_2C\!\!-\!\!CH_2\ \longrightarrow\ \sim\!\!\overset{H}{\underset{H}{C}}\!-\!\overset{H}{\underset{H}{C}}\!-\!\overset{H}{\underset{H}{C}}\!-\!O\!:]^-\ [Li]^+
$$

$$
\sim\!\!\overset{H}{\underset{H}{C}}\!-\!\overset{H}{\underset{H}{C}}\!-\!\overset{H}{\underset{H}{C}}\!-\!O\!:]^-\ [Li]^+\ +\ H_2O\ \longrightarrow\ \sim\!\!\overset{H}{\underset{H}{C}}\!-\!\overset{H}{\underset{H}{C}}\!-\!\overset{H}{\underset{H}{C}}\!-\!OH\ +\ LiOH
$$

Note that if these reagents are added to a lithium-initiated dianion chain, *both* ends of the chain will be capped with the functional group. Such chains are macro diacids or diols. Carboxyl- or hydroxyl-terminated chains may then participate in the usual step-growth reactions.

When monomer is exhausted, if a tetrafunctional monomer such as divinyl benzene (DVB), $H_2C=CH\phi HC=CH_2$, is added to a batch of living chains, it couples to itself and to the living chains. If it is assumed that a DVB can react with another DVB only at one end (there are probably good steric reasons why this should be so), the following type of *star* structures result, with linear branches radiating from a DVB core:

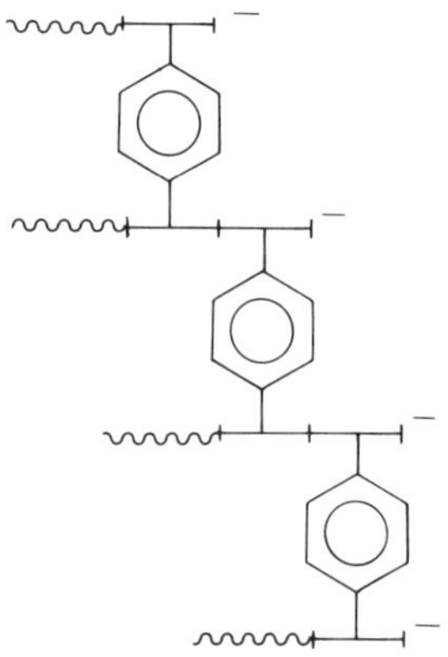

It is conceivable that at this point, a second difunctional monomer could be added, giving a star polymer with two different kinds of branches, but normally, the reactor would be opened and the reaction terminated. From the structure above, there are $(b - 1)$ moles of DVB per mole of star branches, where b is the averge number of branches per star. Since the number of moles of star molecules is equal to the moles of living chains over the average number of branches, and with an initiator such as n-BuLi, each molecule starts one living chain, to make a b-branch star polymer, one must add

$$N_{DVB} = \frac{(b - 1)I_0}{b} \tag{11.1}$$

moles of DVB, where I_0 is the moles initiator charged.

Living chains can also be *linked* by, for example, dichloro compounds, instantly doubling their chain lengths:

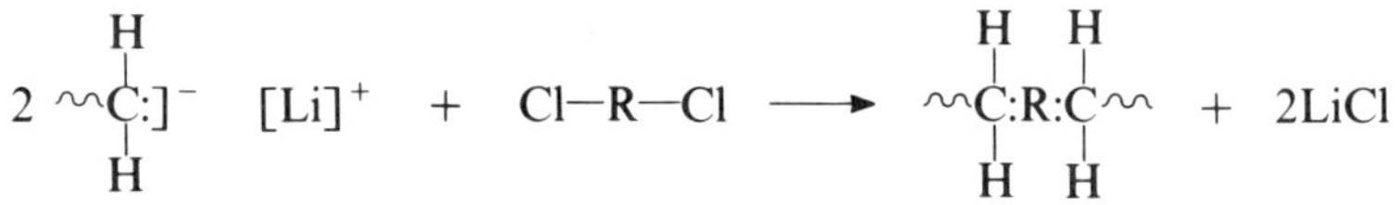

$$2 \; \sim\!\!\overset{\displaystyle H}{\underset{\displaystyle H}{C}}\!:]^- \; [\text{Li}]^+ \; + \; \text{Cl-R-Cl} \; \longrightarrow \; \sim\!\!\overset{\displaystyle H}{\underset{\displaystyle H}{C}}\!:\text{R}:\overset{\displaystyle H}{\underset{\displaystyle H}{C}}\!\!\sim \; + \; 2\text{LiCl}$$

With multifunctional linking agents, this technique can be used to form star polymers with various numbers of branches. It would be nice if all the functional groups on such compounds took part in the linking reaction, but sometimes steric factors lower linking efficiency. Multifunctional initiators can also be used to grow star polymers. Various initiators, linking agents, and other reagents for anionic polymerization have been reviewed.[5-8]

C   Monodisperse Polymers

Consider a batch reactor containing, say, a 20% solution of styrene in tetrahydrofuran (THF). A charge of n-BuLi is suddenly added. In this fairly polar solvent, the n-BuLi ionizes immediately and completely, and with $r_i \approx r_p$, starts chains growing promptly—one chain for each molecule of n-BuLi. In the absence of terminating impurities, the number of growing chains remains constant, and *they compete on an even basis for the available monomer.* They will therefore all have essentially the same length, giving a nearly monodisperse polymer. Not only can anionic polymerization be used to synthesize essentially monodisperse homopolymers, but the blocks in the block copolymers formed this way will also be monodisperse.

As with other modes of polymerization, the number-average chain length is given by the moles of monomer polymerized over the moles of chains present. With an initiator like n-butyllithium, each molecule starts a single chain, so

$$\bar{x}_n = \frac{M_0 - M}{I_0} = \frac{M_0 X}{I_0} \tag{11.2}$$

(If the volume is constant, the moles may be bracketed to give concentrations.) Statistically, the distribution of chain lengths is obtained by answering the following question: Given $M_0 - M$ marbles and I_0 buckets, what will be the distribution of marbles among the buckets if the marbles are thrown completely randomly into the buckets? The result is the Poisson distribution:

$$\left(\frac{n_x}{N}\right) = \frac{(\bar{x}_n)^x \exp(-\bar{x}_n)}{x!} \approx \frac{\exp(x - \bar{x}_n)}{\sqrt{(2\pi x)}} \left(\frac{\bar{x}_n}{x}\right)^x \tag{11.3}$$

(as written above, the distribution does not include the initiator residue). To a good approximation, the polydispersity index of this distribution is*

$$\frac{\bar{x}_w}{\bar{x}_n} \approx 1 + \frac{1}{\bar{x}_n} \tag{11.4}$$

from which it is seen that even at moderate $\bar{x}_n$ values, the polymer is essentially monodisperse. Figure 11.1 compares the Poisson distribution with the most probable distribution, (9.3) and (10.42), for $\bar{x}_n$ values of 100. Because of their sharpness, the number and weight distributions are almost identical.

As in the case of step-growth polymerization, the average chain length increases continuously with conversion (11.2). Unlike step-growth, the reaction mass consists only of monomer molecules and polymer chains of essentially a single length. Equations 11.2 and 11.3 characterize the polymer present.

Example 1. Starting with a batch reactor containing 1×10^{-3} moles of n-BuLi in dilute solution, suggest *two* methods for making the block copolymer

---

* There are the important practical limitations of not being able to add and mix reagents instantaneously and uniformly, and the almost inevitable presence of some terminating impurities. In practice, these always cause some additional broadening of the distribution.

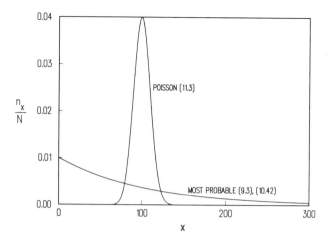

Figure 11.1   Comparison of Poisson and most-probable number-fraction distributions of chain lengths for $\bar{x}_n = 100$.

$[S]_{200}[B]_{1000}[S]_{200}$, where $[S]$ represents the styrene repeating unit and $[B]$ the butadiene repeating unit.

Solution. If we let all reactions go to completion, $X = 1$, then from (11.2), $M_0 = I_0 x$ (since the blocks will be essentially monodisperse, the bar and n have been left off x).

a. Add $200 \times 10^{-3} = 0.2$ mol of styrene to the reactor. When it has completely reacted, living polystyrene chains of $x = 200$ have been formed. Then add $1000 \times 10^{-3} = 1.0$ mol butadiene, and when it has completely reacted, add another 0.2 mol styrene and react to completion before opening the reactor and terminating the chains.

b. As before, start with 0.2 mol of styrene. When it has completely reacted, add 0.5 mol of butadiene, and react to completion. Then, link the resulting 1×10^{-3} mol of $[S]_{200}[B]_{500}^{-}$ living chains with 0.5×10^{-3} moles of a difunctional linking agent Cl–R–Cl to give 0.5×10^{-3} mol of $[S]_{200}[B]_{500}-R[B]_{500}[S]_{200}$. Normally the single R in the middle is insignificant.

Commercially, procedure (*b*) is preferred. To obtain the desired rubbery properties in the central polybutadiene block, nonpolar hydrocarbon solvents are used to promote 1,4 addition. These solvents slow down the crossover reaction from butadiene back to styrene in procedure (*a*), leading to increased polydispersity in the second polystyrene block. The linking reaction is imperfect, and commerical materials contain some diblock chains along with the desired triblock. Keep in mind also that each time material is added to a reactor, the probability of some terminating impurities getting in is increased.

It will be left as an end-of-chapter exercise to show how the polymer in Example 1 could be made by starting with a dianionic initiator.

Example 2. Consider the anionic batch polymerization of 1.0 mol of styrene in tetrahydrofuran solution with 1.0×10^{-3} mol of n-BuLi initiator. Assume that $r_i \approx r_p$ and that mixing is perfect and instantaneous. At 50% conversion, 0.5×10^{-3} mol of water are added to the reaction mass, and the reaction is allowed to continue.

a. At 100% conversion, what chain lengths will be present in the reaction mass?

b. Calculate $\bar{x}_n$ at 100% conversion.

Solution. a. The initiator starts 1×10^{-3} mol of chains growing. At 50% conversion, the water terminates half the chains. Thus, 1.0/4 mol of monomer are present in the 0.5×10^{-3} mol of terminated chains:

$$x(\text{terminated chains}) = \frac{1.0}{4 \times 0.5 \times 10^{-3}} = 500$$

The remaining 1.0(3/4) mol of monomer continue to grow in 0.5×10^{-3} mol of unterminated chains:

$$x(\text{unterminated chains}) = \frac{1.0(3/4)}{0.5 \times 10^{-3}} = 1500$$

b. Since each initiator molecule starts a chain, at $X = 1$,

$$\bar{x}_n = \frac{M_0}{I_0} = \frac{1.0}{1 \times 10^{-3}} = 1000$$

or, using (6.2a),

$$\bar{x}_n = \frac{\sum n_x x}{\sum n_x} = \frac{\left(0.5 \times 10^{-3}\right)(500) + \left(0.5 \times 10^{-3}\right)(1500)}{0.5 \times 10^{-3} + 0.5 \times 10^{-3}} = 1000$$

11.3 ANIONIC KINETICS[5, 6, 9]

A general description of anionic polymerization kinetics is complicated by the associations that may occur, particularly in nonpolar (hydrocarbon) solvents. The rate of propagation is proportional to the product of the monomer concentration and the concentration of *active* living chains $[P_x^-]$

$$r_p = -\left(\frac{1}{V}\right)\left(\frac{dM}{dt}\right) = k_p[M][P_x^-] \tag{11.5}$$

With negligible association (as in tetrahydrofuran solvent, for example, or hydrocarbons at BuLi concentrations less than 10^{-4} molar), each initiator molecule starts a growing chain, and in the absence of terminating impurities, the number of active living chains equals the number of initiator molecules added,

$$P_x^- = I_0 \quad \text{(constant)} \tag{11.6}$$

$$r_p = -\left(\frac{1}{V}\right)\left(\frac{dM}{dt}\right) = k_p\left(\frac{M}{V}\right)\left(\frac{I_0}{V}\right) \tag{11.7}$$

or, making use of the definition of conversion X (10.14) and assuming that the volume of the reaction mass is linear in conversion (10.17), we get

$$\frac{dX}{dt} = k_p[I]_0 \frac{1-X}{1+\varepsilon X} \tag{11.8}$$

Unlike the case for free-radical addition, this can be readily integrated for a batch reactor ($[I]_0$ is constant) to give

$$\varepsilon X + (1+\varepsilon)\ln(1-X) = -k_p[I]_0 t \tag{11.9}$$

but it's still no easier to get X as an explicit function of t.

Because anionic polymerizations are generally carried out in rather dilute solutions in inert solvents, volume changes with conversion tend to be much smaller than when undiluted monomer is polymerized. This often justifies the neglect of volume change ($\varepsilon = 0$), for which (11.9) becomes

$$X = 1 - \exp(-k_p[I]_0 t) \quad (\varepsilon = 0) \tag{11.10}$$

Example 3. For an isothermal reaction (constant k_p) subject to the assumptions in Example 2, obtain an expression that relates $\bar{x}_n$ to time. Neglect volume change.

Solution. Combining (11.2) and (11.10) gives

$$x = \frac{M_0}{I_0}(1 - \exp\{-k_p[I]_0 t\})$$

Since the polymer will be essentially monodisperse, the bar and subscript have been left off x. To use this expression, it would be necessary to know V_0, the initial volume of the reaction mass to get $[I]_0 = I_0/V_0$.

In BuLi polymerizations at high concentrations in nonpolar solvents, the chain ends are present largely as inactive dimers, which dissociate *slightly* according to the equilibrium

$$(P_x^-)_2 \quad \overset{K}{\rightleftharpoons} \quad 2P_x^-$$

Inactive dimer   Active chains

where

$$K = \frac{[P_x^-]^2}{[(P_x^-)_2]} \ll 1 \tag{11.11}$$

The concentration of active chains is then

$$[P_x^-] = K^{1/2}[(P_x^-)_2]^{1/2} \tag{11.12}$$

Now it takes two initiator molecules to make one inactive dimer, so

$$\underbrace{[P_x^-]}_{\text{Negligible}} + 2[(P_x^-)_2] = [I]_0 \tag{11.13}$$

The rate of polymerization then becomes

$$r_p = -\left(\frac{1}{V}\right)\left(\frac{dM}{dt}\right) = k_p K^{1/2}\left(\frac{[I]_0}{2}\right)^{1/2}[M] \tag{11.14}$$

The low value of K, reflecting the presence of most chain ends in the inactive associated state (dimers), gives rise to low rates of polymerization in nonpolar solvents. At very high concentrations, association may be even greater, and the rate essentially independent of $[I]_0$.

Example 4. Which type of isothermal reactor will produce the narrowest possible distribution of chain lengths in an anionic polymerization—batch, continuous stirred tank (backmix) (assume both perfectly stirred), plug-flow tubular, or laminar flow tubular?

Solution. As demonstrated in Example 3, the chain length depends on how long a chain is allowed to grow. In a batch reactor and an *ideal* plug-flow reactor, all chains react for the same length of time; hence, the product will be essentially monodisperse. In a CSTR and a laminar-flow tubular reactor, the residence time of chains in the reactor varies, causing a spread in the distribution. Keep in mind, however, that ideal pluf flow is a practical impossiblity, particularly with highly viscous polymer solutions. Compare these conclusions with those of Chapter X, Example 13.

11.4 GROUP-TRANSFER POLYMERIZATION

In the last decade, the DuPont Company has developed and patented[10] a new type of polymerization that mechanistically is similar to anionic polymerization. *Group-transfer polymerization* (GTP) has been defined as "polymerization of α,β-unsaturated esters, ketones, nitriles, or amides, initiated by silyl ketene acetals."[11] It has most commonly been used to polymerize acrylate and methacrylate monomers with the aid of anionic catalysts (they are true catalysts here) such as the bifluoride ion, $[FHF]^-$. GTP is illustrated below for the polymerization of methyl methacrylate (MMA) with silyl ketene acetal (SKA):

Initiation

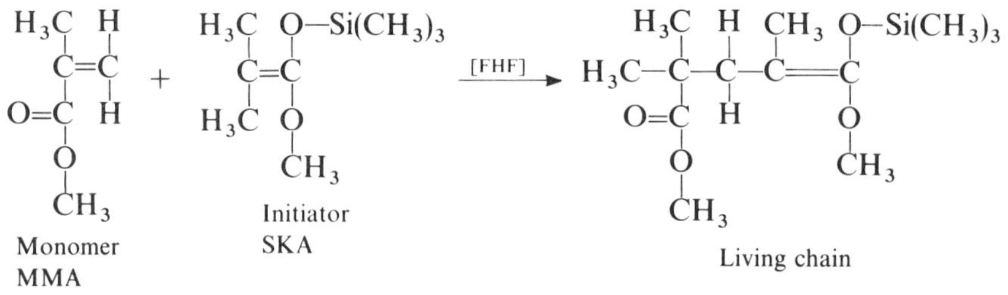

Propagation

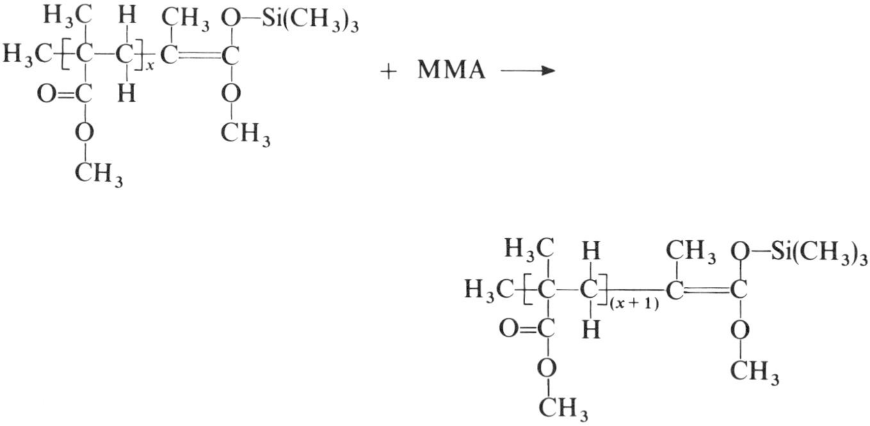

The initiating functionality is transferred to the growing end of the chain as each new monomer unit is added. They are living chains as in anionic addition, and can likewise be used to produce monodisperse polymers, block copolymers, and with the addition of appropriate reagents, chains with desired terminal groups. They can also control stereoregularity in the chain.

Unlike with anionic addition, chain transfer can occur:

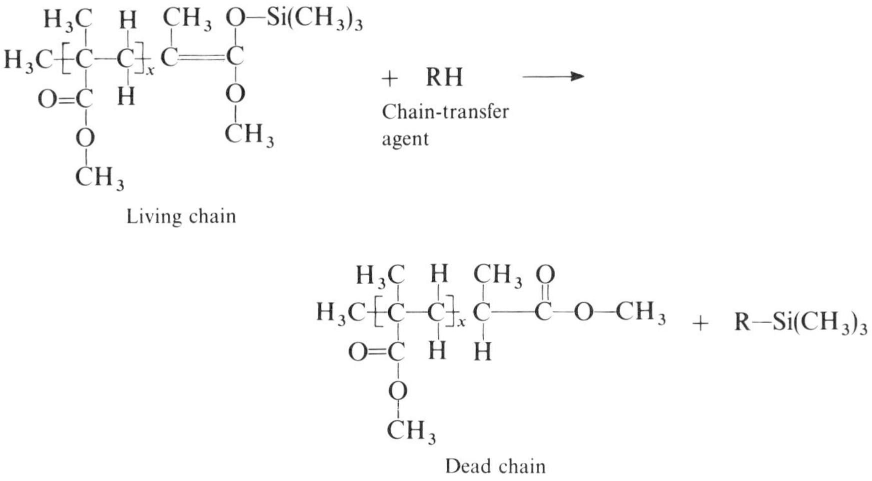

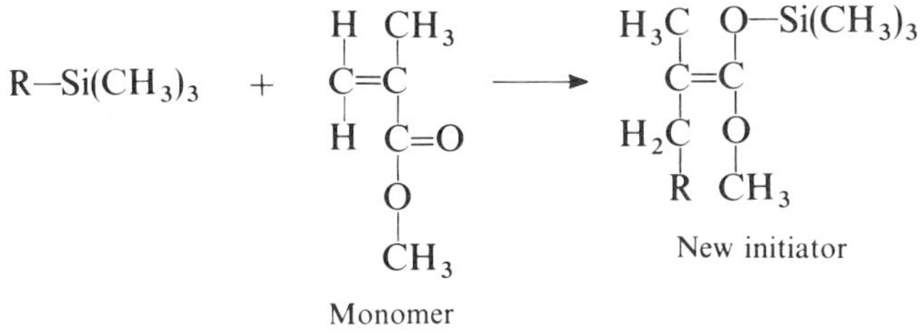

Living chain

+ RH
Chain-transfer
agent

Dead chain

Monomer

New initiator

As in anionic addition polymerization, chain length can be reduced by using more initiator, but because these initiators are rather expensive, it is often preferable to use a chain-transfer agent instead. Also, at low monomer concentrations, termination can occur through cyclization of the chain end.

11.5 HETEROGENEOUS STEREOSPECIFIC POLYMERIZATION[12-14]

In the early 1950s, Karl Ziegler in Germany observed that certain heterogeneous catalysts based on transition metals would polymerize ethylene to a linear, high-density material (high-density polyethylene, Chapter V) at low pressures and temperatures (compared to the existing free-radical process for low-density polyethylene). Giulio Natta in Italy showed that these catalysts would produce highly stereospecific poly α-olefins ($\{H_2C-CHR\}_x$, where $R = C_nH_m$) (notably polypropylene) and polydienes. Ziegler and Natta shared the 1963 Nobel Prize in chemistry for their work. Almost simultaneously, scientists at Amoco and

Phillips in the United States developed heterogeneous catalysts based on Mo and Cr that also produced linear polyethylene under mild conditions.

One example of a *Ziegler–Natta Catalyst* system is titanium tetrachloride, $TiCl_4$, and aluminum triethyl, $Al(C_2H_5)_3$. Vanadium and cobalt chlorides are also used, as in $Al(C_2H_5)_2Cl$. When these substances are mixed in an inert solvent, a crystalline solid is obtained along with a highly colored supernatant (deep violet or brown). It is known that the reaction involves the reduction of the titanium to a lower valence state, probably $+2$, since $TiCl_3$ also forms an effective catalyst. The supernatant liquid alone will polymerize α-olefins, but the resulting polymers show little stereospecificity. Commercial catalyst systems are based on the solid, either alone or on a support such as SiO_2 or $MgCl_2$. A "Phillips" catalyst typically consists of CrO_3 supported on silica or alumina.

Polymerization with Ziegler–Natta catalysts is thought to occur at active sites formed by interaction of the metal alkyl with metal chloride on the surface of the metal chloride crystals. Monomer is chemisorbed at the site (thus accounting for its specific orientation when added to the chain), and propagation occurs by *insertion* of the chemisorbed monomer into the metal-chain bond at the active site.

$$\text{site--}P_x + M \rightarrow \text{site--}P_{x+1}$$

The chain thus grows out from the site like a hair from the scalp.

Hydrogen is used as a chain-transfer agent in these reactions:

$$\text{site--}P_x + H_2 \rightarrow \text{site--H} + \text{H--}P_x$$

Chain transfer with the metal alkyl component in Ziegler–Natta systems has been identified:

$$\text{site--}P_x + Al(R)_3 \rightarrow \text{site--R} + Al(R)_2\text{--}P_x$$

and transfer to monomer may also occur.

Nowadays, most commercial processes based on these catalysts are carried out with the monomers either in the gas phase or in the liquid phase at temperatures such that the polymer precipitates as it is formed. Either way, solid polymer coats the catalyst particles on which it is formed. Solid, porous catalyst particles in the range of 10–100 μm are introduced to the reactor. As polymer begins to form, they rapidly break up into many smaller (0.01–1 μm) fragments (primary crystallites) on which the polymer continues to grow. The reaction mass therefore consists of a suspension of macroparticles, which are in turn made up of an agglomeration of microparticles, each of which surrounds a primary crystallite (Fig. 11.2). According to Ray and co-workers,[15] the macroparticle usually remains intact so that one is generated from each catalyst particle fed to the reactor. The primary catalyst crystallites are distributed more or less uniformly throughout the ultimate macroparticle.

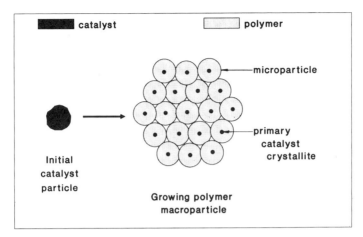

Figure 11.2   Mechanism of particle formation in Ziegler–Natta polymerization.[15]

If you are not yet convinced that these are complex systems, consider the following. The catalysts consist of two or more components, the nature and relative amounts of which influence the rate of polymerization. Each catalyst particle contains a multiplicity of surface sites on which polymer can grow and they may differ in activity (due to different crystal faces, for example). Sites may be deactivated so that their number decreases with time. The monomer must be chemisorbed on the catalyst surface before it can be added to a chain. And before it even gets to the catalyst surface, it must diffuse from the bulk fluid to the surface of the macroparticle, through the interstices (pores) between the microparticles and finally through the layer of solid polymer that coats the catalyst fragment within each microparticle. Furthermore, the heat of polymerization is liberated at the surface of the catalyst fragments and must flow outward. This could give rise to significant temperature gradients, complicating the interpretation of experimental data and limiting the value of isothermal rate expressions. It shouldn't be surprising, therefore, that we can't yet write simple and universally accepted kinetic equations as we could for free-radical and anionic addition polymerizations despite the fact that tens of billions of pounds per year of polymer are produced with these catalyst systems. Nevertheless, the following discussion illustrates some of the approaches that have been taken to describe the kinetics in systems with two-component Ziegler–Natta catalysts such as the $TiCl_4$–$Al(C_2H_5)_3$ system mentioned above.

The mechanism developed to describe the reaction is based on four assumptions[16]:

1. The propagation reaction occurs on active sites on the surface of the metal halide crystals.

2. Active sites are formed by the reaction of chemisorbed metal alkyls with metal halide on the crystal surface.

3. The propagation reaction occurs between a growing chain on the active site and an adjacent chemisorbed monomer molecule.

4. The metal alkyl competes with monomer for adsorption on the crystal surface.

Let C represent the number of catalytic sites on the catalyst surface, presumably associated with the Ti. Before a chain can grow on a site, however, the site must be *activated* by chemisorption of the $Al(C_2H_5)_3$ on an adjacent site. Therefore,

$$C_p^* = C\theta_A \tag{11.15}$$

where C_p^* is the number of growing chains stuck to the catalyst surface and θ_A is the fraction of the surface covered by chemisorbed alkyl.

The rate of polymerization is given by

$$r_p = k_s C_p^* \theta_M = k_s C\theta_M \theta_A \tag{11.16}$$

where k_s is the insertion rate constant, which characterizes the rate of insertion of chemisorbed monomer into the Ti-chain bond, and θ_M is the fraction of the surface covered by chemisorbed monomer.

Monomer and metal alkyl compete for chemisorption on the *uncovered* fraction of the active surface $\theta_0 = (1 - \theta_M - \theta_A)$

$$M + S \underset{k_{M-1}}{\overset{k_{M+1}}{\rightleftharpoons}} M \cdot S \quad \text{(chemisorbed monomer)}$$

$$A + S \underset{k_{A-1}}{\overset{k_{A+1}}{\rightleftharpoons}} A \cdot S \quad \text{(chemisorbed alkyl)} .$$

where S represents a surface site:

$$\text{Rate of monomer adsorption} = k_{M+1}(1 - \theta_M - \theta_A)[M_s] \tag{11.17}$$

$$\text{Rate of monomer desorption} = k_{M-1}\theta_M \tag{11.18}$$

$$\text{Rate of alkyl adsorption} = k_{A+1}(1 - \theta_M - \theta_A)[A_s] \tag{11.19}$$

$$\text{Rate of alkyl desorption} = k_{A-1}\theta_A \tag{11.20}$$

$[M_s]$ and $[A_s]$ are the concentrations of monomer and alkyl *at the surface*. If we assume a steady state for both monomer and alkyl and equate the rates of adsorption and desorption of each we get

$$\theta_M = \frac{K_M[M_s]}{1 + K_M[M_s] + K_A[A_s]} \tag{11.21}$$

and

$$\theta_A = \frac{K_A[A_s]}{1 + K_M[M_s] + K_A[A_s]} \tag{11.22}$$

where $K_M = k_{M+1}/k_{M-1}$ and $K_A = k_{A+1}/k_{A-1}$ are the equilibrium constants for monomer and alkyl adsorption. Equations 11.21 and 11.22 are Langmuir–Hinshelwood absorption isotherms, which when inserted into (11.16) gives[17,*]

$$r_p = k_s C \frac{K_A[A_s]K_M[M_s]}{(1 + K_A[A_s] + K_M[M_s])^2} \tag{11.23}$$

Equation 11.23 appears to give a reasonable description of experimental results. For example, the activity of a catalyst can be greatly enhanced by ball milling and/or dispersing it on the surface of a support such as $MgCl_2$, both of which increase C. A maximum in rate is generally observed as the alkyl concentration is increased. Actually, almost all experiments show the rate to be first order in monomer concentration, and (11.23) is not linear in monomer. This simply means that $K_M[M_s] \ll 1$, which is generally agreed to be the case.

With a little creative algebra, (11.23) can be put into a simpler and more convenient form:

$$r_p = k_s K_M C_p^* \theta_0[M_s] = k_p C^*[M_s] \tag{11.24}$$

where $k_p = k_s K_M$ is a propagation rate constant and $C^* = C_p^* \theta_0$ is interpreted as the number of *active* growing chains, that is, those that possess an open adjacent site for the chemisorption of monomer or alkyl. Again, because

$$\theta_0 = (1 - \theta_M - \theta_A) = \frac{1}{1 + K_A[A_s] + K_M[M_s]} \tag{11.25}$$

is a function of monomer concentration, (11.24) is not first order in monomer unless $K_M[M_s] \ll 1$, which generally seems to be the case. Equation 11.24 is the form usually used by chemical engineers to describe Ziegler–Natta polymerizations.

---

* It should be noted that when dealing with heterogeneous reactions such as these, rates are usually given per unit moles or mass of catalyst rather than per unit of reactor volume, as is the case for homogeneous reactions. In (11.23), for example, the rate would typically be in mol monomer/mol Ti (or $TiCl_4$), second.

Example 5. Estimate (*a*) the average lifetime of a growing chain and (*b*) the rate of polymerization for the polymerization of propylene in a heptane slurry with a modern, high-activity Ziegler–Natta catalyst under condition such that $k_p = 660$ L/mol·s, $C^* = 10^{-5}$ mol/g catalyst, $[M_s] = 4.0$ mol/L,[18] and $\bar{x}_n = 5000$. (*c*) Also estimate how long the reaction must be carried out under these conditions to achieve a catalyst yield of 10 000 g polymer/g catalyst (which is desirable to avoid a costly catalyst removal step). $[M_s]$ is kept constant by bubbling monomer into the reactor as polymerization proceeds.

Solution. ***a.*** By analogy to Example 11, Chapter X, $\bar{t}$ is given by the average number of monomer units in a chain ($\bar{x}_n$) times the number of growing chains (C^*) divided by the rate at which monomer is added to all chains (r_p):

$$\bar{t} = \frac{\bar{x}_n C^*}{r_p} = \frac{\bar{x}_n}{k_p[M_s]}$$

$$= \frac{5000 \text{ mol mon/mol chains}}{(660 \text{ L/mol chains} \cdot \text{s})(4.0 \text{ mol mon/L})} = 1.9 \text{ s}$$

b.

$$r_p = -\left(\frac{1}{m_{\text{cat}}}\right)\left(\frac{dM}{dt}\right) = k_p[M_s]C^*$$

$$= \left(660 \frac{\text{L}}{\text{mol chains} \cdot \text{s}}\right)\left(4.0 \frac{\text{mol mon}}{\text{L}}\right)\left(\frac{10^{-5} \text{ mol chains}}{\text{g cat}}\right)$$

$$= 0.0264 \text{ mol mon/g cat} \cdot \text{s}$$

or, because the molecular weight of propylene is 42

$$r_p = 0.0264 \times 42 \times 3600 = 4000 \text{ g polymer/g cat} \cdot \text{h}$$

c. So, to get the specified catalyst yield, the reaction must be carried out for $10\,000/4000 = 2.5$ h. Note that the rate of polymerization is based on the mass of catalyst rather than the volume. The reaction time here is much greater than the average chain lifetime (like free-radical addition and unlike anionic addition), but the ratio is not as great as in a typical free-radical process.

Unfortunately, things aren't always that simple in real life. For one thing, catalyst activity usually drops with time even in the absence of polymerization. Catalyst half-lives typically are on the order of minutes to hours. This is explained by a progressive deactivation of the sites responsible for polymerization, C^*, and is described quantitatively with an *n*th-order deactivation law:

$$\frac{dC^*}{dt} = -k^*(C^*)^n \tag{11.26}$$

Currently, second-order deactivation, $n = 2$, seems to be the most popular, but $n = 1$ and $n = 3$ have also been used to fit data.

Example 6. Rework Example 5(c), but now assume that the catalyst deactivates by a second-order mechanism with a half-life of 1 h.

Solution. Integration of (11.26) with $n = 2$ and $C^* = C_0^*$ at $t = 0$ gives

$$C^* = \frac{C_0^*}{1 + C_0^* k^* t}$$

Because $C^*/C_0^* = 0.5$ at $t = t_{1/2}$, $t_{1/2} = 1/C_0^* k^*$. By inserting the above into (11.24), we get

$$r_p = -\frac{1}{m_{\text{cat}}} \frac{dM}{dt} = \frac{k_p C_0^* [M_s]}{1 + C_0^* k^* t}$$

Separating variables (remember that $[M_s]$ is constant here) gives

$$-\frac{1}{m_{\text{cat}}} \int_{M_0}^{M} dM = k_p C_0^* [M_s] \int_0^t \frac{dt}{1 + C_0^* k^* t}$$

and integrating gives

$$\frac{M_0 - M}{m_{\text{cat}}} = \frac{k_p C_0^* [M_s]}{C_0^* k^*} \ln(1 + C_0^* k^* t) = k_p C_0^* [M_s] t_{1/2} \ln\left(1 + \frac{t}{t_{1/2}}\right)$$

Now $(M_0 - M)/m_{\text{cat}} = 10\,000$ g/g cat/42 g/mol $= 238$ mol/g cat. Using this value, $t_{1/2} = 1$ h, the parameters from Example 5 and solving for t gives $t = 11.3$ h. Compared with the result of Example 5(c), this illustrates the practical consequences of a deactivating catalyst.

Another complication is the possibility of mass-transfer limitations on the reaction rate. To polymerize, monomer must first diffuse from the bulk fluid surrounding the macroparticles, where its concentration is $[M_b]$, to the catalytic surface within the microparticles, where its concentration is $[M_s]$. Where the inherent rate of reaction on the catalytic surface and/or the resistance to diffusion are high, the overall reaction rate can be determined by the physics of diffusion rather than the chemistry of the reaction.

Another way of putting this is that $[M_s] < [M_b]$. To use (11.23) or (11.24), $[M_s]$ must be known. Relating $[M_s]$ to $[M_b]$ is a complex problem in mass transfer and is beyond the scope of this chapter. It is treated in the literature.[18,19] Suffice it to say that to make calculations, you need to know, in addition to the kinetic parameters above, the geometry of the particles, the mass-transfer coefficient at the surface of the macroparticles, the effective diffusivity for the monomer in the interstices (pores) between the microparticles, and the diffusivity of the monomer through the layer of polymer coating the primary catalyst particles.

Calculations show that rates can be mass-transfer limited, depending on the activity of the catalyst and the physical properties of the system. One interesting result is that the resistance to mass transfer can decrease with time, causing an increase in rate (at constant $[M_b]$). This increase, when combined with a deactivating catalyst, gives rise to a maximum in the polymerization rate with time. The parameters in Example 5 represent a case that is not mass-transfer limited, so $[M_s] \approx [M_b]$.

If all the polymer-producing sites on a catalyst surface had the same activity and were exposed to the same monomer concentration, and if chain transfer occurred randomly, we'd expect the polymer to have a most-probable distribution of chain lengths, (10.47), with a polydispersity index of 2.0.[20] Polymers produced with these heterogeneous catalysts typically have polydispersity indexes from 3 to 20, however. Two reasons have been advanced for these broad distributions: (1) mass-transfer resistance, which causes the monomer concentration to vary with location and perhaps time in the particles, and (2) a range of site activities (k_p's) on the catalyst surface. Different types of sites have been identified on the surfaces.[20]

Recent calculations show that while mass-transfer limitations can contribute somewhat to a broadening of the distribution, a range of the site activities is needed to account for the observed polydispersities. Furthermore, the different sites must have a pretty wide range of activities.[18] To examine the effects of site heterogeneity, (11.24) must be generalized to

$$r_p = [M_s] \sum_{i=1}^{N} k_{pi} C_i^*$$

(11.27)

where N is the number of different types of sites and

$$k_p C^* = \sum k_{pi} C_i^*$$

(11.28)

Example 7. Consider a catalyst that has four types of sites. The fractions of each on the surface are $\theta_i = (C_i^*/\Sigma C_i^*) = 0.519, 0.333, 0.111,$ and 0.037. The k_{pi} are in the ratio $1:3:9:27$.[18] Each site produces polymer with a polydispersity index of 2.0. Assume that the rate of chain transfer is independent of site activity so that $\bar{x}_{ni}$ is proportional to k_{pi}. Assume further that each site is exposed to the same monomer concentration $[M_s]$ (this would not be true with mass-transfer limitations). Calculate the polydispersty index that results from this heterogeneity of sites.

Solution. If we divide (6.6) by (6.5) and convert from molecular weights to chain lengths with (6.4b) and (6.4c) we get

$$\left\langle \frac{\bar{x}_w}{\bar{x}_n} \right\rangle = \sum (w_i \bar{x}_{wi}) \sum \left(\frac{w_i}{\bar{x}_{ni}} \right)$$

(11.29)

where w_i here represents the weight fraction of polymer produced by each type of site. If the ratio $\bar{x}_w/\bar{x}_n$ is the same for each type of site, this becomes

$$\left\langle \frac{\bar{x}_w}{\bar{x}_n} \right\rangle = \left(\frac{\bar{x}_w}{\bar{x}_n} \right) \sum (w_i \bar{x}_{ni}) \sum \left(\frac{w_i}{\bar{x}_{ni}} \right) \tag{11.30}$$

and with $\bar{x}_{ni}$ proportional to k_{pi}:

$$\left\langle \frac{\bar{x}_w}{\bar{x}_n} \right\rangle = \left(\frac{\bar{x}_w}{\bar{x}_n} \right) \sum (w_i k_{pi}) \sum \left(\frac{w_i}{k_{pi}} \right) \tag{11.31}$$

Now with $[M_s]$ common to all sites, the weight of polymer produced at each type of site will be proportional to $k_{pi}C_i^*$, so

$$w_i = \frac{k_{pi}C_i^*}{\sum k_{pi}C_i^*} = \frac{k_{pi}\theta_i}{\sum k_{pi}\theta_i} \tag{11.32}$$

Using the given θ_i and the k_{pi} ratios, 1, 3, 9, and 27 (the proportionality constant between these and the actual k_{pi} cancels out) in (11.32) gives $w_i = 0.148, 0.284,$ 0.284, and 0.284. When these are inserted in (11.31) with $\bar{x}_w/\bar{x}_n = 2.0$ we get for the overall polydispersity index $\langle \bar{x}_w/\bar{x}_n \rangle = 6.4$.

You can get large polydispersity indexes with only two types of sites, provided they have a large enough difference (at least an order of magnitude) in activities. However, this results in a bimodal (two-peaked) molecular weight distribution, which is not generally seen in practice. The parameters in Example 7 were chosen to demonstrate that as few as four types of sites could not only account for the large polydispersity indexes but also produce the unimodal type of molecular weight distributions normally observed.[18]

In some applications, narrow molecular weight distributions are desirable. Polydispersity indexes in the range of 3–4 have been reported for systems in which the catalyst is supported on $MgCl_2$ (selective promotion of certain types of sites?) or when additives are used (selective poisoning of certain types of sites?).[19] Single-site (EXXPOL) catalysts have recently been announced.[20] They are reportedly capable of producing polyolefins with a minimum polydispersity index of 2. As this is written, the first is being commercialized.[21]

The question of temperature gradients in the particles has also been addressed. The conclusion is that they are generally not significant, except in the very early stages of gas-phase polymerizations, where temperatures may get high enough to melt the polymer.[22]

In practice, Ziegler–Natta catalyst systems are difficult to work with. Great care must be exercised in their preparation and use, since they are easily poisoned by water, among other things. They are pyrophoric (spontaneously burst into flame on contact with oxygen) and are used in close proximity to large amounts of flammable monomers and solvents, and so can present a significant safety hazard, both in the laboratory and the plant.

REFERENCES

1. Plesch, P. H. (ed.), *The Chemistry of Cationic Polymerization*, MacMillan, New York, 1963.

2. Russell, K. E., and G. J. Wilson, Cationic polymerizations, in *Polymerization Processes*, C. E. Schildknecht and I. Skeist (eds.), Wiley, New York, 1977, Chapter 10.

3. Percec, V., Recent developments in cationic polymerization, in *Applied Polymer Science*, 2nd ed., R. W. Tess and G. W. Poehlein (eds.), American Chemical Society, Washington, DC, 1985, Chapter 5.

4. Odian, G., *Principles of Polymerization*, 2nd ed., Wiley-Interscience, New York, 1981, Chapter 5-2.

5. Morton, M., *Anionic Polymerization: Principles and Practice*, Academic, New York, 1983.

6. Swarc, M., *Living Polymers and Mechanisms of Anionic Polymerization*, Springer, Berlin, 1983.

7. McGrath, J. E. (ed.), *Anionic Polymerization: Kinetics, Mechanisms and Synthesis*, American Chemical Society, Washington, DC, 1981.

8. Franta, E., and P. Rempp, *Polym. Prepr.* **20**(1), 5 (1979).

9. Odian, G., *op. cit.*, Chapter 5-3.

10. Unites States patents 4,414,372 and 4,417,034.

11. Stinson, S. C., *Chem. Eng. News*, April 27, 1983, p. 43.

12. Keii, T., *Kinetics of Ziegler–Natta Polymerization*, Chapman Hall, London, 1972.

13. Chien, J. C. W. (ed.), *Coordination Polymerization*, Academic, New York, 1975.

14. Karol, F. J., Coordinated anionic polymerizaton and polymerization mechanisms, in *Applied Polymer Science*, 2nd ed., R. W. Tess and G. W. Poehlein (eds.), American Chemical Society, Washington, DC, 1985, Chapter 4.

15. Floyd, S., T. Heiskanen, and W. H. Ray, *Chem. Eng. Progr.* **85**(11), 56 (1988).

16. Tait, P. J. T., *Chem. Technol.* **5**, 688 (1975).

17. Keii, T., et al., *Makromol. Chem.* **183**, 2285 (1982).

18. Floyd, S., et al., *J. Appl. Polym. Sci.* **33**, 1021 (1987).

19. Floyd, S., et al., *J. Appl. Polym. Sci.* **32**, 2935 (1986).

20. Anon,, *Mod. Plast.* **68**(7), 61 (1991): Leaversuch, R. D., *Mod. Plast.* **68**(10), 46 (1991): Anon., *Chem. Eng. News*, Dec. 23, 1991, p. 16.

21. Anon., *Chem. Eng. Prog.* **87**(10), 21 (1991).

22. Zucchini, U., and G. Cecchin, *Adv. Polym. Sci.* **51**, 101 (1983).

PROBLEMS

1. Given a reactor that contains 4.000 mol styrene in solution and supplies of n-BuLi and a tetrafunctional linking agent $R(Cl)_4$, list the amounts and order of addition of reagents necessary to produce a tetra-star polymer with branches of length $x = 1000$. Assume complete and instantaneous ionization, perfect linking, and no terminating impurities.

2. A reactor initially contains 1.00×10^{-3} mol of n-BuLi in a dry, inert solvent. 1.00 mol of styrene (S) is added to the reactor. At 50% conversion,

0.250×10^{-3} mol of a difunctional linking agent, Cl–R–Cl, is added to the reactor. The reaction is then allowed to go to completion. The assumptions in Problem 1 apply.

a. Describe quantitatively the reactor product (how many moles of what).

b. Calculate $\bar{x}_n$ for the product.

3. Repeat Problem 2, but replace the linking agent with an additional 1.00×10^{-3} mol of *n*-BuLi.

4. Repeat Problem 2, but replace the linking agent with 1.00×10^{-3} mol of a dianionic initiator, $^+[Li]\ ^-[:R:]^-\ [Li]^+$.

5. Consider anionic addition polymerization in which most of the chain ends are associated in inactive aggregates of *n* chain ends. The aggregates dissociate slightly to active chain ends according to the equilibrium relation

$$(P_x^-)_n \overset{K}{\rightleftharpoons} nP_x^-$$

where $K \ll 1$. Obtain an expression for the rate of polymerization.

6. Describe quantitatively how the polymer of Example 1 could be made with the dianionic initiator of Problem 4.

7. Describe how you would make a block copolymer of butadiene and nylon 6/6 (Chapter II, Example 4E).

8. Obtain an expression for the number-average chain length at complete conversion in a group-transfer polymerization batch that contains I, M, and RH moles of initiator, monomer, and chain-transfer agent, respectively.

9. The following mechanism has been proposed to explain cationic polymerization (don't bet the farm on it):

Ionization $\qquad\qquad\qquad$ $GH \overset{K}{\rightleftharpoons} G^-H^+$

Initiation $\qquad\qquad\qquad$ $G^-H^+ + M \overset{k_a}{\to} P_1^+G^-$

Propagation $\qquad\qquad\quad$ $P_x^+G^- + M \overset{k_p}{\to} P_{x+1}^+G^-$

Termination $\qquad\qquad\quad$ $P_x^+G^- \overset{k_t}{\to} P_x + GH$

Here, GH is the catalyst and G^- the gegen ion. The ionization step is assumed to be essentially instantaneous, and always at equilibrium, with K being the equilibrium constant. Obtain expressions for the rate of polymerization and the number-average chain length according to this mechanism.

10. Consider an ideal anionic polymerization, the type that when carried out in a batch reactor would produce monodisperse polymer. Here, however, the reaction is to be carried out in a perfectly micromixed CSTR in which the steady-state monomer concentration is $[M]$. The average residence time $\tau = V/Q$, where V is the reactor volume and Q is the volumetric flow rate. The residence-time distribution for such a reactor is

$$E(t) = \frac{1}{\tau} \exp\left(-\frac{t}{\tau}\right)$$

where $E(t)dt$ is the fraction of the material in the product stream that has resided in the reactor for time t. Assume that chains grow only while in the reactor.

In terms of parameters defined above, obtain expressions for $\bar{x}_n$, (n/N), (w/W) (the number- and weight-fraction distributions of chain lengths) and the polydispersity index $\bar{x}_w/\bar{x}_n$. Does any of this look familiar?

11. Consider the reaction in Problem 10 above carried out in a laminar-flow tubular reactor of length L and inner radius R, in which the velocity profile is

$$u(r) = \frac{2Q}{\pi R^2}\left[1 - \left(\frac{r}{R}\right)^2\right]$$

where $u(r)$ is the local velocity at radius r. For this type of reactor, the residence-time distribution is

$$E(t) = 0 \quad \text{for } t < \frac{\tau}{2} \qquad E(t) = \frac{\tau^2}{2t^3} \quad \left(\text{for } t > \frac{\tau}{2}\right)$$

The concentrations of monomer and initiator in the feed are $[M]_0$ and $[I]_0$.

a. First consider the (unrealistic) completely segregated case, in which no radial or axial diffusion is permitted. Obtain expressions for the maximum and minimum values of x and sketch the number distribution.

b. Now consider the more realistic case where diffusion, most importantly radial diffusion of monomer, is permitted. Show how this would alter the distribution you sketched in (**a**).

12. Write the expression giving conversion as a function of time for the reaction in Example 2b.

13. Consider a Ziegler–Natta polymerization in which chains are terminated only by chain transfer with hydrogen. The rate of transfer is given by $r_{tr} = k_{tr} C^* [H_2]$.

a. Obtain an expression for $\bar{x}_n$ in terms of concentrations and rate constants. Assume sites of uniform activity.

b. Obtain the expression for the number-fraction distribution of chain lengths, $(n/N)(x)$ for case (*a*). What is the polydispersity index for this distribution?

14. A gas-phase Ziegler–Natta polymerization is carried out in a constant-volume, isothermal batch reactor. Assume that surface monomer concentration is always proportional to bulk monomer concentration, $[M_s] = k[M_b]$. Obtain expressions for conversion as a function of time for two cases:

    a. With a constant-activity catalyst.

    b. With a catalyst that deactivates according to a second-order mechanism.

    Define any parameters you need that are not defined in the chapter.

15. Consider the polymer in Example 7. Assume that $\bar{x}_{n1} = 1000$, $\bar{x}_{n2} = 3000$, $\bar{x}_{n3} = 9000$, and $\bar{x}_{n4} = 27\,000$.

    a. Calculate $\bar{x}_n$ and $\bar{x}_w$ for the product.

    b. Plot the weight-fraction distribution of chain lengths.

16. A heterogeneous catalyst has two different types of surface sites. They produce equal weights of polymer, but one produces chains of $\bar{x}_n = 1000$ and the other chains of $\bar{x}_n = 10\,000$. The chains produced at each site follow the most-probable distribution.

    a. Calculate $\bar{x}_w$, $\bar{x}_n$, and the polydispersity index for the product.

    b. Plot the weight-fraction distribution of chain lengths.

17. Define in terms of parameters in (11.23) the optimum alkyl concentration for maximum polymer production.

CHAPTER XII

Copolymerization

12.1 MECHANISM

We have seen in previous chapters how the composition of a random copolymer can influence many of its important properties, including solubility, degree of crystallinity, T_g, and T_m. The control of copolymer composition is therefore of great practical importance.

A quantitative treatment of random copolymerization is based on the assumption that the reactivity of a growing chain depends only on its active terminal unit. Therefore, when two monomers, M_1 and M_2 are copolymerized, there are four possible propagation reactions:*

Reaction   Rate equation

$$P_1\cdot + M_1 \xrightarrow{k_{11}} P_1\cdot \qquad k_{11}[P_1\cdot][M_1] \tag{12.1}$$

$$P_1\cdot + M_2 \xrightarrow{k_{12}} P_2\cdot \qquad k_{12}[P_1\cdot][M_2] \tag{12.2}$$

$$P_2\cdot + M_2 \xrightarrow{k_{22}} P_2\cdot \qquad k_{22}[P_2\cdot][M_2] \tag{12.3}$$

$$P_2\cdot + M_1 \xrightarrow{k_{21}} P_1\cdot \qquad k_{21}[P_2\cdot][M_1] \tag{12.4}$$

---

* The notation has been changed a bit here to conform to the existing literature. The subscripts 1 and 2 designate the two monomers being copolymerized, not the number of repeating units. Thus $P_1\cdot$ represents a growing chain (of any length) with a terminal unit of monomer 1, that is, a chain to which a monomer M_1 was the last added. While the mathematics of copolymerization looks the same for the various types of addition polymerization, there are some important mechanistic differences which will be discussed later.

The first subscript on the rate constants designates the nature of the chain end and the second identifies the monomer being added to the chain.

Application of the steady-state assumption to $P_1\cdot$ and $P_2\cdot$ requires that they be generated and consumed at equal rates. $P_1\cdot$'s are generated in reaction 12.4 and consumed in reaction 12.2. Note that (12.1) just converts one $P_1\cdot$ into another one, with no net change in their number. Therefore,

$$k_{12}[P_1\cdot][M_2] = k_{21}[P_2\cdot][M_1] \tag{12.5}$$

The rates of consumption of monomers M_1 and M_2 are

$$-\left(\frac{1}{V}\right)\left(\frac{dM_1}{dt}\right) = k_{11}[P_1\cdot][M_1] + k_{21}[P_2\cdot][M_1] \tag{12.6}$$

$$-\left(\frac{1}{V}\right)\left(\frac{dM_2}{dt}\right) = k_{12}[P_1\cdot][M_2] + k_{22}[P_2\cdot][M_2] \tag{12.7}$$

Dividing (12.6) by (12.7), eliminating the $[P\cdot]$'s with (12.5), and recalling that $M_i = [M_i]V$ gives

$$\frac{dM_1}{dM_2} = \frac{M_1}{M_2}\left[\frac{r_1 M_1 + M_2}{M_1 + r_2 M_2}\right] \tag{12.8}$$

where r_1 and r_2 are the *reactivity ratios*.

$$r_1 = \frac{k_{11}}{k_{12}} = \text{relative preference of } P_1\cdot \text{ for } M_1/M_2$$

$$r_2 = \frac{k_{22}}{k_{21}} = \text{relative preference of } P_2\cdot \text{ for } M_2/M_1$$

Reactivity ratios are experimentally determined[1] or may be estimated.[2] In organic free-radical copolymerizations, for a given monomer pair they are pretty much independent of initiator and solvent, and are only weakly temperature dependent. In ionic copolymerizations, however, they depend strongly on the gegen ion and solvent.

This relation may be put in a more convenient form by defining

f_1 = mole fraction of monomer 1 in the reaction mass *at any instant*
F_1 = mole fraction of monomer 1 in the copolymer formed *at that instant*.

$$F_1 = (1 - F_2) = \frac{dM_1}{d(M_1 + M_2)} \tag{12.9}$$

$$f_1 = (1 - f_2) = \frac{M_1}{M_1 + M_2} \tag{12.10}$$

Combination of (12.8)–(12.10) (if you enjoy algebra, you really ought to try this one) gives

$$F_1 = \frac{r_1 f_1^2 + f_1 f_2}{r_1 f_1^2 + 2f_1 f_2 + r_2 f_2^2} = \frac{(r_1 - 1)f_1^2 + f_1}{(r_1 + r_2 - 2)f_1^2 + 2(1 - r_2)f_1 + r_2} \quad (12.11)$$

$$F_1 = F_1(r_1, r_2, f_1)$$

The quantity F_1, the *instantaneous copolymer composition*, is analogous to $\bar{x}_n$, the instantaneous number-average chain length in free-radical addition polymerization. Like $\bar{x}_n$, it depends on the conditions in the reactor at a particular instant. It, too, is really an average, since not all the copolymer formed at a particular instant has exactly the same composition. However, the instantaneous distribution of compositions is normally much narrower than the instantaneous distribution of chain lengths, and because the fact that it is an average is not normally of great practical significance and can't be controlled anyhow, the overbar is left off F_1.

12.2 SIGNIFICANCE OF REACTIVITY RATIOS

To gain an appreciation for the physical significance of (12.11), let's look at some special cases of the reactivity ratios.

Case 1: $r_1 = r_2 = 0$. With both reactivity ratios zero, neither type of chain end can add its own monomer, so a *perfectly alternating* copolymer results, $F_1 = 0.5$ regardless of f_1, until one of the monomers is used up, at which point, polymerization stops.

Case 2: $r_1 = r_2 = \infty$. Here, $P_1 \cdot$'s can add only M_1 monomer, and $P_2 \cdot$'s only M_2, so the polymer formed will be a physical mixture of homopolymer 1 and homopolymer 2 chains.

Case 3: $r_1 = r_2 = 1$. Under these conditions, the growing chains find the monomers equally attractive, so the addition depends only on the ratio of monomers in the vicinity of the chain ends, $F_1 = f_1$.

Case 4: $r_1 r_2 = 1$. This is the so-called ideal copolymerization, where each chain displays the same preference for one of the monomers over the other: $k_{11}/k_{12} = k_{21}/k_{22}$, so it doesn't matter what's on the end of the chain. In this case, (12.11) reduces to

$$F_1 = \frac{r_1 f_1}{(r_1 - 1)f_1 + 1} \quad (12.12)$$

The reader familiar with distillation theory will note here the exact analogy between (12.12) and the vapor–liquid equilibrium composition relation for ideal solutions with a constant relative volatility.

Case 5: $r_1 < 1, r_2 < 1$. This common situation corresponds to an azeotrope in vapor–liquid equilibrium. At the azeotrope,

$$F_{1az} = f_{1az} = \frac{(1 - r_2)}{(2 - r_1 - r_2)} \tag{12.13}$$

Systems for which $r_1 > 1$, $r_2 > 1$ also form azeotropes, but have been reported only rarely.

12.3 VARIATION OF COMPOSITION WITH CONVERSION

In general, $F_1 \neq f_1$, that is, the composition of the copolymer formed at any instant will differ from that of the monomer mixture from which it is being formed. Thus, as the reaction proceeds, the unreacted monomer mixture will be depleted in the more reactive monomer, and as the composition of the unreacted monomer changes, so will that of the polymer being formed, in accordance with (12.11).

Example 1. Draw curves of instantaneous copolymer composition, F_1, vs. monomer composition, f_1, for the following systems, and indicate the direction of composition drift as the reaction proceeds in a batch reactor.

a. Butadiene (1), styrene (2), 60°C; $r_1 = 1.39$, $r_2 = 0.78$.
b. Vinyl acetate (1), styrene (2), 60°C; $r_1 = 0.01$, $r_2 = 55$.
c. Maleic anhydride (1), isopropenyl acetate (2), 60°C; $r_1 = 0.002$, $r_2 = 0.032$.

Solution. Application of (12.11) gives the plots in Fig. 12.1. Note that system (*a*) approximates ideal copolymerization, case 4 above. In system (*b*), styrene is the preferred monomer, regardless of the terminal radical; hence, the copolymer is largely styrene until styrene monomer is nearly used up. System (*c*) approximates case 1 above.

The direction of composition drift with conversion is indicated by arrows. Note that system (*c*) forms an azeotrope at $F_1 = f_1 = 0.493$. To the left of the azeotrope, M_1 is the more reactive monomer ($F_1 > f_1$), but to the right, M_2 is more reactive ($F_1 < f_1$). Therefore, with an initial monomer composition $f_{10} < 0.493$, F_1 and f_1 will decrease with conversion, but if $f_{10} > 0.493$, they will increase. If the initial monomer charge is exactly at the azeotropic composition, there will be no composition drift.

Consider a batch consisting of a total of M moles of monomer ($M = M_1 + M_2$). At time t, the monomer has a composition f_1. In the time interval dt, dM moles of monomer polymerize to form copolymer with a composition F_1. Therefore, at time $t + dt$, there are left ($M - dM$) moles monomer whose composition has been changed to ($f_1 - df_1$). Writing a material balance on monomer 1 gives M_1 in monomer at $t = M_1$ in monomer at ($t + dt$) + M_1 in

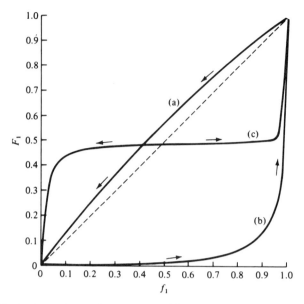

Figure 12.1 Instantaneous copolymer composition (F_1) vs. monomer composition (f_1). (a) Butadiene (1), styrene (2), 60°C; $r_1 = 1.39$, $r_2 = 0.78$. (b) Vinyl acetate (1), styrene (2), 60°C; $r_1 = 0.01$, $r_2 = 55$. (c) Maleic anhydride (1), isopropenyl acetate (2); $r_1 = 0.002$, $r_2 = 0.032$. The direction of composition drift in a batch reaction is indicated by arrows (Example 1).

polymer formed in interval dt:

$$f_1 M = (M - dM)(f_1 - df_1) + F_1 dM \qquad (12.14)$$

Expanding and neglecting second-order differentials gives

$$\frac{dM}{M} = \frac{df_1}{F_1 - f_1} \qquad (12.15)$$

At the start of the reaction, there are M_0 moles monomer present with a composition f_{10}, and at some later time, there are M moles monomer left with a composition f_1. Integrating between these limits gives

$$\ln \frac{M}{M_0} = \int_{f_{10}}^{f_1} \frac{df_1}{F_1 - f_1} \qquad (12.16)$$

This equation is the exact analog of the Rayleigh equation relating the amount and composition of the still-pot liquid in a batch distillation. By choosing values for f_1 and calculating the corresponding F_1's with (12.11), the integral may be evaluated to obtain a relation between the monomer composition and conversion $(X = 1 - (M/M_0))$. An analytic solution to (12.11) and (12.16) has been obtained.[3] For $r_1 \neq 1$, $r_2 \neq 1$ (if either is equal to 1, see the original reference),

$$\frac{M}{M_0} = \left[\frac{f_1}{f_{10}}\right]^\alpha \left[\frac{f_2}{f_{20}}\right]^\beta \left[\frac{f_{10} - \delta}{f_1 - \delta}\right]^\gamma = 1 - X \qquad (12.17)$$

where $\alpha = \dfrac{r_2}{1 - r_2}$

$\beta = \dfrac{r_1}{1 - r_1}$

$\gamma = \dfrac{1 - r_1 r_2}{(1 - r_1)(1 - r_2)}$

$\delta = \dfrac{1 - r_2}{2 - r_1 - r_2}$

Knowing the monomer composition f_1 as a function of conversion immediately gives the instantaneous copolymer composition F_1 as a function of conversion through (12.11). This is important to know, because if there is a large variation in the composition of the copolymer formed from the beginning of the reaction to high conversions, there may be a wide variation in its properties, also.

Example 2. Discuss the possible influence of a wide variation in F_1 on the transparency of amorphous copolymers.

Solution. Amorphous, random copolymers of fairly uniform composition are normally transparent because on a macroscopic scale, they are homogeneous systems in which the discontinuities (groups of a few repeating units) are very much smaller than the wavelengths of visible light. Because of the general mutual insolubility of two different polymers, copolymers of widely differing F_1 values may actually form a heterogeneous system in which the discontinuities, globules of one phase in a matrix of another, will be larger than the wavelength of light. Because of the compositional difference of the phases, they almost certainly will differ in refractive index as well, leading to a system that scatters light and is therefore not transparent.

In addition to the instantaneous copolymer composition F_1, another quantity of interest is $\langle F_1 \rangle$, the *cumulative composition* of the copolymer that has been formed up to a particular conversion. $\langle F_1 \rangle$ is exactly analogous to $\langle \bar{x}_n \rangle$, the cumulative number-average chain length. For a batch reactor, it is obtained through a material balance:

moles M_1 charged = moles M_1 in copolymer + moles M_1 left in monomer

$$f_{10}M_0 = \langle F_1 \rangle (M_0 - M) \qquad\qquad + f_1 M \qquad (12.18)$$

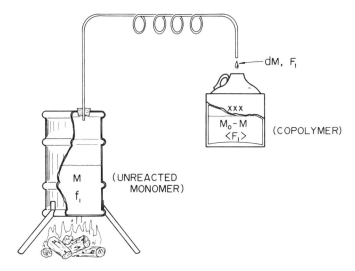

Figure 12.2   Batch distillation analogy to copolymerization. Holdup in vapor space and lines is nelgected.

Rearranging gives

$$\langle F_1 \rangle = \frac{f_{10} - f_1(M/M_0)}{1 - (M/M_0)} = \frac{f_{10} - f_1(1 - X)}{X} \qquad (12.19)$$

The distillation analog of this equation tells the well-educated bootlegger how much of his 20-proof sour mash he must distill over to have 120-proof white lightning in the jug under his condenser. Figure 12.2 illustrates the quantities defined in terms of the distillation analogy. The "well-stirred, adiabatic bathtub" analogy developed in Section 10.9 to visualize the relation between $\bar{x}_n$ and $\langle \bar{x}_n \rangle$ is equally applicable to the relation between F_1 and $\langle F_1 \rangle$. Here, F_1 is analogous to the temperature of the water leaving the spigot and $\langle F_1 \rangle$ to the temperature of the water in the bathtub.

Calculational procedures for a batch reactor are summarized in the following diagram for a system of known r_1, r_2, and f_{10}:

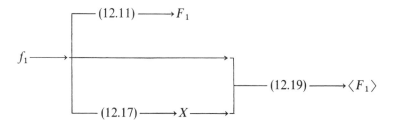

Example 3. For the styrene–butadiene system of Example 1, plot instantaneous copolymer composition F_1 and cumulative copolymer composition

$\langle F_1 \rangle$ vs. conversion for a batch reaction starting with a 50–50 (mol%) initial monomer charge.

Solution. As illustrated in the preceding diagram, calculations are facilitated by choosing f_1 as the independent variable. From Example 1, butadiene is the more reactive monomer, so as the reaction proceeds, both polymer and monomer will be enriched in styrene, and f_1 will decrease from 0.5 to 0. For a given value of f_1, the conversion X is calculated from (12.17) with $\alpha = 3.55$, $\beta = -3.57$, $\gamma = 0.983$, and $\delta = -1.29$. F_1 is obtained through (12.11) as in Example 1. $\langle F_1 \rangle$ is calculated from (12.19) using the chosen f_1 and the values of X obtained from (12.17).

The results are plotted in Fig. 12.3. Note that in this particular case, $\langle F_1 \rangle$ does not vary much between 0 and 100% conversion, but F_1 changes considerably, particularly at high conversions. Note also that $\langle F_1 \rangle$ at 100% conversion must always equal f_{10}. Because the entire initial monomer charge has been converted to copolymer, their compositions must be the same.

The preceding mathematical developments apply equally to copolymers formed by free-radical addition and anionic addition (though the numerical values of r_1 and r_2 would, in general, be different for the same monomer pair in each of the modes). The nature of the molecules formed is quite different, however. Recall that in free-radical addition, the average chain lifetime is an infinitesimal fraction of the reaction time, while in anionic addition, the chains continue to grow throughout the reaction. Thus, in free-radical addition, F_1 is the composition of entire chains formed during a conversion increment dX. $\langle F_1 \rangle$ represents the average composition of the mixture of chains of different F_1 formed up to conversion X. In classic anionic addition, all the chains are

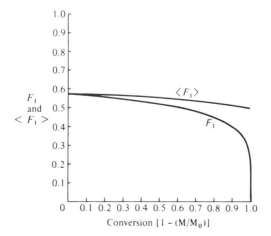

Figure 12.3 Instantaneous (F_1) and cumulative ($\langle F_1 \rangle$) copolymer composition vs. conversion, $r_1 = 1.39$, $r_2 = 0.78$, $f_{10} = 0.50$ (Example 2).

essentially identical. As f_1 changes with conversion, we get a *composition gradient* in each chain as the reaction proceeds. F_1 is the composition of the portion of each chain formed over a conversion increment dX and $\langle F_1 \rangle$ is the average composition of each chain (as well as the entire copolymer) up to conversion X. Copolymerization with typical Ziegler–Natta catalysts should be more like that in free-radical systems.

Example 4. Suggest three techniques for producing copolymers of fairly uniform composition, that is, those in which F_1 does not vary much.

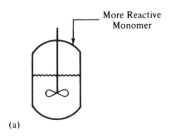

(a)

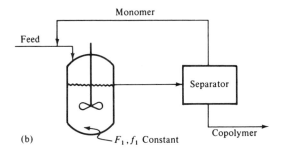

(b)

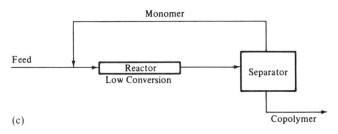

(c)

Figure 12.4 Techniques for minimizing the spread of copolymer composition: (*a*) semibatch reactor; (*b*) continuous stirred-tank reactor; (*c*) tubular reactor.

Solution. Three possibilities are sketched in Fig. 12.4. With a semibatch reactor, the more reactive monomer is replenished as the reaction proceeds to maintain f_1 (and therefore F_1) constant. A method for calculating the appropriate rate of addition has been described.[4] In a continuous stirred tank (backmix) reactor, both f_1 and F_1 are constant with time. In a continuous plug-flow reactor, the variation in F_1 can be kept small by limiting the conversion per pass in the reactor. Note that the last two techniques require facilities for separating unreacted monomer from the polymer, and in most cases, recycling it.

Example 5. Consider the semibatch reactor illustrated in Fig. 12.4*a*.

Let $P(t)$ = moles monomer (both) in polymer formed to time t

$A(t)$ = moles more reactive monomer added to time t

In this setup, if the relation between $A(t)$ and $P(t)$ is arranged properly, not only will the copolymer be of uniform composition, but both monomers will be used up simultaneously. Obtain an expression that relates $A(t)$ to $P(t)$ and system constants.

Solution. First, make a total material balance on monomer: moles initially charged + moles added = moles unreacted + moles in copolymer:

$$M_0 + A(t) = M(t) + P(t)$$

Then, make a material balance on the more reactive monomer 1:

$$f_{10}M_0 + A(t) = M(t)f_1 + P(t)F_1$$

By eliminating $M(t)$ between the two, keeping in mind that if F_1 is to be constant, $\langle F_1 \rangle = F_1$ and $f_1 = f_{10}$ (constant), we find

$$A(t) = \frac{F_1 - f_{10}}{1 - f_{10}} P(t) = \text{constant} \times P(t)$$

Thus, the more reactive monomer is added in direct proportion to the amount of polymer formed (in terms of moles monomer polymerized). In practice, the real trick is keeping track of $P(t)$. One approach is to make a dynamic heat balance on the reactor. The liberated heat of polymerization is proportional to $P(t)$.

12.4 COPOLYMERIZATION KINETICS

The problem of free-radical copolymerization kinetics is not nearly in such good shape. In addition to the four propagation reactions, there are three possible termination reactions ($P_1\cdot + P_2\cdot$, $P_1\cdot + P_1\cdot$, $P_2\cdot + P_2\cdot$), each with its own rate constant. A general rate equation has been developed,[5] but because of a lack of independent knowledge of the constants involved and the mathematical complexity, it hasn't been used much. An approximate integrated expression including first-order initiator decay has been presented.[6] It involves a single, average termination rate constant. This "constant" would not be expected to remain constant as composition varies. Nevertheless, there appears to be some experimental verification of the relation. In certain instances where the compositions aren't varying too much (for example, in the control of a continuous backmix reactor), simplification of (12.6) and (12.7) to

$$-\left(\frac{1}{V}\right)\left(\frac{dM_1}{dt}\right) = K_1[M_1] \qquad (12.20)$$

$$-\left(\frac{1}{V}\right)\left(\frac{dM_2}{dt}\right) = K_2[M_2] \qquad (12.21)$$

where the "constants" $K_1 = (k_{11}[P_1\cdot] + k_{21}[P_2\cdot])$ and $K_2 = (k_{12}[P_1\cdot] + k_{22}[P_2\cdot])$ are determined experimentally may prove satisfactory.

12.5 PENULTIMATE EFFECTS AND CHARGE-TRANSFER COMPLEXES

There are certain cases in which the equations developed in Section 12.1 do not adequately describe copolymerization. Two approaches have been taken to remedy these deficiencies: invoking penultimate effects and, more recently, postulating the formation of charge-transfer complexes. In the former, the next-to-last monomer unit in a growing chain also exerts an influence on the addition of the next monomer molecule.[7-10] In the latter, a 1:1 complex forms reversibly between electron-donating and electron-accepting comonomers (introducing an equilibrium constant to the analysis). This complex may then polymerize (from either end—introducing four more reactivity ratios) with itself or with the uncomplexed monomers.

Complex formation is particularly helpful in explaining free-radical copolymerizations in systems such as styrene–maleic anhydride. This system forms a 1:1 copolymer over most of the range of monomer composition, and the addition of maleic anhydride greatly enhances the rate of polymerization over that of pure styrene, despite the fact that maleic anhydride will not homopolymerize at a noticeable rate. These observations are consistent with the formation of a strong, readily polymerized complex between the monomers. The general equations to describe such copolymerizations have been presented by Seiner and Litt[11] and applied in a number of special cases.[12-15]

REFERENCES

1. Greenly, R. Z., *Polymer Handbook*, 3rd ed., J. Brandrup and E. H. Immergut (eds.), Wiley, New York, 1989, Chapter II, p. 153.

2. Alfrey, T., Jr., and C. C. Price, *J. Polym. Sci.* **2**, 101 (1947).

3. Meyer, V. E., and G. G. Lowry, *J. Polym. Sci.* **A3**, 2843 (1965).

4. Hanna, R. J., *Ind. Eng. Chem.* **49**, 208 (1957).

5. DeButts, E. H., *J. Am. Chem. Soc.* **72**, 411 (1950).

6. O'Driscoll, K. F., and R. S. Knorr, *Macromolecules* **1**, 367 (1968).

7. Merz, E. T., T. Alfrey, and G. Goldfinger, *J. Polym. Sci.* **1**, 75 (1946).

8. Barb, W. G., *J. Polym. Sci.* **11**, 117 (1953).

9. Ham, G. E., *J. Polym. Sci.* **54**, 1 (1961).

10. Ham, G. E., *J. Polym. Sci.* **61**, 9 (1962).

11. Seiner, J. A., and M. Litt, *Macromolecules* **4**, 308 (1971).

12. Litt, M., *Macromolecules* **4**, 312 (1971).

13. Litt, M., and J. A. Seiner, *Macromolecules* **4**, 314 (1971).

14. Litt, M., and J. A. Seiner, *Macromolecules* **4**, 316 (1971).

15. Spencer, H. G., *J. Polym. Sci., Polym. Chem. Ed.* **13**, 1253 (1975).

PROBLEMS

1. *a.* Plot F_1 and $\langle F_1 \rangle$ vs. conversion in a batch reactor for systems (*b*) and (*c*) of Example 1. $f_{10} = 0.50$.

   b. If you were going to produce a copolymer of uniform composition $F_1 = \langle F_1 \rangle = 0.5$ in a semibatch reactor (Fig. 12.4*a*), which monomer would have to be added to the reactor for systems (*a*), (*b*), and (*c*)?

2. Consider a copolymerization system in which $r_1 = r_2 = 0.5$. Sketch F_1 and $\langle F_1 \rangle$ vs. conversion in a batch reactor for f_{10}'s of 0.25, 0.50, and 0.75.

3. Crud Chemicals wishes to make a copolymer for beverage bottles using styrene for low cost and processability and acrylonitrile (AN) for its barrier (to CO_2 and O_2, mainly) properties. They want a uniform composition of 75 wt % acrylonitrile (component 1) and 25% styrene (component 2). This is 86%(1), 14%(2) on a mole basis. With more AN, the copolymer becomes difficult to process (recall from Chapter III that homopolyacrylonitrile, though linear, is not thermoplastic). With less, barrier properties suffer. For this system, $r_1 = 0.040$, $r_2 = 0.40$. Crud proposes to use the semibatch technique of Fig. 12.4*a*.

   For a product that is to contain 100 total moles monomer at complete conversion, address the following:

   a. Which monomer must be added over the course of the reaction?

   b. How many moles of the monomer in (*a*) must be added over the course of the reaction?

   c. Calculate the initial reactor charge.

   d. Explain why this is a difficult copolymer to make.

   e. Suppose Crud were just to dump 86 mol of (1) and 14 mol of (2) into the reactor and let it react to completion. Plot F_1 and $\langle F_1 \rangle$ vs. conversion. Why isn't this a good idea?

4. Most commercial copolymers of styrene and acrylonitrile, known as SAN, are about 75% styrene and 25% acrylonitrile, just the reverse of the composition in Problem 3. Why is this copolymer much easier to make at approximately uniform composition?

5. We wish to make an acrylonitrile (1)–styrene (2) copolymer (see Problem 3) by the technique shown in Fig. 12.4c. The range of instantaneous copolymer composition must be limited to $0.40 < F_1 < 0.60$.

   a. What is the feed composition to the reactor?

   b. What should the maximum conversion be?

   c. What will $\langle F_1 \rangle$ be?

6. Crud Chemicals wishes to make a pilot-plant batch of block copolymer $[M_2]_{500}[M_1]_{500}$ by anionic polymerization. They start with 5.00 mol of M_2 and an appropriate quantity of n-BuLi in their reactor. They intend to let the M_2 react to completion and then add 5.00 mol of M_1. Unfortunately, they get impatient (as often happens at Crud), and dump in the 5.00 mol of M_1 after only 2.50 mol of M_2 have polymerized. For this system, $r_1 = 2.00$, $r_2 = 0.50$.

   Plot F_1 and $\langle F_1 \rangle$ vs. total moles monomer polymerized, from 0 to 10. Describe the molecules produced. Make the usual assumptions of complete ionization: no termination, instantaneous addition, and perfect mixing of reagents, etc.

7. For a copolymerizing system of known r_1, r_2 and f_1, obtain an expression for the probability that a $P_1 \cdot$ will add M_1 (rather than M_2).

8. Crud Chemicals is producing a copolymer of uniform composition in a semibatch reactor, as illustrated in Fig. 12.4a. For their system, $r_1 = 2.0$, $r_2 = 0.5$, and $f_{10} = 0.5$. They adjust the rate of addition of the more reactive monomer according to the heat evolution in the reactor. Normally, the system works quite well. One day, however, the feed valve sticks full open at 50% conversion, suddenly dumping the remaining monomer into the reactor. Nevertheless, they let the reaction go to completion. Plot F_1 and $\langle F_1 \rangle$ vs. conversion (of all monomer) for this screwed-up reaction.

9. Consider the use of a CSTR to produce a copolymer of uniform composition, Fig. 12.4b. Here, f_{10} is the composition of the gross feed (fresh feed + recycle) to the reactor and F_1 and f_1 are the steady-state compositions in the reactor and its effluent stream.

   a. Obtain the expression that relates f_{10} to conversion X in such a reactor.

   b. How is the recycle ratio (moles recycled per mole of fresh feed) related to X in such a process, assuming complete separation and recycling of monomer?

   c. What must the mole fraction of monomer 1 be in the fresh feed to such a process, again assuming complete separation and recycling of monomer?

d. See Problem 3. Calculate the gross feed composition and the recycle ratio to make that copolymer in a CSTR operating at 50% conversion.

10. Your boss calls you in to tell you that you're going to be in charge of implementing a control system to produce a copolymer of uniform composition in a semibatch reactor (Fig. 12.4a). The reactor is fitted with a jacket to which steam can be fed to heat the reactants to the desired reaction temperature (during which conversion is negligible). Once the reaction has begun, the steam is shut off and replaced by cooling water, which is throttled by a feedback control system to maintain the reactor temperature fairly constant. What information would you need to know, and how would you go about designing a control system to feed the more reactive monomer at the correct rate?

11. We have seen how the compositions in a copolymerizing system can be described in terms of two parameters, r_1 and r_2. What does it take to describe the *terpolymerization* of monomers M_1, M_2, and M_3? If you're feeling particularly masochistic, you might try to work out the terpolymerization equivalents of (12.11).

12. Crud Chemicals is producing a copolymer of $F_1 = 0.5$ in a CSTR (Fig. 12.4b). For their system $r_1 = 2.0, r_2 = 0.5$. Things are going along fine until the feed pump for monomer 1 suddenly quits. It takes the maintenance people a couple of hours to get it back on line. Sketch qualitatively F_1 and $\langle F_1 \rangle$ vs. time for the period from immediately before the breakdown to the restoration of steady-state operation.

13. Write the equations equivalent to (10.1) through (10.4b) that are necessary to describe rigorously the kinetics of copolymerization.

CHAPTER XIII

Polymerization Practice

13.1 BULK POLYMERIZATION

The simplest and most direct method of converting monomer to polymer is known as *bulk* or *mass* polymerization. A typical charge for a free-radical bulk polymerization might consist of a liquid monomer, a monomer-soluble initiator, and perhaps a chain-transfer agent.

As simple as this seems, some serious difficulties can be encountered, particularly in free-radical bulk polymerizations. One of them is illustrated in Fig. 13.1,[1] which indicates the course of polymerization for various concentrations of methyl methacrylate in benzene, an inert solvent. The reactions were carefully maintained at constant temperature. At low polymer concentrations, the conversion vs. time curves are described by (10.19). As polymer concentrations increase, however, a distinct acceleration of the rate of polymerization is observed which does not conform to the classical kinetic scheme. This phenomenon is known variously as *autoacceleration*, the *gel effect*, or the *Tromsdorff effect*.

The reasons for this behavior lie in the difference between the propagation reaction (10.3) and the termination steps (10.4*a* and *b*), and the extremely high viscosities of concentrated polymer solutions (10^4 poise might be a ballpark figure). The propagation reaction involves the approach of a small monomer molecule to a growing chain end, whereas termination requires that the ends of two growing chains get together. At high concentrations of polymer, it becomes exceedingly difficult for the growing chain ends to drag their chains through the entangled mass of dead polymer chains. It is nowhere near as difficult for a monomer molecule to pass through the reaction mass. Thus, the rate of the termination reaction is limited not by the nature of the chemical reaction, but by the rate at which the reactants can diffuse together; that is, it is *diffusion controlled*. This lowers the effective termination rate constant k_t, and since k_t appears in the denominator of (10.13), the net effect is to increase the rate of

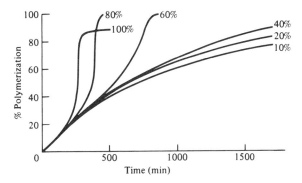

Figure 13.1 Polymerization of methyl methacrylate at 50°C in the presence of benzoyl peroxide at various concentrations of monomer in benzene.[1]

polymerization. At very high polymer concentrations and below the temperature at which the chains become essentially immobile—T_g of the monomer-plasticized polymer—even the propagation reaction is diffusion limited, hence the leveling off of the 100% curve.

The difficulties are compounded by the inherent nature of the reaction mass. Vinyl monomers have rather large exothermic heats of polymerization, typically between -10 and -21 kcal/mol. Organic systems also have low heat capacities and thermal conductivities, about half those of aqueous systems. To top it all off, the tremendous viscosities prevent effective convective (mixing) heat transfer. As a result, overall heat-transfer coefficients are very low, making it difficult to remove the heat generated by the reaction. This raises the temperature, further increasing the rate of reaction (Example 3, Chapter X) and heat evolution, and can ultimately lead to disaster. To quote Schildknecht[2] on laboratory bulk polymerizations, "If a complete rapid polymerization of a reactive monomer in large bulk is attempted, it may lead to loss of the apparatus, the polymer or even the experimenter."

Example 1. The *maximum possible* temperature rise in a polymerizing batch may be calculated by assuming that no heat is transferred from the system. Estimate the adiabatic temperature rise for the bulk polymerization of styrene, $\Delta H_p = -16.4$ kcal/mol, molecular weight = 104.

Solution. The polymerization of 1 mol of styrene liberates 16 400 cal (assuming complete conversion). In the absence of heat transfer, all this energy heats up the reaction mass. The heat capacities of organic compounds are often difficult to find, and since the reaction mass is going from monomer to polymer, which in general have different heat capacities, the heat capacity of the reaction mass changes with conversion and probably also with temperature. To a reasonable approximation, however, the heat capacity of most liquid organic systems may be taken as 0.5 cal/g°C. Thus,

$$\Delta T_{max} = 16\,400\,\frac{cal}{mol} \times \frac{1\,mol}{104\,g} \times \frac{g\,°C}{0.5\,cal} \approx 315°C(!)$$

(The normal boiling point of styrene is 146°C.)

These problems are circumvented in several ways:

1. By keeping at least one dimension of the reaction mass small, permitting heat to be conducted out. Polymethyl methacrylate sheets are cast between glass plates at a maximum thickness of 5 in. or so.

2. By maintaining low reaction rates through low temperatures, low initiator concentrations, and initiators that have relatively large energies of activation. The polymerization times for the sheets in (1) are on the order of 30 to 100 h and the temperatures are raised slowly as the monomer concentration drops. This approach has obvious economic disadvantages.

3. By starting with a *sirup* instead of the pure monomer. A sirup is a solution of the polymer in the monomer. It can be made in either of two ways: (i) by carrying the monomer to partial conversion in a kettle, or (ii) by dissolving preformed polymer in monomer. Starting off with a sirup means that some of the conversion has already been accomplished, cutting heat generation and monomer concentration in the final polymerization. Since the density of a typical reaction mass increases on the order of 10–20% between 0 and 100% conversion in a bulk polymerization, the use of a sirup has the added advantage of cutting shrinkage when casting a polymer.

4. By carrying out the reaction continuously, with a lot of heat-transfer surface per unit conversion.

Batch bulk polymerization is often used to make objects with a desired shape by polymerizing in a mold. Examples are casting, potting, and encapsulation of electrical components and impregnation of reinforcing agents followed by polymerization. Continuous bulk polymerization is used for the production of thermoplastics by both step-growth and free-radical addition mechanisms. A continuous bulk process is outlined in Fig. 13.2.[3] Conversion is carried to about 40% in a stirred tank. The reaction mass then passes down a tower with the temperature increasing to keep the viscosity to a manageable level and to drive up the conversion. The tower may be a simple gravity-flow device, or it may contain slowly rotating spiral blades that scrape the walls, promoting heat transfer and conveying the reaction mass downward. The reaction mass is fed from the tower to a vented extruder at better than 95% conversion. Some additional conversion takes place in the extruder, and a vacuum sucks off unreacted monomer, which is recycled. The extruded strands of molten polymer are then water cooled and chopped to form the roughly $\frac{1}{8} \times \frac{1}{8} \times \frac{1}{8}$-in. ($3 \times 3 \times$ 3-mm) pellets, which are sold to processors as "molding powder." Sheets of polymethyl methacrylate are also continuously cast from sirup between polished sheet-metal belts.

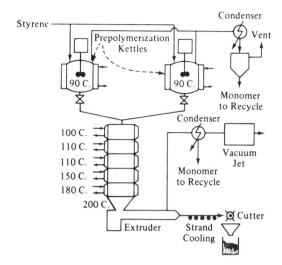

Figure 13.2 Continuous bulk polymerization of styrene. Reprinted by special permission from *Chemical Engineering*, August 1, 1962. Copyright © 1962, by McGraw-Hill, Inc., New York, NY 10036.

Most bulk polymerizations are homogeneous. However, if the polymer is insoluble in its monomer and precipitates as the reaction proceeds, the process is sometimes known as *heterogeneous bulk* or *precipitation polymerization*. Two examples of such polymers are polyacrylonitrile and polyvinyl chloride (PVC). The latter is produced commercially by a heterogeneous bulk process, which allows control of particle size and porosity for optimum plasticizer absorption.[4] Such heterogeneous polymerizations do not follow the kinetic scheme developed in Chapter X for homogeneous reactions.

Conventional low-density polyethylene (LDPE) is produced at pressures in the range of 15 000 to 50 000 psi (1000 to 3400 atm), and temperatures of about 150 to 300°C.[5] These pressures are well above the critical pressure of ethylene (731 psi), so the monomer can be considered to be in the "fluid" phase. Depending on the temperature, pressure, and polymer molecular weight, one or two phases may be present as polymer forms. Peroxide and azo free-radical initiators are used, as are small amounts of oxygen. Oxygen reacts with monomer to form peroxide initiators in situ.

Two types of reactors are used for high-pressure ethylene polymerization: autoclaves and tubular. The former are basically high-pressure CSTRs and the latter are continuous plug-flow reactors. Figure 13.3 illustrates a high pressure tubular process.[6] The reactors are made of thick-walled pipe, typically from 1 to 3 in. in diameter and from 800 to 2500 ft long. They're heat exchangers as well, with heat-transfer fluid circulating around the pipe, first to heat the ethylene to reaction temperature and then to remove the large heat of polymerization. Residence times are on the order of a minute, with single-pass conversions of

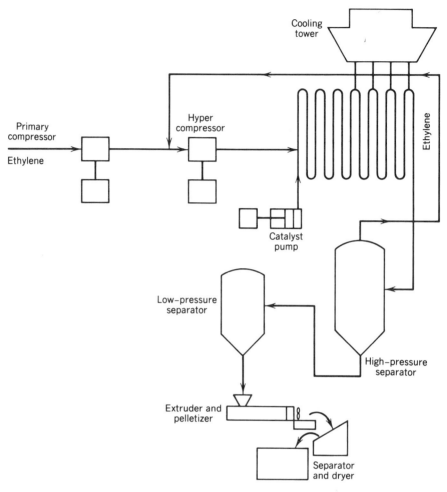

Figure 13.3   Tubular low-density polyethylene process. Reprinted with permission from *CHEMTECH* **13**(4), 223 (1983). Copyright 1983 American Chemical Society.

20% or less. Production rates are in the range of 8 to 125 million lb/yr per reactor.[5]

Separation of polymer from unreacted monomer is basically a matter of reducing the pressure and allowing the monomer to flash off. Given the pressures involved, it's no surprise that the compressors represent a significant fraction of the capital and operating costs of these processes.

Bulk polymerization has several advantages:

1. Since only monomer, initiator, and perhaps chain-transfer agent are used, the purest possible polymer is obtained. This can be important in electrical and optical applications.

2. Objects may be conveniently cast to shape. If the polymer is one that is crosslinked in the synthesis reaction, this is the only way of obtaining such objects short of machining from larger blocks.

3. Bulk polymerization provides the greatest possible polymer yield per reactor volume.

Among its disadvantages are the following:

1. It is often difficult to control.

2. To keep it under control, it may have to be run slowly, with the attendant economic disadvantages.

3. As indicated by Eqs. 10.13 and 10.27, it may be difficult to get both high rates and high average chain lengths because of the opposing effects of $[I]$.

4. It can be difficult to remove the last traces of unreacted monomer. This can be important, for example, if the polymer is intended for use in food-contact applications.

13.2 GAS-PHASE OLEFIN POLYMERIZATION

Gas-phase polyolefin processes based on Ziegler–Natta catalysts have undergone tremendous growth in the past decade. The popular UNIPOL® fluidized-bed process is shown in Fig. 13.4.[6] It typically operates at pressures in the range of 100–300 psi (7–20 atm) and at temperatures under 100°C.[6] Monomer, catalyst particles, and hydrogen chain-transfer agent (see Section 11.5) are fed to the fluidized-bed reactor. The upward-circulating monomer suspends the growing polymer particles. The reactors have a characteristic bulge at the top which is a disengaging section. It lowers the velocity of the circulating monomer, keeping the growing polymer particles in the lower reaction zone. The monomer is circulated through a heat exchanger to remove the heat of polymerization. Polymer is discharged from the reactor in granular form, looking for all the world like powdered laundry detergent. It may be used as is or extruded and pelletized to form the more conventional molding powder.

Impact-modified polypropylene is produced by using two reactors in series.[7] The first produces polypropylene particles, which are fed to the second reactor. A mixture of ethylene and propylene monomers (but no additional catalyst) is fed to the second reactor. The monomers diffuse to the still-active catalyst within the polypropylene particles from the first reactor, where they copolymerize to form a core of ethylene–propylene rubber within a polypropylene shell. This morphology greatly enhances the impact strength over that of polypropylene alone.

Compared to older processes, a number of significant advantages are claimed for this one. With modern high-yield catalysts (10 000 g polymer/g catalyst), there is no need to remove the catalyst from the product; thus a difficult and

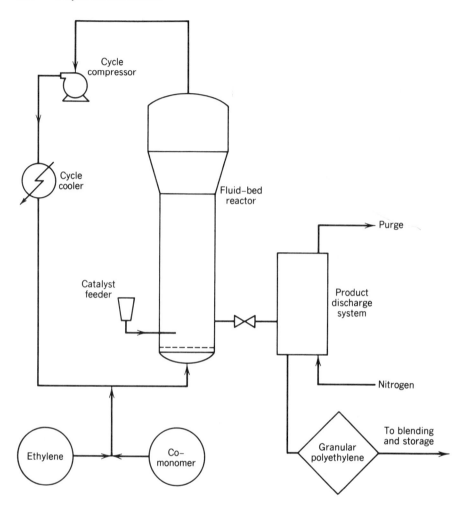

Figure 13.4 UNIPOL® gas-phase polyolefin process. Reprinted with permission from *CHEMTECH* **13**(4), 225 (1983). Copyright 1983 American Chemical Society.

costly step is eliminated. Modern catalysts also produce polypropylene with such a high isotactic content that removal of an atactic fraction, a common step in older processes, is not required. With low pressures, equipment and compression costs are low. No solvent is required. Separation of unreacted monomer is simple. With different catalysts and minor variations in process conditions, virtually the same equipment can be used to manufacture isotactic polypropylene (PP), high-density polyethylene (HDPE), and the so-called linear low-density polyethylene (LLDPE). Overall, major advantages in capital and operating costs, energy requirements, pollution, and superior product properties are cited by the licensors of this process.

13.3 SOLUTION POLYMERIZATION

The addition of an inert solvent to a bulk polymerization mass minimizes many of the difficulties encountered in bulk systems. As shown in Fig. 13.1, it reduces the tendency toward autoacceleration in free-radical addition. The inert diluent adds its heat capacity without contributing to the evolution of heat, and it cuts the viscosity of the reaction mass at any given conversion. In addition, the heat of polymerization may be conveniently and efficiently removed by refluxing the solvent. Thus, the danger of runaway reactions is minimized.

Example 2. Estimate the adiabatic temperature rise for the polymerization of a 20% (by weight) solution of styrene in an inert organic solvent.

Solution. In 100 g of the reaction mass, there are 20 g of styrene, so the energy liberated on its complete conversion to polymer is

$$(20 \text{ g}) \frac{(1 \text{ mol})}{(104 \text{ g})} \frac{(16\,400 \text{ cal})}{(\text{mol})} = 3150 \text{ cal}$$

The adibatic temperature rise is then

$$\Delta T_{\text{max}} = (3150 \text{ cal}) \frac{(\text{g}^{\circ}\text{C})}{(0.5 \text{ cal})} \frac{1}{(100 \text{ g})} \approx 63^{\circ}\text{C}$$

The advantages of solution polymerization are as follows:

1. Heat removal and control are easier than with bulk polymerization.
2. Since the reactions are more likely to follow known theoretical kinetic relations, the design of reactor systems is facilitated.
3. For some applications (lacquers, for example) the desired polymer solution is obtained directly from the reactor.

Among its disadvantages are the following:

1. Since both rate and average chain length are proportional to $[M]$ the use of a solvent lowers them. Additional lowering of x will occur if the solvent acts as a chain-transfer agent.
2. Large amounts of expensive, flammable, and perhaps toxic solvent are generally required.
3. Separation of the polymer and recovery of the solvent require additional technology.
4. Removal of the last traces of the solvent and monomer may be difficult.
5. Use of an inert in the reaction mass lowers the yield per volume of reactor.

Solution polymerizations are often used for the production of thermosetting condensation polymers, which are carried to a conversion short of the gel point in the reactor. The crosslinking is later completed in a mold. Such reactions may be carried out in a refluxing organic solvent. The water of condensation is carried overhead along with the solvent vapors. When the vapors are condensed, the water forms a second phase which is decanted before the solvent is returned to the reactor. Not only does this drive the reaction toward higher conversions, but the amount of water evolved provides a convenient measure of conversion (see Chapter IX, Example 1).

Ionic polymerizations are almost exclusively solution processes. Many Ziegler–Natta polymerizations are also.[5] They can be run under conditions such that the polymer product stays in solution, as in the production of stereospecific rubbers. The crystalline polymers polyethylene and isotactic polypropylene are commonly produced at temperatures sufficiently below T_m so that the polymer product is a solid that grows on the catalyst particles as in gas-phase polymerizations. Such processes are known as *slurry* polymerizations.

Figure 13.5 sketches a generic process utilizing a Ziegler–Natta catalyst system. Heat removal from the reactor(s) may be accomplished by refluxing the solvent, cooling jackets, external pumparound heat exchangers, or combinations of these. Catalyst deactivation with methanol or acid and separation by filtration or centrifugation are shown, although with modern high-yield catalysts these difficult and expensive steps can be eliminated. Some polypropylene processes also required a solvent-washing step to remove minor amounts of atactic material, but again, with newer catalysts, this may no longer be necessary.

Solvent and unreacted monomer are stripped with hot water and steam and recovered, leaving water slurry of polymer, which is then dried to form a "crumb." With rubbers, the crumb is compacted and baled; with plastics, it is normally extruded and pelletized. Reactor designs for these processes are interesting and varied.[8] Most of the processes are continuous.

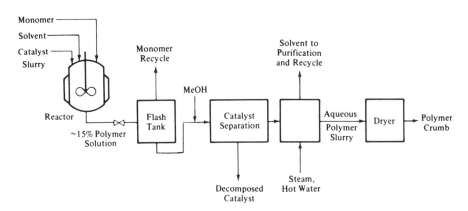

Figure 13.5   Ziegler–Natta solution process.

13.4 INTERFACIAL POLYCONDENSATION

A variation of solution polymerization known as interfacial polycondensation has been used in the laboratory for a long time,[9] and rumor has it that it is now being applied commercially. One monomer of a condensation pair is dissolved in one solvent, and the other member of the pair in another solvent. The two solvents are insoluble in one another. The polymer is soluble in neither, and forms at the interface between them. One of the phases generally also contains an agent that reacts with the molecule of condensation to drive the reaction to completion.

An example of such a process is the preparation of nylon 6/10 from hexamethylene diamine and sebacoyl chloride (the acid chloride form of sebacic acid):

$$H_2N\text{-}(CH_2)_6\text{-}NH_2 + Cl\text{-}\overset{O}{\overset{\|}{C}}\text{-}(CH_2)_8\text{-}\overset{O}{\overset{\|}{C}}\text{-}Cl \longrightarrow$$

$$\text{-}[\overset{H}{\overset{|}{N}}\text{-}(CH_2)_6\text{-}\overset{H}{\overset{|}{N}}\text{-}\overset{O}{\overset{\|}{C}}\text{-}(CH_2)_8\text{-}\overset{O}{\overset{\|}{C}}]_x\text{-} + HCl$$

The acid chloride is dissolved in, for example, CCl_4, and the diamine in water, along with some NaOH to soak up the HCl. In the classic "rope trick" demonstration, the aqueous layer is gently floated on top of the organic layer in a beaker. The reactants diffuse to the interface, where they react rapidly to form a polymer film. With care, the film can be withdrawn from the interface in the form of a continuous, hollow strand which traps considerable liquid. New polymer forms at the interface as the old is withdrawn. Commercially, it is probably easier simply to stir the phases together.

A major advantage of this technique is that these reactions usually proceed very rapidly at room temperature and atmospheric pressure, in contrast to the long times, high temperatures, and vacuums usually associated with polycondensations. This must be balanced against the cost of preparing the special monomers, such as the acid chloride above, and the need to separate and recycle solvents and unreacted monomers.

13.5 SUSPENSION POLYMERIZATION

Under the discussion of bulk polymerization, it was mentioned that one of the ways of facilitating heat removal was to keep one dimension of the reaction mass small. This is carried to its logical extreme in suspension polymerization by suspending the monomer in the form of droplets 0.01 to 1 mm in diameter in an inert, nonsolvent liquid (almost always water). In this way, each droplet becomes an individual bulk reactor with dimensions small enough that heat removal is no problem. The heat can easily be soaked up by and removed from the low-viscosity, inert suspension medium.

An important characteristic of these systems is that the suspensions are *thermodynamically unstable*, and must be maintained with agitation and suspending agents. A typical charge might consist of

Monomer (water insoluble)   ⎫
Initiator (monomer soluble)   ⎬   Monomer phase
Chain-transfer agent (monomer soluble) ⎭
Water

Suspending agent ⎰ Protective colloid
                  ⎱ Insoluble inorganic salt

Two types of suspending agent are used. A protective colloid is a water-soluble polymer whose function is to increase the viscosity of the continuous water phase. This hydrodynamically hinders coalescence of monomer drops, but is inert with regard to the polymerization. A finely divided insoluble inorganic salt such as $MgCO_3$ may also be used. It collects at the droplet–water interface by surface tension, and prevents coalescence of the drops upon collision. A pH buffer is sometimes also used to help stability.

The monomer phase is suspended in the water at about a $\frac{1}{2}$ to $\frac{1}{4}$ monomer/water volume ratio. The reactor is purged with nitrogen and heated to start the reaction. Once underway, temperature control in the reactor is facilitated by the added heat capacity of the water and the low viscosity of the reaction mass—essentially that of the continuous phase—allowing easy heat removal through a jacket.

The size of the product beads depends on the strength of agitation, as well as the nature of the monomer and suspending system. Between about 20 and 70% conversion, agitation is critical. Below this range, the organic phase is still fluid enough to redisperse, and above it, the particles are rigid enough to prevent agglomeration; but if agitation stops or weakens between these limits, the sticky particles will coalesce or agglomerate in a large mass and finish polymerization that way. Again quoting Schildknecht,[10] "After such uncontrollable polymerization is completed in an enormous lump, it may be necessary to resort to a compressed air drill or other mining tools to salvage the polymerization equipment."

Since any flow system is bound to have some relatively stagnant corners, it has been impractical to run suspension polymerization continuously on a commercial scale. Figure 13.6[11] shows a typical process. The reactors are usually jacketed, stainless or glass-lined steel kettles of up to 50 000-gal capacity. The polymer beads are filtered or centrifuged and water washed to remove the protective colloid and/or rinsed with a dilute acid to decompose the $MgCO_3$. The beads are quite easy to handle when wet, but they tend to pick up a static charge when dry, making them cling to each other and everything else. The

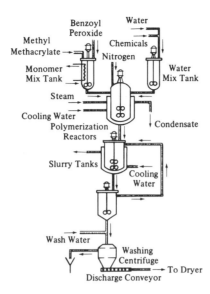

Figure 13.6 Suspension polymerization of methyl methacrylate. Excerpted by special permission from *Chemical Engineering*, June 6, 1966. Copyright © 1966 by McGraw-Hill, Inc., New York, NY 10036.

beads can be molded directly, extruded and chopped to form molding powder, or used as is, for example, as ion-exchange resins or the beads from which polystyrene foam cups and packing supports are made.

Ion-exchange resins are basically suspension beads of polystyrene cross-linked by copolymerization with a few percent divinyl benzene, which are then treated chemically to provide the necessary functionality. To reduce mass-transfer resistance in the ion-exchange process, an inert solvent may be incorporated in the organic suspension phase. When polymerization is complete, the solvent is removed, leaving a highly porous bead with a large internal surface area ("macroreticular"). Foam beads are linear polystyrene containing an inert liquid blowing agent, usually pentane. Pentane may be added to the monomer prior to polymerization, but more commonly, it is added to the reactor after polymerization and is absorbed by the polystyrene beads. When exposed to steam in a mold, the beads soften and are foamed and expanded by the volatilized blowing agent to form the familiar cups and other foam items.

Commercially, suspension polymerization has been limited to the free-radical addition of water-insoluble liquid monomers. With a volatile monomer such as vinyl chloride, moderate pressures are required to maintain it in the liquid state. It is possible, however, to perform inverse suspension polymerizations, with a hydrophilic monomer or an aqueous solution of a water-soluble monomer suspended in a hydrophobic continuous phase.

The major advantages of suspension polymerization, then, are the following:

1. Easy heat removal and control.
2. The polymer is obtained in a convenient, easily handled, and often directly useful form.

Disadvantages include the following:

1. Low yield per reactor volume.
2. A somewhat less pure polymer than from bulk polymerization, since there are bound to be remnants of the suspending agent(s) adsorbed on the particle surface.
3. The inability to run the process continuously, although if several batch reactors are alternated, the process may be continuous from that point on.
4. It can't be used to make condensation polymers or for ionic or Ziegler–Natta polymerizations.

13.6   EMULSION POLYMERIZATION

When the supply of natural rubber from the East was cut off by the Japanese in World War II, the United States was left without an essential material. The success of the Rubber Reserve Program in developing a suitable synthetic substitute and the facilities to produce it in the necessary quantities is one of the all-time outstanding accomplishments of chemists and engineers, comparable to the Manhattan (atomic bomb) project and the production of penicillin in its significance at the time. The styrene–butadiene copolymer rubber GR-S (government rubber–styrene) or SBR (styrene–butadiene rubber) as it is now called is still the most important synthetic rubber, and much of it is still produced, along with a variety of other polymers, by the emulsion polymerization process developed during the war.

The theory behind the emulsion reaction is discussed in Chapter 10. A generic commercial process is shown in Fig. 13.7. The reactors are usually stainless or glass-lined steel tanks, similar to those used for suspension polymerization. In contrast to suspension polymerization, however, a proper emulsion is thermodynamically stable, and therefore emulsion polymerization can be run continuously. Newer processes often have several continuous stirred-tank reactors in series.

The product of an emulsion polymerization is a *latex*—polymer particles on the order of 0.05 to 0.15 μm stabilized by the soap. These latices are often important items of commerce in their own right. Two familiar examples are "white glue" and latex paints.

Where the polymer must be mixed with other materials, the process of *master batching* sometimes allows this to be done conveniently and uniformly. In

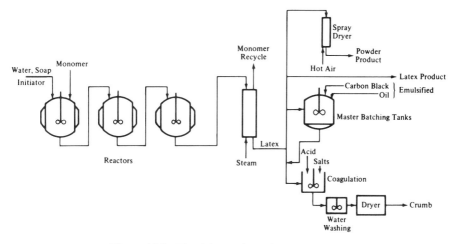

Figure 13.7 Emulsion polymerization process.

rubber technology, carbon black and oil are emulsified and mixed with the rubber latex and then the mixture is coagulated together, giving a uniform and intimate dispersion of the additives in the rubber.

The increased recognition of possible adverse physiological effects of even small amounts of residual monomer makes emulsion polymerization attractive in certain applications. The extremely small size of the latex particles provides a very short diffusion path for the removal of small molecules from the polymer, by steam stripping, for example, permitting very low levels of residual monomer to be obtained.

For many applications, the solid polymer must be recovered from the latex. The simplest method is spray drying, but since no attempt is made to remove the soap, the product is an extremely "dirty" polymer. A latex may be "creamed" by adding a material such as acetone that is at least a partial solvent for the polymer. This makes the particles sticky and causes some agglomeration. The latex is then coagulated by adding an acid, usually sulfuric, which converts the soap to the insoluble hydrogen form, and/or by adding an electrolyte salt, which disrupts the stabilizing double layer on the particles, causing them to agglomerate through electrostatic attraction. The former method leaves much insoluble material adsorbed on the particle surfaces, but in some applications, this may even be beneficial. For example, the fatty acid acts as a lubricant in tire manufacturing. The coagulated polymer "crumb" is then washed, dried and either baled or processed further.

As with suspension polymerization, commercial emulsion polymerization has pretty much been restricted to the free-radical addition of water-insoluble, liquid monomers (with volatile monomers such as butadiene and vinyl chloride, moderate pressures are required to keep them in the liquid phase). Inverse emulsion polymerizations, with a hydrophylic monomer phase dispersed in a continuous hydrophobic phase, are possible, however.

In summary, the advantages of emulsion polymerization are as follows:

1. Ease of control. The viscosity of the reaction mass is much less than that of a true solution of comparable polymer concentration, the water adds its heat capacity, and the reaction mass may be refluxed.
2. It is possible to obtain both high rates of polymerization and high average chain lengths through the use of high soap and low initiator concentrations.
3. The latex product is often directly valuable or aids in obtaining uniform compounds through master batching.
4. The small size of the latex particles allows the attainment of low residual monomer levels.

Its disadvantages are the following:

1. It is difficult to get pure polymer. The tremendous surface area of the tiny particles provides plenty of room for adsorbed impurities. This includes water attracted by residual soap, traces of which can cause problems in certain applications.
2. Considerable technology is required to recover the solid polymer.
3. The water in the reaction mass lowers the yield per reactor volume.
4. It can't be used to make condensation polymers, or for ionic or Ziegler–Natta polymerizations.

REFERENCES

1. Schulz and G. Harborth, *Makromol. Chem.* **1**, 106 (1947).
2. Schildknecht, C. E., *Polymer Processes*, Interscience, New York, 1956, p. 38.
3. Wohl, M. H., *Chem. Eng.* Aug. 1, 1962, p. 60.
4. Thomas. J. C., *Soc. Plast. Eng. J.* **23**(10), 61 (1967).
5. Albright, L. F., *Processes for Major Addition-Type Plastics and their Monomers*, McGraw-Hill, New York, 1974.
6. Karol, F. J., *CHEMTECH* **13**(4), 222 (1983).
7. Haggin, J., *Chem. Eng. News* March 31, 1986, p. 15; Bisio, A., *Chem. Eng. Progr.* **85**(5), 76 (1989).
8. Gerrens, H., *CHEMTECH*, **12**(7), 434 (1982).
9. Morgan, P. W., *Soc. Plast. Eng. J.* **15**(6), 485 (1959).
10. Schildknecht. C. E., *Polymer Processes*, Interscience, New York, 1956, p. 94.
11. Guccione, E., *Chem. Eng.* June 6, 1966, p. 138.

PROBLEMS

1. What difficulties would you foresee in carrying out anionic or Zeigler–Natta polymerizations by suspension or emulsion processes?

2. Suspension polymerization is routinely used to produce ion-exchange beads from copolymers of styrene and divinyl benzene. Why isn't it used to make polymer for ion-exchange membranes of similar composition?

3. Emulsion polymerization is used extensively for the production of synthetic rubbers, but suspension polymerization is not. Why?

4. *a*. Autoacceleration is common in the free-radical polymerization of vinyl monomers. Why isn't it observed in the anionic addition polymerization of the same monomers?

   b. What do you think would be a distinguishing molecular characteristic of polymer formed in a strongly autoaccelerated reaction, and why?

5. The semibatch method of producing copolymers of uniform composition (Fig. 12.3*a*) can be used in conjunction with all modes of polymerization but suspension. Why?

6. Describe a suspension-polymerization formulation for the production of polyacrylic acid. This vinyl monomer is similar to acetic acid in most properties.

7. The 100% monomer curve in Fig. 13.1 levels off at about 90% conversion. What would you do to push the conversion to 99%?

8. Why do you suppose such high pressures are needed for LDPE production? They aren't needed for the other free-radial reactions we've seen.

9. To be useful in paint formulations, a polymer must form a continuous film on the surface to which it is applied.

   a. What key properties must the polymer phase in a latex-paint formulation have?

   b. One of the major advantages of latex paints over the older oil-base paints is easy cleanup with water. You may have noticed, however, that if you let your brushes sit around too long before rinsing them, you can no longer get the paint out with water. Why?

10. In the two-reactor, impact-modified polypropylene process described in Section 13.2, it was noted that no new catalyst was introduced to the second reactor. Suppose it was. How would that alter the product?

11. Crud Chemicals is preparing a batch of glyptal resin by refluxing 2 kmol of glycerine and 3 kmol of phthalic anhydride (see Chapter IX, Example 2) in toluene. The water of condensation is decanted before returning the toluene to the reactor. Calculate the maximum liters of water they can collect before shutting down the reaction.

PART III

POLYMER PROPERTIES

CHAPTER **XIV**

Rubber Elasticity

14.1 INTRODUCTION

Natural and synthetic rubbers possess some interesting, unique, and useful mechanical properties. No other materials are capable of reversible extension of 600–700%. No other materials exhibit an increase in modulus with increasing temperature. It was recognized long ago that vulcanization was necessary for rubber deformation to be completely reversible. We now know that this is a result of the crosslinks so introduced preventing the bulk slippage of the molecules past one another, eliminating flow (irrecoverable deformation). More recently, this function of the covalent crosslinks has been assumed by rigid domains (either glassy or crystalline) within some linear polymers (see Section 21.2). Thus, when a stress is applied to a sample of crosslinked rubber, equilibrium is established fairly rapidly. Once at equilibrium, the properties of the rubber can be described by thermodynamics.

14.2 THERMODYNAMICS OF ELASTICITY

Consider an element of material with dimensions $a \times b \times c$, as sketched in Fig. 14.1. Applying the first law of thermodynamics to this system yields

$$dU = dQ - dW \tag{14.1}$$

where dU is the change in the system's *internal energy*, and dQ and dW are the heat and work exchanged between system and surroundings as the system undergoes a differential change. (We have adopted the convention here that work done by the system on the surroundings is positive.)

We will consider three types of mechanical work:

1. Work done by a uniaxial tensile force f:

$$dW(\text{tensile}) = -f\,dl \tag{14.2}$$

233

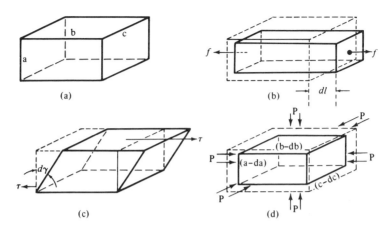

Figure 14.1 Types of mechanical deformation: (*a*) unstressed; (*b*) uniaxial tension; (*c*) pure shear; (*d*) isotropic compression.

where dl is the differential change in the system's length arising from the application of the force f. This is the fundamental definition of work. The negative sign arises from the need to reconcile the mechanical convention of treating a tensile force (which does work on the system) as positive with the thermodynamic convention above.

2. Work done by a shear stress τ:

$$dW(\text{shear}) = (\text{force})(\text{distance}) = -(\tau bc)(a\,d\gamma) = -\tau V\,d\gamma \qquad (14.3)$$

where γ is the shear strain (Fig. 14.1*c*) and $V = abc =$ the system volume.

3. Work done by an isotropic pressure in changing the volume:

$$dW(\text{pressure}) = P(cb)da + P(ac)db + P(ab)dc = P\,dV \qquad (14.4)$$

Note that no minus sign is needed here. A positive pressure causes a decrease in volume (negative dV) and does work on the system.

If the deformation process is assumed to occur *reversibly* (in a thermodynamic sense), then

$$dQ = T\,dS \qquad (14.5)$$

where S is the system's entropy.

Combining the preceding five equations gives a general relation for the change of internal energy of an element of material undergoing a differential deformation:

$$dU = T\,dS - P\,dV + f\,dl + V\tau\,d\gamma \qquad (14.6)$$

Now, let's consider three individual types of deformation:

1. Uniaxial tension at constant volume and temperature. Under these conditions, $dV = \tau = 0$. Dividing the remaining terms in (14.6) by dl, restricting to constant T and V, and solving for f gives

$$f = \left(\frac{\partial U}{\partial l}\right)_{T,V} - T\left(\frac{\partial S}{\partial l}\right)_{T,V} \tag{14.7}$$

2. Pure shear at constant volume and temperature. Here, $dV = f = 0$. Dividing the remaining terms in (14.6) by $d\gamma$, restricting to constant T and V, and solving for τ gives

$$\tau = \frac{1}{V}\left(\frac{\partial U}{\partial \gamma}\right)_{T,V} - \frac{T}{V}\left(\frac{\partial S}{\partial \gamma}\right)_{T,V} \tag{14.8}$$

3. Isotropic compression only, at constant temperature:

$$P = -\left(\frac{\partial U}{\partial V}\right)_{T} + T\left(\frac{\partial S}{\partial V}\right)_{T} \tag{14.9}$$

It is very difficult to carry out tensile experiments at constant volume to obtain the partial derivatives in (14.7). Most tests are carried out at constant pressure (atmospheric), and in general, there is a change in volume with tensile straining. Fortunately, Poisson's ratio is approximately 0.5 for rubbers, so this change in volume is small, and (14.7) is approximately valid for tensile deformation at constant pressure, also. For precise work, the hydrostatic pressure must be varied to maintain V constant, or theoretical corrections applied to the constant-pressure data to obtain the constant-volume coefficients.[1,2] In pure shear, V should be constant and (14.8) valid.

$$\nu = \left|\frac{lateral\ strain}{Axial\ strain}\right| = -\frac{\varepsilon_y}{\varepsilon_x} = -\frac{\varepsilon_z}{\varepsilon_x}$$

A Types of Elasticity

Equations 14.7–14.9 reveal that there are *energy* (the first term on the right) and *entropy* (the second term on the right) contributions to the tensile force, shear stress, or isotropic pressure. In polymers, *energy elasticity* represents the storage of energy resulting from rotation about bonds (Fig. 8.4) and the straining of bond angles and lengths from their equilibrium values. Interestingly, energy elasticity is almost entirely intramolecular rather than intermolecular in origin; that is, there is no change in the energy interaction between different molecules with deformation.[2]

Entropy elasticity is caused by the decrease in entropy upon straining. This can be visualized by considering a single polymer molecule subjected to a tensile stress. In an unstressed state, the molecule is free to adopt an extremely large

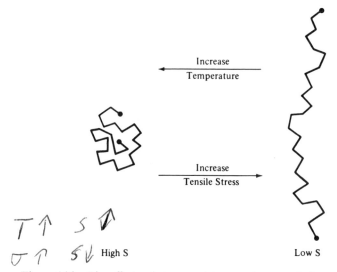

Figure 14.2   The effects of stress and temperature on chain configurations.

number of random, "balled-up" configurations (Fig. 14.2a) switching from one to another through rotation about the bonds. Now imagine the molecule to be stretched out under the application of a tensile force (Fig. 14.2b). It is obvious that there are far fewer configurational possibilities, and the more it is stretched, the fewer there are. Now, $S = k \ln \Omega$, where k is Boltzmann's constant and Ω is the number of configurational possibilities, so *stretching decreases the entropy* (increases the order). Raising the temperature has precisely opposite effect. The added thermal energy of the chain segments increases the intensity of their lateral vibrations, favoring a return to the more random or higher entropy state. This tends to pull the extended chain ends together, giving rise to a retractive force.

It should be obvious from the discussion above that to exhibit significant entropy elasticity, the material must be above its glass-transition temperature and cannot have appreciable amounts of crystallinity.

B   The "Ideal" Rubber

In a gas subjected to an isotropic pressure, the energy term in (14.9) arises from the change in intermolecular forces with volume, and the entropy term from the increased room (and therefore greater "disorder") the molecules gain with volume. In an ideal gas, there are no intermolecular forces, $(\partial U/\partial V)_T = 0$. By analogy, in an ideal rubber, $(\partial U/\partial l)_{T,V} = (\partial U/\partial \gamma)_{T,V} = 0$, and elasticity arises only from entropy effects. For many gases around room temperature and above, and around atmospheric pressure and below, $(\partial U/\partial V)_T < T(\partial S/\partial V)_T$, and the ideal gas law is a good approximation. Similarly, as illustrated in Fig. 14.3,[3]

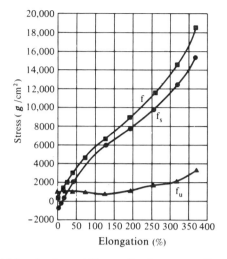

Figure 14.3   Energy (f_U) and entropy (f_S) contributions to tensile stress in natural rubber at 20°C. Reprinted from Anthony et al.[3] Copyright 1942 by the American Chemical Society. Reprinted by permission of the copyright owner.

under some circumstances, $(\partial U/\partial l)_{T,V} < T(\partial S/\partial l)_{T,V}$ for rubbers, and they behave approximately as ideal rubbers.

C   Effects of Temperature at Constant Force

Now let's consider what happens to the length of a piece of rubber when its temperature is changed while a weight is suspended from it, that is, when it is maintained at a constant tensile force. Assuming constant volume (an approximation, in the usual constant-pressure experiment), $dU = T\,dS + f\,dl$. Solving for dl, dividing by dT, and restricting to constant f as well as V gives

$$\left(\frac{\partial l}{\partial T}\right)_{f,V} = \frac{1}{f}\left(\frac{\partial U}{\partial T}\right)_{f,V} - \frac{T}{f}\left(\frac{\partial S}{\partial T}\right)_{f,V} \tag{14.10}$$

As before, the first term on the right represents energy elasticity and the second represents entropy elasticity. Since internal energy generally increases with temperature, the partial derivative in the energy term is positive, as is f. The energy term, therefore, causes an increase in length with temperature (positive contribution to $(\partial l/\partial T)_{f,V}$). This is the normal thermal expansion observed in all materials (metals, ceramics, glasses, etc.), reflecting the increase in the average distance between atomic centers with temperature. All factors in the entropy term are positive, however, and since it is preceded by the negative sign, it gives rise to a *decrease* in length with increasing temperature. In rubbers, where the entropy effect overwhelms the normal thermal expansion, this is what is actually

$T \uparrow$ Length $\downarrow$ Const Stress

observed. In all other materials, where the structural units are confined to a single arrangement (e.g., the atoms in a crystal lattice cannot readily interchange) the entropy term is small.

The magnitude of the entropy contraction in rubbers is typically much greater than the thermal expansion of other materials. An ordinary rubber band will contract an inch or so when heated to 300°F under stress, while the expansion of a piece of metal of similar length over a similar temperature range would not be noticeable to the naked eye. According to a well-known metallurgist,[4] "Polymers are all entropy."

D   Effects of Temperature at Constant Length

It is interesting to consider what happens to the force in a piece of rubber when it is heated while stretched to a constant length. Use the exact thermodynamic Maxwell relation

$$-\left(\frac{\partial S}{\partial l}\right)_{T,V} = \left(\frac{\partial f}{\partial T}\right)_{l,V} \tag{14.11a}$$

to describe approximately the usual experiment conducted at constant pressure. A better approximation to the constant pressure experiment is

$$-\left(\frac{\partial S}{\partial l}\right)_{T,V} \approx \left(\frac{\partial f}{\partial T}\right)_{P,\alpha} \tag{14.11b}$$

where $\alpha = l/l_0$ is the extension ratio, the ratio of stretched to unstretched length at a particular temperature. Combining (14.11a) and (14.7) gives

$$\left(\frac{\partial f}{\partial T}\right)_{l,V} = \frac{f}{T} - \frac{1}{T}\left(\frac{\partial U}{\partial l}\right)_{T,V} \tag{14.12}$$

Now, both f and T are positive, so the first term on the right causes the force to increase with temperature, a result of the greater thermal agitation (tendency toward higher entropy) of the extended chains. The partial derivative in the second term is usually (but not always!) positive, as energy is stored spring-like with extension. With the negative sign in front, this term predicts a relaxation of the tensile force with increasing temperature. Again, the second term reflects the ordinary thermal expansion obtained with all materials, but in rubbers, at reasonably large values of f, it is overshadowed by the first (entropy) term, and the force increases with temperature. For an ideal rubber, $(\partial U/\partial l)_{T,V} = 0$, and integration of (14.12) at constant volume gives

$$f = (\text{constant})\,T \qquad (\text{ideal rubber}) \tag{14.13}$$

This is analogous to the linearity between P and T in an ideal gas at constant V.

$T \uparrow \quad F \uparrow$

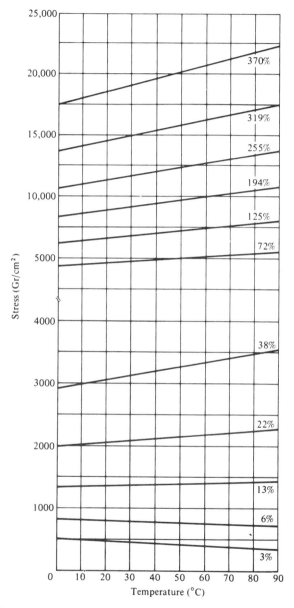

Figure 14.4 Force vs. temperature in natural rubber maintained at constant extension (% relative to length at 20°C).[3] Copyright 1942 by the American Chemical Society. Reprinted by permission of the copyright owner.

These observations are confirmed in Fig. 14.4. The negative slope at low elongations arises from the predominance of thermal expansion when elongation, and hence f, is low. Note that there is an intermediate elongation, the *thermoelastic inversion* point, at which force is essentially independent of temperature, where thermal expansion and entropy contraction balance.

14.3   STATISTICS OF IDEAL RUBBER ELASTICITY[1, 2, 5]

A typical rubber consists of long chains connected by short crosslinks every few hundred carbon atoms. The chain segments between crosslinks are known as *network chains*. The change in entropy upon stretching a sample containing N moles of network chains is

$$S - S_0 = NR \ln \frac{\Omega}{\Omega_0} \qquad (14.14)$$

where the subscript 0 refers to the unstretched state, Ω is the number of configurations available to the N moles of network chains, and R is the gas constant. By statistically evaluating the Ω's, it is possible to show that for constant-volume stretching,

$$S - S_0 = -\frac{1}{2} NR \left[\left(\frac{l}{l_0} \right)^2 + 2 \left(\frac{l_0}{l} \right) - 3 \right] \qquad (14.15)$$

$$= -\frac{1}{2} NR (\alpha^2 + 2\alpha^{-1} - 3)$$

where $\alpha = l/l_0$ is the extension ratio. (Newer theories give somewhat different results,[2] but this is adequate for our purposes).

For an ideal rubber, in which the tensile force is given by

$$f = -T \left(\frac{\partial S}{\partial l} \right)_{T, V} \qquad \text{(ideal rubber)} \qquad (14.16)$$

differentiation of (14.15) and insertion into (14.16) then gives

$$f = \frac{NRT}{l_0} (\alpha - \alpha^{-2}) \qquad (14.17)$$

Also,

$$N = \frac{\text{mass}}{\bar{M}_c} = \frac{\rho V}{\bar{M}_c} = \frac{\rho l_0 A_0}{\bar{M}_c} = \frac{\rho l A}{\bar{M}_c} \qquad (14.18)$$

where $\bar{M}_c$ is the number-average molecular weight of the network chains, ρ the density, and A the cross-sectional area of the sample (since the volume change in stretching a piece of rubber is negligible, $A_0 l_0 \approx Al$). Therefore,

$$f = \frac{\rho A_0 RT}{\bar{M}_c}(\alpha - \alpha^{-2}) \qquad (14.19)$$

The *engineering tensile stress* σ_e is defined as the tensile force divided by the initial cross-sectional area of the sample A_0, and is, therefore,

$$\sigma_e \equiv \frac{f}{A_0} = \frac{\rho RT}{\bar{M}_c}(\alpha - \alpha^{-2}) \qquad (14.20)$$

and the *true tensile stress* σ_t, the tensile force over the actual area A at length l, is

$$\sigma_t \equiv \frac{f}{A} = \frac{\rho RT}{\bar{M}_c}(\alpha^2 - \alpha^{-1}) \qquad (14.21)$$

Since the tensile strain ε is

$$\varepsilon = \frac{l - l_0}{l_0} = \alpha - 1 \qquad (14.22)$$

the slope of the true stress–strain curve (tangent Young's modulus) is

$$E = \left(\frac{\partial \sigma_t}{\partial \varepsilon}\right)_T = \left(\frac{\partial \sigma_t}{\partial l}\right)_T \left(\frac{\partial l}{\partial \varepsilon}\right)_T = \frac{\rho RT}{\bar{M}_c}(2\alpha + \alpha^{-2}) \qquad (14.23)$$

and the initial modulus (as $\alpha \to 1$) for an ideal rubber becomes

$$E(\text{initial}) = \frac{3\rho RT}{\bar{M}_c} \qquad (14.24)$$

Equations 14.19–14.24 point out two important concepts: (1) the force (or modulus) in an ideal rubber sample held at a particular strain increases in porportion to the absolute temperature (in agreement with Eq. 14.13), and (2) the force is *inversely* proportional to the molecular weight of the chain segments between crosslinks. Thus, increased crosslinking, which reduces $\bar{M}_c$, is an effective means of stiffening a rubber. Equation 14.24 is often used to obtain $\bar{M}_c$ from mechanical tests and thereby evaluate the efficiency of various crosslinking procedures.

Even non-crosslinked polymers exhibit rubbery behaviour above their T_g values for limited periods of time. This is due to mechanical entanglements acting as temporary crosslinks; $\bar{M}_c$ then represents the average length of chain segments between entanglements.

When compared with experimental data, (14.19) does a reasonably good job in compression, but begins to fail at extension ratios α greater than about 1.5, where the experimental force becomes greater than predicted. There are a number of reasons for this. First, (14.15) is based on the assumption of a Gaussian distribution of network chains. This assumption fails at high elongations, and it is also in error if crosslinks are formed when chains are in a strained configuration. Second, it does not take into account the presence of chain end segments, which do not contribute to the support of stress. Third, some rubbers (natural, in particular) begin to crystallize as a result of chain orientation at high elongations. (Orientation reduces ΔS_m and raises T_m above the test temperature. See Chapter VIII, Example 6.) This causes the stress–strain curve to shoot up markedly. Theoretical modifications to the theory are available for the first two factors, which improve things considerably in the absence of crystallinity, but there is as yet no satisfactory treatment of the effect of crystallinity on stress–strain properties.[2]

It is important to keep in mind also, that in practice, rubbers are rarely used in the form of pure polymer. They are almost always reinforced with carbon black, and often contain other fillers, plasticizing and extending oils, etc., all of which influence the stress–strain properties, and are not considered in the theories discussed here.

REFERENCES

1. Flory, P. J., *Principles of Polymer Chemistry*, Cornell UP, Ithaca, NY, 1953, Chapter 11.
2. Mark, J. E., and B. Erman, *Rubberlike Elasticity: A Molecular Primer*, Wiley, New York, 1988.
3. Anthony, R. L., R. H. Caston, and E. Guth, *J. Phys. Chem.* **46**, 826 (1942).
4. Paxton, H. W., personal communication.
5. Tobolsky, A. V., *Properties and Structure of Polymers*, Wiley, New York, 1960, Chapter 2.

PROBLEMS

1. Recall the definition of enthalpy: $H \equiv U + PV$. Write the enthalpy analogs of (14.7) and (14.8) for experiments carried out at constant pressure. Sort of makes you wonder why we bother with internal energy at all, doesn't it?

2. Derive (14.11a). *Hints*: Start with (14.6), with $dV = d\gamma = 0$. Make use of the definition of Helmholtz free energy: $A \equiv U - TS$. Remember that second mixed partial derivatives of thermodynamic state functions are independent of the order in which you take them.

3. The following data were obtained in a classroom demonstration in which a rubber band with inked-on gage marks was looped over the hook of

a spring balance:

f (lb)	l (cm)
0	2.42
0.33	2.80
0.54	3.08
0.79	3.58
1.10	3.97
1.36	4.70
1.66	5.60
1.89	6.42
2.20	7.28
2.68	8.45
3.07	9.30

The cross section of the rubber band measured 1 mm × 6 mm. The density of the rubber was probably around 1 g/cm^3 and the temperature around 25°C (remember, this isn't rocket science). Assume that the band is made of natural rubber, $[C_5H_8]_x$. How well does (14.19) fit these data? Estimate the average number of repeating units between crosslinks.

4. Show that the fraction of the energy contribution (f_U) to the total force (f) in a uniaxially stretched rubber is given by

$$\frac{f_U}{f} = 1 - \frac{T}{f}\left(\frac{\partial f}{\partial T}\right)_{l,V} = -T\left(\frac{\partial \ln (f/T)}{\partial T}\right)_{l,V}$$

5. Use the equation of Problem 4 and the data in Figure 14.4 to calculate the fractional energy contribution to the force in the material.

6. Sketch qualitatively the expected variation in f_U/f with degree of crosslinking in a rubber.

7. It was mentioned that $(\partial U/\partial l)_{T,V}$ is not always positive (though it's still smaller than the entropy term). It is negative for linear polyethylene, at least at low elongations, for example. Why? *Hint*: Recall Figure 8.4.

8. This chapter has emphasized the differences between the behavior of polymers and nonpolymers in uniaxial tension. How would you expect them to compare in isotropic compression? Why?

9. Repeat Problem 6 for rubbers in isotropic compression.

CHAPTER XV

Purely Viscous Flow

15.1 INTRODUCTION

*Rheology** is the science of the deformation and flow of materials. Much contemporary rheological work is concerned directly with polymer systems because they exhibit such interesting, unusual, and difficult-to-describe (at least from the standpoint of traditional materials) deformation behavior. The simple and traditional linear engineering models, Newton's law (for flow) and Hooke's law (for elasticity), often just aren't reasonable approximations. Not only are the elastic and viscous properties of polymer melts and solutions usually nonlinear, but they exhibit a combination of viscous and elastic response, the relative magnitudes of which depend on the temperature and the time scale of the experiment. This *viscoelastic* response is dramatically illustrated by Silly Putty (a silicone polymer). When bounced (stressed rapidly), it is highly elastic, recovering most of the potential energy it had before being dropped. If stuck on the wall (stressed over a long time period), however, it will slowly flow down the wall, albeit with a high viscosity, and will show little tendency to recovery any deformation.

We limit ourselves here to two-dimensional deformations. A detailed three-dimensional treatment of rheology is beyond the scope of this book. Several excellent treatises are available.[1-8]

15.2 BASIC DEFINITIONS

We begin our treatment of rheology with a discussion of *purely viscous flow*. For our purposes, this will be defined as a deformation process in which all the

---

* This word has been misprinted with a "t" in place of the "r". After reading some of the current literature, one isn't always sure that this is an error.

applied mechanical energy is nonrecoverably dissipated as heat in the material through molecular friction. This process is known as *viscous energy dissipation.* Purely viscous flow is in most cases a good approximation for dilute polymer solutions, and often for concentrated solutions and melts where the stress on the material is not changing too rapidly; that is, where equilibrium flow is approached.

The *viscosity* of a material expresses its resistance to flow. It is defined quantitatively in terms of two basic parameters; the *shear stress* τ and the *shear rate* (more correctly, the rate of shear straining) $\dot{\gamma}$. These quantities are defined in Fig. 15.1. Consider a point in a laminar flow field (Fig. 15.1). A rectangular coordinate system is established with the x axis (sometimes designated the 1 coordinate direction) in the direction of flow, and the y axis (the 2 direction) perpendicular to surfaces of constant fluid velocity, that is, parallel to the *velocity gradient.* The z axis (3 or neutral direction) is mutually perpendicular to the others. This is known as *simple shearing* flow. It is one example of a *viscometric flow,* a flow field in which the velocity and its gradient are everywhere perpendicular, with a third neutral direction mutually perpendicular to the others (see Bird et al.[2] for a more rigorous definition). Other examples of viscometric flows are given in Chapters XVI and XVII. Viscometric flows can often be treated analytically. Fortunately, many laminar-flow situations encountered in practice either are viscometric flows or least can reasonably be approximated by them.

A fluid layer at y moves with a velocity $u = dx/dt$ in the x direction, while the layer at $y + dy$ has a velocity $u + du$. The displacement gradient, dx/dy, is

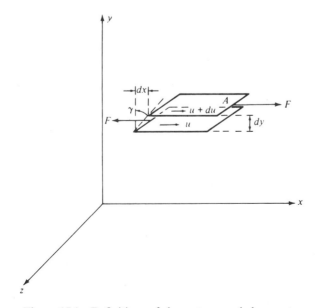

Figure 15.1   Definitions of shear stress and shear rate.

known as the *shear strain*, and is given the symbol γ.

$$\gamma = \frac{dx}{dy} = \text{shear strain} \qquad \text{(dimensionless)} \qquad (15.1)$$

The *time rate of change* of shear strain, $\dot{\gamma}$ (the dot is Newton's notation for the time derivative), is the so-called *shear rate*. Since the order in which the mixed second derivative is taken is immaterial (note that by sticking to two dimensions, we can write total rather than partial derivatives), then

$$\dot{\gamma} = \frac{d}{dt}(\gamma) = \frac{d}{dt}\left(\frac{dx}{dy}\right) = \frac{d}{dy}\left(\frac{dx}{dt}\right) = \frac{du}{dy} \quad (\text{time}^{-1}) \qquad (15.2)$$

Thus, in simple shearing flow, the shear rate is identical to the velocity gradient, du/dy. (This is not the case for all viscometric flows.)

The shear stress is the force (in the direction of flow) per unit area normal to the y axis:

$$\tau_{yx} = \frac{F \text{ (in } x \text{ direction)}}{A \text{ (normal to } y \text{ direction)}} \left(\frac{\text{force}}{\text{length}^2}\right) \qquad (15.3)$$

The subscript yx will henceforth be dropped unless specifically needed.

The *viscosity* η is defined as the ratio of shear stress to shear rate:

$$\eta \equiv \frac{\tau}{\dot{\gamma}} \qquad (15.4)$$

Traditionally, most rheological work has been done in the cgs system, with force in dynes, mass in grams, length in centimeters, and time in seconds. In this system, the unit of viscosity is $\text{dyne} \cdot \text{s/cm}^2$ or *poise* (P). Now SI units are *de rigueur*, with force in newtons (N) and length in meters. SI viscosities are in $\text{N} \cdot \text{s/m}^2 = \text{Pa} \cdot \text{s}$ (1 $\text{Pa} \cdot \text{s} = 10$ P). Equations here will be written with either of these systems in mind. When using the English system of pound force, pound mass, feet, and seconds, each stress or pressure as written here must be multiplied by the dimensional constant $g_c = 32.2 \text{ ft} \cdot \text{lb}_m/\text{lb}_f \cdot \text{s}^2$, and viscosities are in $\text{lb}_m/\text{ft} \cdot \text{s}$.

15.3 RELATIONS BETWEEN τ AND $\dot{\gamma}$: FLOW CURVES

When most materials are subjected to a constant shear rate (or constant shear stress) at a fixed temperature, a corresponding steady-state value of shear stress (or shear rate) is soon established. The steady-state relation between shear stress and shear rate at constant temperature is known as a *flow curve*.

Newton's "law" of viscosity states that the shear stress is linearly proportional to the shear rate, the proportionality constant being the viscosity η:

$$\tau = \eta\dot{\gamma} \tag{15.5}$$

Fluids that obey this hypothesis are termed Newtonian. The hypothesis holds quite well for many nonpolymer fluids, such as gases, water, and toluene. This type of flow behavior would be expected for small, relatively symmetrical molecules, where the structure and/or orientation do not change with the intensity of shearing.

An arithmetic flow curve (τ vs. $\dot{\gamma}$) for a Newtonian fluid is a straight line through the origin with a slope η (Fig. 15.2a). Because τ and $\dot{\gamma}$ often cover very wide ranges, it is usually preferable to plot them on log–log coordinates. Taking logarithms of both sides of Eq. 15.5 yields

$$\log \tau = \log \eta + 1 \log \dot{\gamma}$$

Hence, a log–log plot of τ vs. $\dot{\gamma}$, a logarithmic flow curve, will be a line of slope unity for a Newtonian fluid (Fig. 15.2b).

Unfortunately, many fluids do not obey Newton's hypothesis. Both *dilatant* (shear-thickening) and *pseudoplastic* (shear-thinning) fluids have been observed (Fig. 15.2). On log–log coordinates, dilatant flow curves have a slope greater than 1 and pseudoplastics have a slope less than 1. Dilatant behavior is reported for certain slurries and implies an increased resistance to flow with intensified shearing. *Polymer melts and solutions are invariably pseudoplastic*, that is, their resistance to flow decreases with the intensity of shearing.

For non-Newtonian fluids, since τ is not directly proportional to $\dot{\gamma}$, the viscosity is not constant. Plots (or equations) giving η as a function of $\dot{\gamma}$ (or τ) are an equivalent method of representing a material's equilibrium viscous shearing

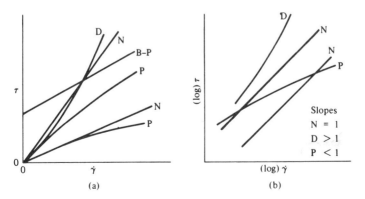

Figure 15.2 Types of flow curves: (a) arithmetic; (b) logarithmic. N, Newtonian; P, pseudoplastic; B–P, Bingham plastic (infinitely pseudoplastic); D, dilatant.

properties. A knowledge of the relation between any two of the three variables (τ, η, and $\dot{\gamma}$) completely defines the equilibrium viscous shearing behavior, since they are related by (15.4).

15.4 TIME-DEPENDENT BEHAVIOR

The types of non-Newtonian flow just described, though shear dependent, are time independent. As long as a constant shear rate is maintained, the same shear stress or viscosity will be observed at equilibrium. Some fluids exhibit *reversible* time-dependent properties, however. When sheared at a constant rate or stress, the viscosity of a *thixotropic* fluid will decrease over a period of time (Fig. 15.3), implying a progressive breakdown of structure. If the shearing is stopped for a while, the structure reforms, and the experiment may be duplicated. The ketchup that splashes all over after a period of vigorous tapping is a classic example. Thixotropic behavior is important in the paint industry, where smooth, even application with brush or roller is required, but it is desirable for the paint on the surface to "set up" to avoid drips and runs after application. The opposite sort of behavior is manifested by *rheopectic* fluids, for example, certain drilling muds used by the petroleum industry. When subjected to continuously increasing and then decreasing shearing, time-dependent fluids give flow curves as in Fig. 15.4.

Once in a while, polymer systems will appear to be thixotropic or rheopectic. Careful checking (including before and after molecular weight determinations) invariably shows that the phenomenon is not reversible and is due to degradation or crosslinking of the polymer when in the viscometer for long periods of time, particularly at elevated temperatures. Other transient time-dependent

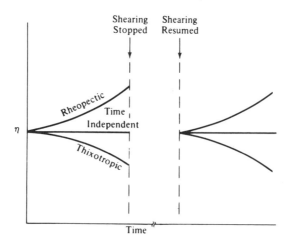

Figure 15.3 Time-dependent fluids.

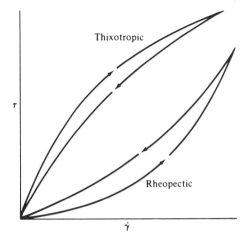

Figure 15.4 Flow curves for time-dependent fluids under continuously increasing and then decreasing shear.

effects in polymers are due to elasticity, and will be considered later, but for chemically stable polymer melts or solutions, the *equilibrium* viscous properties are time independent. We treat only such systems from here on.

15.5 POLYMER MELTS AND SOLUTIONS

When the flow properties of polymer melts and solutions can be measured over a wide enough range of shearing, the logarithmic flow curves appear as in Fig. 15.5. It is generally observed that:

1. At low shear rates (or stresses), a "lower Newtonian" region is reached with a so-called *zero-shear viscosity* η_0.
2. Over several decades of intermediate shear rates, the material is pseudoplastic.
3. At very high shear rates, an "upper Newtonian" region, with viscosity η_∞ is attained.

This behavior can be rationalized in terms of molecular structure. At low shear, the randomizing effect of the thermal motion of the chain segments overcomes any tendency toward molecular alignment in the shear field. The molecules are thus in their most random and highly entangled state, and have their greatest resistance to slippage (flow). As the shear is increased, the molecules will begin to untangle and align in the shear field, reducing their resistance to slippage past one another. Under severe shearing, they will be pretty much

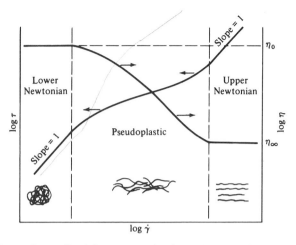

Figure 15.5   Generalized flow properties for polymer melts and solutions.

completely untangled and aligned, and reach a state of minimum resistance to flow. This is illustrated schematically in Fig. 15.5.

Intense shearing eventually leads to extensive breakage of main-chain bonds, that is, mechanical degradation. Furthermore, differentiation of (14.3) with respect to time reveals that the rate of viscous energy dissipation per unit volume is equal to $\tau\dot{\gamma}$. It thus becomes exceedingly difficult to maintain the temperature constant under intense shearing, so good data that illustrate the upper-Newtonian region are relatively rare, particularly for polymer melts.

It is worthwhile to consider here what happens to the highly oriented molecules when the shear field is removed. The randomizing effect of thermal energy tends to return them to their low-shear configurations, giving rise to an elastic retraction.

Some actual flow curves for polymer melts are shown in Fig. 15.6.[9] The data cover only a portion of the general range described above, because very few instruments are capable of obtaining data over the entire range. The polyisobutylene data cover the transition from lower Newtonian to pseudoplastic regions, but the polyethylene data are confined to the pseudoplastic region. No trace of the upper Newtonian region is seen for either material at the shear rates investigated.

15.6   QUANTITATIVE REPRESENTATION OF FLOW BEHAVIOR

To handle non-Newtonian flow analytically, it is desirable to have a mathematical expression relating τ, η, and $\dot{\gamma}$, as Newton's law does for Newtonian fluids. A wide variety of such *constitutive relations* has been proposed, both theoretical and empirical.[2, 5, 10] All appear to fit at least some experimental data over

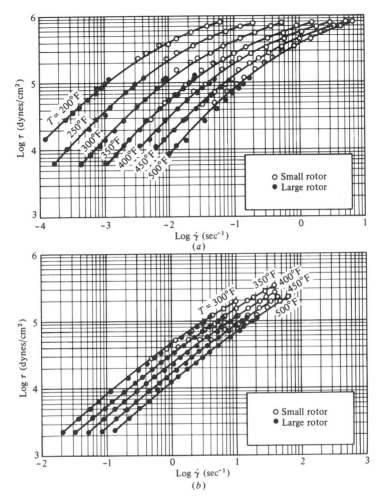

Figure 15.6 Flow curves for polymer melts[9]: (a) L-80 polyisobutylene; (b) low-density polyethylene.

a limited range of shear rates, but in general, the more adjustable parameters in the equation, the better fit it provides. (There's an old saying that with six constants you can draw an elephant, and with a seventh make his trunk wave.) The mathematical complexity of the equations increases greatly with the number of parameters, soon outstripping the data available to establish the parameters, and making the equations impractical for engineering calculations.

The traditional engineering model for purely viscous non-Newtonian flow is the so-called *power law*:

$$\tau = K(\dot{\gamma})^n \tag{15.6a}$$

This is a two-parameter model, the adjustable parameters being the *consistency* K and the *flow index* n. As written above, the dimensions of K depend on the magnitude of n (dimensionless), so if you're a dimensional purist, the power law may be written

$$\tau = K|\dot{\gamma}|^{n-1}\dot{\gamma} \qquad (15.6b)$$

This way, K has the usual viscosity units.

On log τ vs. log $\dot{\gamma}$ coordinates, a power-law fluid is represented by a straight line with slope n. Thus, for $n = 1$, it reduces to Newton's law, for $n < 1$, the fluid is pseudoplastic, and for $n > 1$, the fluid is dilatant. The power law can reasonably approximate only portions of actual flow curves over one or two decades of shear rate (see Fig. 15.6), but it does so with fair mathematical simplicity and has been adequate for many engineering purposes. Many useful relations have been obtained simply by replacing Newton's law with the power law in the usual fluid-dynamic equations.

Example 1. Determine the equation relating viscosity to shear rate for a power-law fluid.

Solution.

$$\eta \equiv \frac{\tau}{\dot{\gamma}} = \frac{K(\dot{\gamma})^n}{\dot{\gamma}} = K(\dot{\gamma})^{n-1} \qquad (15.7a)$$

or, if you prefer (15.6b) to (15.6a),

$$\eta = K|\dot{\gamma}|^{n-1} \qquad (15.7b)$$

Thus, a log–log plot of η vs. $\dot{\gamma}$ for a power-law fluid is linear with a slope of $n-1$ (Fig. 15.7). This points up an interesting limitation of the power law. Where the shear rate goes to zero (e.g., for Poiseuille flow at the center line of a cylindrical tube) the viscosity approaches infinity for a pseudoplastic power-law fluid, for which $n < 1$.

The Carreau model[2]

$$\frac{\eta - \eta_\infty}{\eta_0 - \eta_\infty} = [1 + (\lambda_c\dot{\gamma})^2]^{(n-1)/2} \qquad (15.8)$$

is a four-parameter model capable of representing all the features of the general flow curve in Fig. 15.5. It has been quite successful in fitting data for polymer melts and solutions over at least three or four decades of shear rate. The parameter λ_c is a time constant or characteristic time. For $(\lambda_c\dot{\gamma})^2 \ll 1, \eta \to \eta_0$ and for $(\lambda_c\dot{\gamma})^2 \gg 1, \eta \to \eta_\infty$. In between, it generates a power-law region with a log–log slope of $n-1$ (Fig. 15.7).

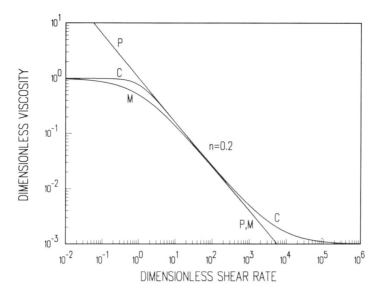

Figure 15.7 The dependence of viscosity on shear rate according to P, the power law (15.7b), η/K vs. $|\dot{\gamma}|$; C, the Carreau model (15.8), η/η_0 vs. $\lambda_c\dot{\gamma}$, $\eta_\infty/\eta_0 = 10^{-3}$; and M, the modified Cross equation (15.9), η/η_0 vs. $C\eta_0\dot{\gamma}$. For all, $n = 0.2$.

Not long ago, an equation with four constants would have been deemed excessively complex for engineering calculations, but the computer doesn't mind a bit. Actually, in many practical applications, the shear rates don't get high enough to approach the upper-Newtonian region, and a truncated (three-parameter) form of the Carreau equation, with $\eta_\infty = 0$, is adequate.

The (much) "modified Cross" model[11] has seen increasing application in recent years:

$$\eta(T, \dot{\gamma}) = \frac{\eta_0(T)}{1 + [C\eta_0(T)\dot{\gamma}]^{1-n}} \tag{15.9}$$

This is a three-parameter model, with the constant C having the dimensions of a reciprocal modulus (area/force). The product $C\eta_0$ has dimensions of time, and may be thought of as a time constant or characteristic time. At low shear rates, $\eta \to \eta_0$, and at high shear rates it gives a power-law region with a log–log slope of $n - 1$ (Fig. 15.7).

Its great virtue is that unlike other models, it explicitly incorporates the dependence of viscosity on temperature as well as shear rate through the temperature dependence of the zero-shear viscosity (to be discussed shortly). Equation 15.9 is written to emphasize that point. This makes it particularly well suited for nonisothermal flow calculations.

Figure 15.8 shows fits of the modified Cross model to the data[12] in Fig. 15.6a. (Note the magnitudes of those viscosities!) The constants C and n are common

Viscosity, poise

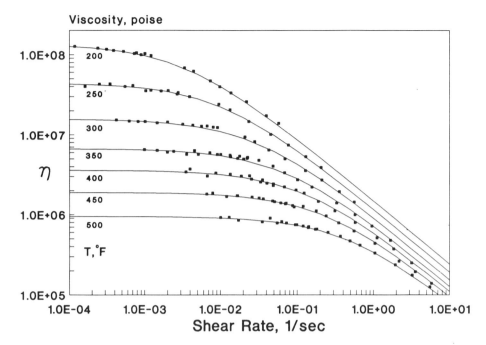

Figure 15.8 The modified Cross equation (15.9) (solid curves) fit to data of Fig. 15.6a.[12] For all isotherms $C = 2.2 \times 10^{-6}$ cm^2/dyn, $n = 0.21$. $\eta_0(200°F) = 1.34 \times 10^8$ P, $\eta_0(250) = 4.40 \times 10^7$, $\eta_0(300) = 1.57 \times 10^7$, $\eta_0(350) = 6.67 \times 10^6$, $\eta_0(400) = 3.60 \times 10^6$, $\eta_0(450) = 1.90 \times 10^6$, $\eta_0(500) = 9.50 \times 10^5$.

to all the isotherms. The fits are quite good. The equation correctly reproduces the fact that isothermal flow curves at different temperatures tend to "come together" at high shear rates, as seen in Figs. 15.6 and 15.8. Keep in mind, however, that each isotherm covers less than three decades of shear rate, so (15.9) is not severely challenged by these data.

As long as you don't approach the upper Newtonian region (generally the case in practice), there is no a priori basis on which to choose between (15.8) with $\eta_\infty = 0$ and (15.9) to fit isothermal data. They differ only in the "sharpness" of the transition from lower Newtonian to power-law regions (Fig. 15.7) and that seems to depend on the material. For a given polymer, it is known that the transition becomes more gradual as the breadth of the molecular weight distribution increases, so polymers with large polydispersity indexes might be better fit by (15.9) and more nearly monodisperse materials by (15.8), but data are not available to support sweeping generalizations.

As will be illustrated in the next chapter, certain important calculations (e.g., the determination of velocity profiles and flow rates in tubes) are facilitated by flow equations that are explicit in τ, that is, have the form $\dot{\gamma} = \dot{\gamma}(\tau)$ or $\eta = \eta(\tau)$. The power law can be written in either form. Some of the equations not given here are in or can be put in one of these forms. Unfortunately, they tend not to fit data as well as the Carreau and modified Cross equations, neither of which can be made explicit in τ. This has led to the practice of fitting flow data with polynomials such as

$$\log \dot{\gamma} = a_0 + a_1 (\log \tau)^1 + a_2 (\log \tau)^2 + \cdots \tag{15.10a}$$

or

$$\log \eta = b_0 + b_1 (\log \tau)^1 + b_2 (\log \tau)^2 + \cdots \tag{15.10b}$$

where the a_i and b_i are functions of temperature.

In principle, this is OK, and it can give good fits, but in practice, two cautions are in order. First, it's easy to get carried away and include more terms in the equation than are justified by the data. This leads to the fitting of experimental scatter. Second, great care must be exercised when extending the equations beyond the range of the data used to determine them. Uncritical extrapolation can lead to artifacts such as shear stresses that *decrease* with increasing shear rate, and negative viscosities, which violate both common sense and the second law of thermodynamics. These problems are less likely to arise when an established flow equation is used.

15.7 TEMPERATURE DEPENDENCE OF FLOW PROPERTIES

Also of engineering interest is the variation of flow properties with temperature. The temperature dependence of the *zero-shear* viscosity can often be represented by the relation

$$\eta_0(T) = Ae^{E/RT} \tag{15.11}$$

where E is the activation energy for viscous flow.

The η_0's from Fig. 15.8 are plotted according to (15.11) in Fig. 15.9. The fit is quite good. The slope is $E/2.303R$, from which $E = 11.4$ kcal/mol. Thus, the combination of (15.11) and (15.9) does a pretty good job of describing both the shear rate and temperature dependence of viscosity, at least for these data.

Example 2. What must the temperature be to reduce the 100°C zero-shear viscosity of L-80 PIB by an order of magnitude?

Solution. The temperature dependence of zero-shear viscosity is given by (15.11). Therefore, between temperatures T_1 and T_2,

$$\frac{\eta_{02}}{\eta_{01}} = \frac{\exp(E/RT_2)}{\exp(E/RT_1)} = \exp\left[\frac{E}{R}\left(\frac{1}{T_2} - \frac{1}{T_1}\right)\right]$$

Here, $T_1 = 100°C = 373$ K and $\eta_{02}/\eta_{01} = 0.1$. With $R = 1.99$ cal/mol·K and

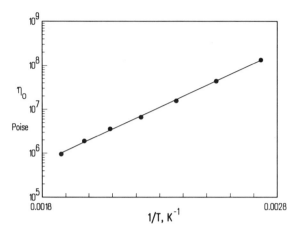

Figure 15.9 Temperature dependence of zero-shear viscosity. Zero-shear viscosities from modified Cross fits to data of Fig. 15.6a plotted according to (15.11).

the activation energy given above, solving for T_2 gives

$$T_2 = 439 \text{ K} = 166°\text{C}$$

The temperature is, therefore, an effective means of controlling melt viscosity in processing operations, but two drawbacks must be kept in mind: (1) it takes time and costs money to put in and take out thermal energy, and (2) excessive temperatures can lead to degradation of the polymer.

The famous Williams–Landell–Ferry or *WLF* equation[13] is useful for describing the temperature dependence of several linear mechanical properties of polymers (see Chapter XVIII). For the zero-shear viscosity, it may be written

$$\log_{10} \frac{\eta_0(T)}{\eta_0(T^*)} = \frac{-C_1(T - T^*)}{C_2 + (T - T^*)} \tag{15.12}$$

where T^* is a reference temperature. If the reference temperature is chosen as the glass-transition temperature, $T^* = T_g$, the "universal" constants $C_1 = 17.44$ and $C_2 = 51.6$ (with T's in K) give a rough fit for a wide variety of polymers. The WLF equation is most useful in this form because T_g is extensively tabulated.[14] Better fits can be obtained by using constants specific to the polymer, but these constants are less readily available. It has also been suggested that fit can be improved by using $C_1 = 8.86$ and $C_2 = 101.6$, with T^* adjusted to fit specific data, if available. When this is done, T^* generally turns out to be $T_g + (50 \pm 5)°\text{C}$.[2] Whichever constants are used, application of the WLF equation should be limited to the range $T_g < T < T_g + 100°\text{C}$. The WLF equation (15.12) can also be used in conjunction with (15.9). In the author's experience, however, (15.11) generally gives a better fit of η_0 vs. T data than (15.12).

We have seen how the modified Cross equation gives the temperature dependence of viscosity at finite shear rates. With other equations, the temperature dependence of all the constants would have to be known. Another approach has used an equation that is independent of any model for the flow curve. Since the viscosity is a function of temperature and shear stress or shear rate

$$\eta = f(\tau, T) \qquad \text{or} \qquad \eta = f'(\dot{\gamma}, T)$$

By analogy to (15.11), these functions are approximated by

$$\eta = B \exp\left(\frac{E_\tau}{RT}\right) \qquad \text{(at constant } \tau) \qquad (15.13a)$$

$$\eta = C \exp\left(\frac{E_{\dot{\gamma}}}{RT}\right) \qquad \text{(at constant } \dot{\gamma}) \qquad (15.13b)$$

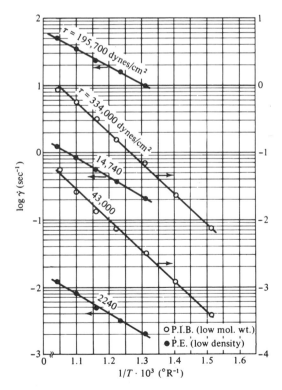

Figure 15.10 Temperature dependence of shear rate at constant shear stress.[9]

where E_τ is the activation energy for flow at constant shear stress and $E_{\dot\gamma}$ is the activation energy for flow at constant shear rate. Figure 15.10 shows plots of $\log \dot\gamma$ vs. $1/T$ at constant τ for the materials in Fig. 15.6.[9] Since $\dot\gamma = \tau/\eta$, the slope of these lines is $-E_\tau/2.303R$.

It will be left as an end-of-chapter exercise to show that $E_\tau > E_{\dot\gamma}$. In the limit $\tau \to 0$, $\dot\gamma \to 0$ and for all Newtonian fluids, $E_\tau = E_{\dot\gamma} = E$.

15.8 INFLUENCE OF MOLECULAR WEIGHT ON FLOW PROPERTIES

It has long been known that a polymer's molecular weight exerts a strong influence on its melt or solution viscosity. Experiments show that

$$\eta_0 \propto \bar M_w^1 \qquad \text{for } \bar M_w < \bar M_{wc} \tag{15.14a}$$

$$\eta_0 \propto \bar M_w^{3.4} \qquad \text{for } \bar M_w > \bar M_{wc} \tag{15.14b}$$

where $\bar M_{wc}$ is a critical average molecular weight, thought to be the point at which molecular entanglements begin to dominate the rate of slippage of the molecules. It depends on the temperature and polymer type, but most commercial polymers are well above $\bar M_{wc}$.

Equation 15.14 holds quantitatively for just about all polymer melts. The addition of a low molecular weight solvent, of course, cuts down entanglements and raises $\bar M_{wc}$. Equation 7.13 suggests that entanglements set in when the dimensionless product of intrinsic viscosity and concentration exceeds one, $[\eta]c > 1$. Thus, if you know $[\eta]$ for your polymer in the solvent, you can get an idea of the concentration above which (15.14b) ought to hold. In practice, even moderately concentrated (say 25% or more) polymer solutions have viscosities proportional to $\bar M_w^{3.4}$, provided $\bar M_w$ is in the range of commercial importance.

Example 3. Obtain an approximate equation that relates $\bar M_{wc}$ to concentration and the MHS constants (6.16) for polymer solutions.

Solution. First, we assume that entanglements begin when the Berry number, $[\eta]c$, equals 1. Because all we're after here is a rough estimate, we can assume that $\bar M_w \approx \bar M_v$ (see Chapter 6, Example 4). Thus,

$$\bar M_{wc} \approx \left(\frac{1}{Kc}\right)^{1/a}$$

As the shear rate is increased, the number of entanglements between chains is reduced, and as expected, the dependence of viscosity on molecular weight decreases (Fig. 15.11).

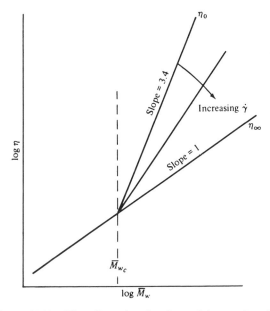

Figure 15.11   The effect of molecular weight on viscosity.

Example 4. By what percentage must $\bar{M}_w$ be changed to cut η_0 in half?

Solution. From (15.14*b*),

$$\frac{\eta_{02}}{\eta_{01}} = \left(\frac{\bar{M}_{w2}}{\bar{M}_{w1}}\right)^{3.4}$$

$$\frac{\bar{M}_{w2}}{\bar{M}_{w1}} = \left(\frac{\eta_{02}}{\eta_{01}}\right)^{1/3.4} = (1/2)^{1/3.4} = 0.816$$

$$\% \text{ change} = 100\left(\frac{\bar{M}_{w2} - \bar{M}_{w1}}{\bar{M}_{w1}}\right) = 100\left(\frac{\bar{M}_{w2}}{\bar{M}_{w1}} - 1\right) = -18.4\%$$

Thus, it takes only an 18% decrease in chain length to halve the zero-shear melt viscosity. Although things won't be quite so dramatic at high shear rates, this illustrates the importance of controlling molecular weight during synthesis to achieve a product with the desired processing properties.

15.9   THE EFFECTS OF PRESSURE ON VISCOSITY

For all fluids, viscosity increases with increasing pressure, as the free volume, and hence the ease of molecular slippage, is decreased. With liquids, including

polymer melts, because of their relative incompressibility, the effect becomes noticeable only at fairly high pressures (hundreds of atmospheres). Nevertheless, such pressures are reached in certain processes (e.g., injection molding) and in certain types of capillary viscometers (Chapter XVI). Available data are fit by

$$\eta_0(P) = \alpha e^{\beta P} \tag{15.15}$$

where α and β are constants. Equation 15.15 may be combined with (15.11) to give an equation that expresses both the temperature and pressure dependence of zero-shear viscosity:

$$\eta_0(T, P) = D \exp\left(\frac{E}{RT} + \beta P\right) \tag{15.16}$$

Presumably, (15.16) could be used in conjuction with (15.9) to express viscosity as a function of $\dot{\gamma}$, T, and P, but for pressure, there is as yet little experimental justification for this.

Carley[15] has critically reviewed work on polymer melts and concludes that pressure effects are of minor significance in most processing situations, provided the temperature is not too close to a transition. High pressures raise both T_g and T_m slightly, and of course the viscosity shoots up tremendously as either is approached.

REFERENCES

1. Barnes, H. A., J. F. Hutton, and K. Walters, *An Introduction to Rheology*, Elsevier, New York, 1989.

2. Bird, R. B., R. C. Armstrong, and O. A. Hassager, *Dynamics of Polymeric Liquids*, Vol. 1, *Fluid Mechanics*, Wiley, New York, 1977.

3. Coleman, B. D., H. Markovitz, and W. Noll, *Viscometric Flows of Non-Newtonian Fluids*, Springer, New York, 1966.

4. Christensen, R. M., *Theory of Viscoelasticity: An Introduction*, Academic, New York, 1971.

5. Darby, R., *Viscoelastic Fluids*, Dekker, New York, 1976.

6. Dealy, J. M., and K. F. Wissbrun, *Melt Rheology and Its Role in Plastics Processing*, Van Nostrand Reinhold, New York, 1990.

7. Middleman, S., *The Flow of High Polymers*, Interscience, New York, 1968.

8. Tanner, R. I., *Engineering Rheology*, Oxford UP, New York, 1985.

9. Best, D. M., and S. L. Rosen, *Polym. Eng. Sci.* **8**(2), 116 (1968).

10. Bird, R. B., W. E. Stewart, and E. N. Lightfoot, *Transport Phenomena*, Wiley, New York, 1960, Chapter 1.

11. Tseng, H. C., et al., *Preprints, Soc. Plast. Engrs. Annual Tech. Conf.*, 1985, p. 716.

12. Best, D. M., Thesis, Carnegie Institute of Technology, Pittsburgh, PA, 1966.

13. Tobolsky, A. V., *Properties and Structure of Polymers*, Wiley, New York, 1960, Chapter II.8.

14. Brandrup, J., and E. H. Immergut (eds.), *Polymer Handbook*, 3d ed., Wiley-Interscience, New York, 1989.

15. Carley, J. F., *Modern Plastics* **39**(4), 123 (1961)

PROBLEMS

1. The data below were taken from a prehistoric Ph.D. thesis (Rosen, S. L., Cornell University, 1964) recently unearthed by archeologists. They were obtained at 100°C on a sample of poly(ethyl acrylate) with a viscosity-average molecular weight of 310 000 and a very broad molecular weight distribution.

   Compare the truncated ($\eta_\infty = 0$) form of the Carreau equation (15.8) and the modified Cross equation (15.9) for fitting these data.

τ (dyn/cm^2)	$\dot{\gamma}$ (s^{-1})	τ	$\dot{\gamma}$
3.15×10^3	0.00578	1.36×10^5	1.01
3.70	0.00696	1.67	1.60
4.53	0.00913	2.44	4.15
5.91	0.0112	2.76	5.08
7.34	0.0145	3.22	9.66
1.00×10^4	0.0213	3.59	12.76
1.47	0.0321	4.29	25.4
2.20	0.0548	5.03	50.8
2.72	0.0727	6.59	127
4.00	0.123	7.96	254
5.23	0.189	9.40	550
6.49	0.262	1.16×10^6	1500
7.73	0.358	1.33	3140
9.00	0.463	1.52	6610

2. Assume that the L-80 PIB in Figs. 15.6a and 15.8 has $\bar{M}_w = 80\,000$. Its $\eta_0 = 6.67 \times 10^6$ P at 350°F.

   a. If L-120 PIB has $\bar{M}_w = 120\,000$, what will its η_0 be at 350°F?

   b. In what proportions should L-80 and L-120 PIBs be mixed to produce a material with $\eta_0 = 8.0 \times 10^6$ P at 350°F? *Hint:* See Equation 6.6.

3. Write the power law in the form $\eta = \eta(\tau)$.

4. The following data were obtained by N. D. Sylvester (Ph.D. Thesis, Carnegie Institute of Technology, Pittsburgh, PA, 1968) at room temperature on a 1.10% aqueous solution of Separan® AP-30, a partially hydrolyzed

polyacrylamide of molecular weight 2–3 million:

τ (dyn/cm^2)	$\dot{\gamma}$ (s^{-1})
365	415
434	748
471	929
536	1360
566	1630
616	2110
653	2360
695	2710
748	3400

Fit these data with one of the equations in the chapter. Obtain all of the constants in the equation.

5. Figure 15.7 illustrates that the modified Cross equation (15.9) reduces to the power law when $(C\eta_0\dot{\gamma})^{(1-n)} \gg 1$. Assume that the temperature dependence of η_0 is given by (15.11). What is the relation between E and $E_{\dot{\gamma}}$ in the power-law region according to the modified Cross equation?

6. Some equations give the viscosity as an explicit function of shear stress, $\eta = \eta(\tau)$. Two equations of this form are

$$\eta = A + B\left[1 + \left(\frac{\tau}{C}\right)^2\right]^{-1} \qquad\qquad \eta = \frac{D}{1 + E(\tau)^m}$$

$$\text{(I)} \qquad\qquad\qquad\qquad \text{(II)}$$

where A, B, C, D, E, and m are constants.

   a. Which of the above models (if either) is capable of showing a zero-shear viscosity? What is the zero-shear viscosity in terms of the constants.
   b. Repeat (a) for an infinite shear rate viscosity.
   c. Which (if either) can exhibit a power-law region? What is the power-law constant n in this region in terms of the parameters in the equation?

7. Prove that for a pseudoplastic fluid, $E_\tau \geq E_{\dot{\gamma}}$, and that in the low-shear limit and for all Newtonian fluids, $E_\tau = E_{\dot{\gamma}}$. Hint: Consider viscosity to be a function of temperature and either shear rate or shear stress, and write an expression for its total differential.

8. What is the relation between E_τ and $E_{\dot{\gamma}}$ for a power-law fluid?

9. R. A. Stratton (J. Colloid Interface Sci. **22**, 517 (1966)) provides τ vs. $\dot{\gamma}$ data for five essentially monodisperse polystyrenes of different molecular weight. Are these data better fit by the Carreau or modified Cross equations?

Stratton obtains η_0 values by extrapolating plots of η vs. τ to $\tau = 0$. Do these equations give substantially different values? Stratton's values at 183°C are as follows:

$\bar{M}_w \times 10^{-5}$	$\eta_0 \times 10^{-4}$ (P)
0.48	0.153
1.17	2.88
1.79	12.4
2.17	21.9
2.42	33.5

Do the above data conform to Eq. 15.14*b*?

10. Another four-parameter model for equilibrium viscous flow is

$$\tau = \left(\frac{A + B}{1 + C\tau^m} \right) \dot{\gamma}$$

where A, B, C, and m are constants.

   a. In terms of model parameters, what is η_0?
   b. In terms of model parameters, what is η_∞?
   c. If the model is to represent a Newtonian fluid, what conclusion(s) can be drawn about the parameters?

11. Equations 15.11 and 15.12 are two different ways of representing the temperature dependence of zero-shear viscosity. For a material that follows the WLF equation (15.12), obtain an expression for the temperature-dependent activation energy E in (15.11).

12. A polymer is being processed at 500 K. Mechanically, things are OK, but discoloration of the parts indicates that some thermal degradation of the material is occurring during processing. A suggestion is made that the temperature be lowered to 450 K to minimize degradation. You point out that this could cause serious mechanical problems unless a switch was made to a different molecular weight polymer.

   Given that the polymer has a flow activation energy of 10 kcal/mol and that the current material has $\bar{M}_w = 400\,000$, what can you say about the $\bar{M}_w$ of a different grade of the same polymer that will allow adequate processing at 450 K?

13. Sometimes even liquids made up of small, symmetrical molecules appear to be psuedoplastic when measurements are made at very high shear rates in certain viscometers. Why?

14. Add one constant to the modified Cross equation (15.9) so that it will reproduce an upper-Newtonian region.

15. Consider the sample of polymethyl methacrylate in Example 3, Chapter VI. What is the minimum concentration of that polymer in acetone at 30°C necessary to assure that (15.14b) is applicable?

16. A Bingham plastic (if such a thing really exists) has a flow curve $\tau - \tau_y = \eta'\dot{\gamma}$, where τ_y is a yield stress which must be exceeded to cause flow. Sketch $\log \tau$ vs. $\log \dot{\gamma}$ and $\log \eta$ vs. $\log \dot{\gamma}$ for a Bingham plastic, showing limiting behavior at low and high $\dot{\gamma}$.

CHAPTER XVI

Viscometry and Tube Flow

16.1 INTRODUCTION

The quantities η, τ, and $\dot{\gamma}$ are defined in terms of viscometric flows. This chapter considers the principles behind the devices used to establish viscometric flows and thereby determine the viscous flow properties discussed in the previous chapter. Details of the instruments and procedures have been reviewed.[1,2] One of the common techniques, Poisueille (laminar) flow in cylindrical tubes, is also important from a technological standpoint, since polymer melts and solutions are often transported and processed in this fashion.

The chapter concludes with a discussion of the turbulent flow of non-Newtonian polymer fluids. Because of the tremendous viscosities of polymer fluids, laminar flow is far more prevalent than in nonpolymer fluids; in fact, turbulence is normally encountered only with dilute polymer solutions (less than a few weight percent, or so). Nevertheless, when turbulent flow does occur, there are some interesting and practically important differences from the turbulent flow of Newtonian fluids.

16.2 VISCOUS ENERGY DISSIPATION

Regardless of the viscometric technique, determination of the true, isothermal flow properties at high shear rates can be complicated by high rates of viscous energy dissipation, which make it hard to maintain isothermal conditions. If we divide both sides of (14.3) by $V\,dt$, we see that the *rate of viscous energy dissipation per unit volume* $\dot{E}$ is

$$\dot{E} = \tau\dot{\gamma} = \eta\dot{\gamma}^2 \tag{16.1}$$

Example 1. Obtain an expression for the adiabatic rate of temperature rise in a polymer sample subjected to a shear stress τ and shear rate $\dot{\gamma}$.

Solution. In the absence of heat transfer, the rate of temperature rise $\dot{T}$ is given by the rate of energy dissipation per unit volume divided by the volumetric heat capacity ρC_p:

$$\dot{T} = \frac{\tau \dot{\gamma}}{\rho C_p} \tag{16.2}$$

Not only is this an important consideration in viscometry, where great care must be taken in the design of viscometers to permit adequate temperature regulation, but it also must be taken into account in the design of processing systems. In the steady-state operation of extruders, for example, virtually all the energy required to melt and maintain the polymer in the molten state is supplied by the mechanical drive. Here, however, we will limit our considerations to isothermal flows.

16.3 POISEUILLE FLOW

Axial, laminar (Poiseuille) flow in a tube of cylindrical cross section is another example of a viscometric flow field. Here, geometry dictates the use of cylindrical coordinates. Fluid motion is in the x ('1') direction along the tube axis, the velocity gradient is directed everywhere in the outward radial r (or '2') direction, and the mutually perpendicular or neutral direction is the tangential θ (or '3') coordinate. Consider a cylindrical fluid element of radius r and length dx (Fig. 16.1). Assume that the pressure in the fluid is a function of the distance along the tube (x coordinate) only. The net pressure force pushing the element in the x direction is the differential pressure drop across the element, $-dP$, times the area of the element's ends, πr^2. The motion of the element is resisted by a shear force at its surface, which is the product of the shear stress, τ_{rx}, and the surface area of the cylinder, $2\pi r\, dx$. In equilibrium flow, these forces must balance, so

$$\underset{\text{Surface shear force}}{2\pi r\, dx\, \tau_{rx}} \quad = \quad \underset{\text{Net pressure force}}{-\pi r^2\, dP} \tag{16.3}$$

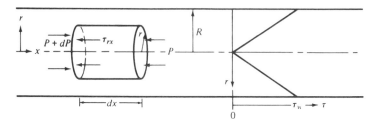

Figure 16.1   Force balance on an element in a cylindrical tube.

Thus, the dependence of shear stress on radius (denoted $\tau(r)$) is

$$\tau_{rx}(r) = -\left(\frac{r}{2}\right)\left(\frac{dP}{dx}\right) \tag{16.4}$$

For a tube with inner radius R (dropping the rx subscript on shear stress)

$$\tau_w = -\left(\frac{R}{2}\right)\left(\frac{dP}{dx}\right) \tag{16.5}$$

where τ_w is the shear stress at the tube wall ($r = R$), and

$$\tau(r) = \frac{r}{R}\tau_w \tag{16.6}$$

So, the shear stress varies linearly with radius from zero at the tube center to a maximum of $\tau_w = -(R/2)(dP/dx)$ at the tube wall. *Note that this result does not depend in any way on the fluid properties.*

Since the axial fluid velocity u is a function of radial position only (denoted $u(r)$), we may write

$$du(r) = \frac{du}{dr}dr \tag{16.7}$$

Integrating with the boundary conditions $u(R) = 0$ (i.e., the fluid sticks to the wall) and u at radius $r = u(r)$ gives

$$u(r) = \int_0^{u(r)} du(r) = \int_R^r \frac{du}{dr}dr \tag{16.8}$$

Realizing that the velocity gradient (du/dr) is the shear rate $\dot{\gamma}$ (just replace y with r in (15.2)), we find

$$u(r) = \int_R^r \dot{\gamma}[\tau(r)]\,dr = \int_R^r \frac{\tau(r)}{\eta[\tau(r)]}\,dr \tag{16.9}$$

where $\tau(r)$ is given by (16.4). The functions $\dot{\gamma}(\tau)$ or $\eta(\tau)$ represent the material's equilibrium viscous properties or flow curve, as discussed in the previous chapter. So, for a given pressure gradient, the velocity profile in the tube may always be calculated from the flow curve by

1. Choosing an r and calculating τ from (16.4)
2. Obtaining $\dot{\gamma}$ (or η) at the τ above from the flow curve
3. Representing numerically or graphically the relation $\dot{\gamma}(r)$ (or $\tau(r)/\eta(r)$)
4. Integrating from R to r, which gives u at r.

$$\tau = \kappa \dot{\gamma}^n$$

$$\dot{\gamma} = \left(\frac{\tau}{\kappa}\right)^{\frac{1}{n}}$$

If an analytic representation of the flow curve is available, (16.9) may be integrated directly (provided the representation is simple enough). For example (verification will be left as an exercise for the reader, as they say in the math books), for a power-law fluid $\dot{\gamma}(\tau) = (\tau/K)^{1/n}$, and

$$u(r) = \frac{[(-dP/dx)/2K]^{1/n}}{(1/n) + 1} [R^{(1/n)+1} - r^{(1/n)+1}] \tag{16.10}$$

Equation 16.10 reduces to the familiar Newtonian parabolic profile for $n = 1$.

Note that the above calculations are facilitated if the flow curve can be written as an explicit function of τ, that is, in the form $\dot{\gamma} = \dot{\gamma}(\tau)$ or $\eta = \eta(\tau)$. While the power law meets that criterion, neither the Carreau (15.8) nor the modified Cross (15.9) equations do. For this reason, it's sometimes viewed as being easier to fit data with a polynomial such as (15.10a) or (15.10b), rather than to use an established constitutive equation to make these calculations (but note the cautions!).

Looking at a differential ring of the tube cross section, with thickness dr, located at radius r where the velocity is $u(r)$ (Fig. 16.2), the differential volumetric flow rate is

$$dQ = u(r) 2\pi r \, dr \tag{16.11}$$

where $u(r)$ is the local velocity and $2\pi r \, dr$ the area of the differential ring. Integrating over the tube cross section gives

$$Q = 2\pi \int_0^R u(r) r \, dr \tag{16.12}$$

So, knowing the velocity profile allows calculation of the volumetric throughput Q. For a power-law fluid, insertion of (16.10) into (16.12) and turning the crank gives

$$Q = \left(\frac{-dP/dx}{2K} \right)^{1/n} \left[\frac{\pi}{(1/n) + 3} \right] R^{(1/n)+3} \tag{16.13}$$

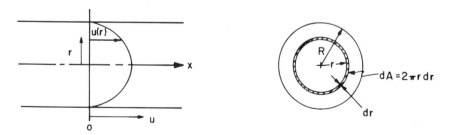

Figure 16.2   Determination of volumetric flow rate.

The *average velocity* V in the tube is defined by

$$V \equiv \frac{Q}{\pi R^2} \qquad (16.14)$$

Velocity profiles for power-law fluids in a tube are plotted in dimensionless form in Fig. 16.3 for several values of n. The greater the degree of pseudoplasticity (i.e., the lower n), the flatter the profile becomes. For the Newtonian fluid, $n = 1$, the profile is parabolic.

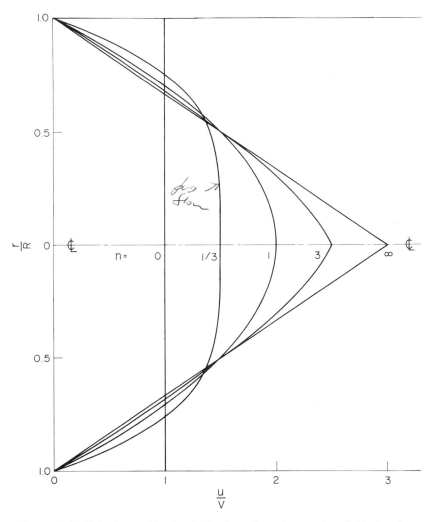

Figure 16.3 Velocity profiles for the laminar flow of power-law fluids in tubes.

16.4   DETERMINATION OF FLOW CURVES

The formulas just developed allow the relation of pressure gradient, velocity profile, and volumetric flow rate *as long as the shear stress–shear rate relation* (flow curve) *for the fluid is known*. The problem in viscometry is just the reverse: How is the flow curve obtained from pressure drop–volumetric flow rate measurements in a cylindrical tube? If the mathematical form of the flow curve, that is, a particular constitutive equation, is assumed a priori, the integrated equations as developed above may be used to establish the parameters in the constitutive relation. For example, if it is assumed that the power law represents the flow curve of a fluid under investigation, two readings of dP/dx versus Q in a tube of known R will allow calculation of K and n from (16.13). However, in the general case, the form of the constitutive equation is not known a priori, and must be established by viscometry. This may be done, first by integrating (16.12) by parts

$$Q = 2\pi \left[\frac{r^2 u(r)}{2} \Big|_0^R - \int_0^R \frac{r^2}{2} \, du(r) \right] \tag{16.15}$$

The first term in the brackets is zero, because at the upper limit $u(R) = 0$, and at the lower limit $r = 0$. Therefore,

$$Q = -\pi \int_0^R r^2 \, du(r) = -\pi \int_0^R r^2 \left(\frac{du(r)}{dr} \right) dr = -\pi \int_0^R r^2 \dot{\gamma}(r) \, dr \tag{16.16}$$

Using (16.6) to change the independent variable from r to τ, and recalling that at $r = 0$, $\tau = 0$ and at $r = R$, $\tau = \tau_w$, we find

$$\frac{\tau_w^3 Q}{\pi R^3} = -\int_0^{\tau_w} \dot{\gamma}(\tau) \tau^2 \, d\tau \tag{16.17}$$

Applying Leibnitz' rule to differentiate both sides of (16.17) with respect to τ_w gives

$$\frac{1}{\pi R^3} \left(\tau_w^3 \frac{dQ}{d\tau_w} + 3\tau_w^2 Q \right) = -\dot{\gamma}_w \tau_w^2 \tag{16.18}$$

or

$$-\dot{\gamma}_w = \frac{1}{\pi R^3} \left(\tau_w \frac{dQ}{d\tau_w} + 3Q \right) \tag{16.19}$$

Since $d \ln \Phi = d\Phi/\Phi$ (Φ = any variable)

$$-\dot{\gamma}_w = \frac{1}{\pi R^3} \left(3Q + \frac{Q \, d \ln Q}{d \ln \tau_w} \right) \tag{16.20}$$

Handwritten annotations in margins: "Apparent shear rate" (pointing to equation), "Might ... huge ... of dict."

$$-\dot{\gamma}_w = \frac{4Q}{\pi R^3}\left(\frac{3}{4} + \frac{1}{4}\frac{d\ln Q}{d\ln \tau_w}\right) \tag{16.21}$$

$$-\dot{\gamma}_w = \Gamma\left(\frac{3}{4} + \frac{1}{4}\frac{d\ln \Gamma}{d\ln \tau_w}\right) \tag{16.22}$$

where

$$\Gamma = \frac{4Q}{\pi R^3} = \frac{8V}{D} \tag{16.23}$$

is known as the *apparent* shear rate.

Equations 16.21 and 16.22 are equivalent forms of the *Rabinowitsch equation*. They allow calculation of the shear rate at the tube wall from experimental data. Note the necessity of obtaining the slope of a log Γ versus log τ_w plot (or otherwise differentiating the data) to obtain $\dot{\gamma}_w$. For a single tube, Γ is proportional to Q and τ_w to (dP/dx), so $(d\ln Q)/[d\ln (dP/dx)]$ may be substituted for $(d\ln \Gamma)/(d\ln \tau_w)$ in (16.22). However, (16.22) allows data from several different tubes, if available, to be combined. A look at (16.17) reveals that for the flow of a given fluid in all cylindrical tubes, τ_w *is a unique function of the apparent shear rate* Γ, because the value of the integral depends only on the upper limit τ_w, regardless of the nature of the $\dot{\gamma}(\tau)$ relation (flow curve) of the fluid. Therefore, for a given fluid, data from various tubes should all fall on a common $\tau_w - \Gamma$ curve. This provides more accurate values for the derivatives. Values of $\dot{\gamma}_w$ are then combined with values of τ_w calculated from (16.5) to give the flow curve.

Neglect of the *Rabinowitsch correction* (the term in brackets in (16.21) and (16.22)) can lead to serious error. To illustrate, differentiation of (16.13) and substitution of (16.5) and (16.23) reveal that for a power-law fluid, $(d\ln \Gamma)/(d\ln \tau_w) = 1/n$, a constant. The Rabinowitsch equation then becomes

$$-\dot{\gamma}_w = \Gamma\left(\frac{3n + 1}{4n}\right) \qquad \text{(power-law fluids)} \tag{16.24}$$

Only for a Newtonian fluid, $n = 1$, is Γ equal to the true shear rate at the tube wall $\dot{\gamma}_w$. For an n of $\frac{1}{3}$, a reasonable value for a polymer melt or solution, the correction term is 1.5, so the apparent shear rate Γ is 50% lower than the true shear rate at the tube wall $\dot{\gamma}_w$. The Rabinowitsch correction accounts for the fact that the velocity gradient (shear rate) at the tube wall is greater for pseudoplastic fluids than for a Newtonian fluid at the same Q, as illustrated in Fig. 16.3.

Much older literature data from capillary rheometers are represented as being τ_w vs. $\dot{\gamma}_w$ or $\eta = \tau_w/\dot{\gamma}_w$ vs. $\dot{\gamma}_w$ (true flow curves), but in reality are τ_w vs. Γ or *apparent viscosity* $\eta_a = \tau_w/\Gamma$ vs. Γ. The fact that the Rabinowitsch correction has not been applied may not be mentioned, so care must be exercised when using such data.

Example 2. The researchers at Crud Chemicals have obtained good data on a polygunk solution from their capillary rheometer. Unfortunately, they don't understand the Rabinowitsch correction, so they don't apply it, and the relation that they report as being viscosity vs. shear rate is really apparent viscosity vs. apparent shear rate:

$$\eta_a \equiv \frac{\tau_w}{\Gamma} = 13.2\Gamma^{-0.650}$$

where η_a is in poise and Γ in s^{-1}. Obtain the true viscosity-shear rate relation for Crud.

Solution. The form of their equation suggests that they've got a power-law fluid, for which

$$\eta \equiv \frac{\tau_w}{\dot{\gamma}_w} = K(\dot{\gamma}_w)^{n-1} \tag{15.7}$$

Inserting (16.24) for $\dot{\gamma}_w$ and rearranging gives

$$\eta_a = K\left(\frac{3n+1}{4n}\right)^n \Gamma^{n-1}$$

Comparing this with their experimental relation, we see that they do indeed have a power-law fluid, with $n = 1 - 0.650 = 0.350$ and $K = 13.2[(3n+1)/4n]^{-n} = 11.6$. The true viscosity–shear rate relation is therefore

$$\eta = 11.6\dot{\gamma}^{-0.650}$$

The preceding analysis of flow in tubes has assumed that the flow properties of the fluid are independent of pressure. This is probably a reasonable assumption in most cases. However, as mentioned in Section 15.7, viscosity increases with pressure, particularly as T_g or T_m is approached, and if the pressure drop across the tube is very large, corrections may be necessary. Such calculations have been discussed.[3]

16.5 ENTRANCE CORRECTIONS

A capillary viscometer is a device in which the fluid under investigation is forced from a reservoir through a cylindrical capillary tube. They are operated with either flow rate or pressure as the independent variable. In the former case, the fluid is usually driven by a piston advancing through a cylindrical reservoir at a known constant rate with the force on the piston recorded; in the latter case, regulated gas pressure drives the fluid and the volumetric flow rate is measured.

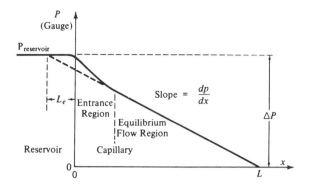

Figure 16.4 Pressure profile in a capillary viscometer.

Unfortunately, calculating the equilibrium pressure gradient from the data so obtained is not always a simple matter of setting $dP/dx = \Delta P/L$, where ΔP is the overall pressure drop across the capillary of length L. The gradient does not become constant until equilibrium flow is reached some distance downstream from the entrance of the capillary, as illustrated in Fig. 16.4, and approximating the true equilibrium gradient $(dP/dx)_{eq}$ with the measured $\Delta P/L$ can cause considerable error. Higher than equilibrium pressure losses are observed in the entry region because (1) kinetic energy must be added to the fluid as it is accelerated from low velocities in the reservoir to higher velocities in the capillary, (2) rearrangement of the velocity profile dissipates additional energy, and (3) a viscoelastic fluid *stores* some energy elastically when going from the low-stress reservoir to the high-stress capillary (think of stretching a rubber band). Recovery of this stored energy at the tube exit leads to such things as die swell, melt fracture, and other anomalies observed in the flow of viscoelastic fluids.

Bagley[4] has developed a method for getting the true pressure gradient from capillary viscometer data. As seen in Fig. 16.4,

$$\left(\frac{dP}{dx}\right)_{eq} = \frac{\Delta P}{L + L_e} = \frac{\Delta P}{L + eR} \tag{16.25}$$

where L_e is a fictitious "entrance length." L_e is commonly expressed as the product of e, the *entrance correction* and the tube radius R. Inserting (16.25) into (16.5), dividing numerator and denominator by R, and recalling that τ_w is a function of Γ only for a given fluid in cylindrical tubes gives

$$\tau_w = \frac{R\Delta P}{2(L + eR)} = \frac{\Delta P}{2[(L/R) + e]} = f(\Gamma) \tag{16.26}$$

where f denotes "function of." Rearranging, we get

$$\Delta P = 2\left(\frac{L}{R} + e\right)f(\Gamma) \tag{16.27}$$

Thus, if a series of experiments is run in which the capillary L/R ratio is varied, a *Bagley plot* of ΔP vs. L/R at constant $\Gamma = 4Q/\pi R^3$ should give a series of straight lines, one for each constant value of Γ. The $\Delta P = 0$ intercept is the entrance correction, $-e$ (Fig. 16.5). The various intercepts at constant Γ give e as a function of τ_w or $\dot\gamma_w$. The need to use several capillaries greatly increases the work, but for melts, it appears that $L/R \gg e$ (i.e., the entrance correction becomes negligible) only for $L/R > 100$ or so. For solutions, at high flow rates, the correction may be significant at much higher L/R values.

The reverse of this procedure may be important in design situations; that is, the entrance correction must be known as a function τ_w and included in the overall pressure drop calculations. In extruder dies and spinerettes for fiber spinning, L/R values rarely exceed 10, and the entrance loss may be the major contribution to the overall pressure drop. Carley[5] provides examples of design calculations involving these concepts.

Once the entrance correction has been established, it can be used to calculate the true equilibrium gradient through (16.25), which in turn is used in (16.5) and (16.21) or (16.22) to obtain the flow curve, relating the shear stress to shear rate, both at the tube wall. In a few instances, other minor corrections have been applied, for example, considering the pressure drop in the viscometer reservoir.[6]

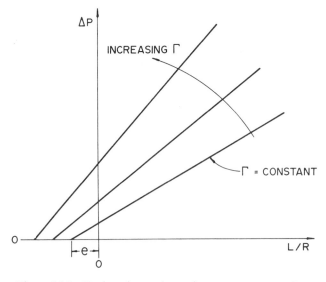

Figure 16.5   Bagley plot to determine entrance corrections.

16.6   SCALEUP FOR LAMINAR FLOW IN CYLINDRICAL TUBES

When comparing viscometric data from tubes with data obtained from other types of viscometers, and when using tube-flow data to predict behavior in other geometries, the use of the Rabinowitsch correction is essential to give the true shear stress–shear rate relation. If all you're concerned with is laminar flow in cylindrical tubes, however, it's not necessary. As pointed out above, τ_w is a unique function of Γ for the laminar flow of a given fluid in all cylindrical tubes. Thus, experimental determinations of τ_w vs. Γ obtained from cylindrical tubes are applicable for scaleup purposes to other cylindrical tubes through which the same fluid is flowing (at the same temperature). For full-scale piping systems, it is usually permissible to write $dP/dx = \Delta P/L$.

Example 3. Consider the equilibrium laminar flow of the same non-Newtonian fluid in two different cylindrical tubes. Available data are as follows:

	R (in.)	Q (gal/min)	ΔP (psi)	L (ft)
Tube 1	1	2	1000	400
Tube 2	2	16	?	200

Can you calculate ΔP in the second tube? If not, what additional information would be required?

Solution. Application of (16.23) reveals that $\Gamma_2 = \Gamma_1$ (conversion factors cancel out, so you can use the given mongrel units directly). Because τ_w is a unique function of Γ, $\tau_{w2} = \tau_{w1}$ also, so from (16.5) with $dP/dx = \Delta P/L$,

$$R_2\left(\frac{\Delta P_2}{L_2}\right) = R_1\left(\frac{\Delta P_1}{L_1}\right)$$

from which $\Delta P_2 = 250$ psi.

We were able to obtain this solution without additional knowledge only because $\Gamma_2 = \Gamma_1$. Had this not been the case, we would have needed the actual relation between τ_w and Γ.

Example 4. Laboratory experiments with a capillary viscometer have established the following empirical relation for a polymer solution at room temperature:

$$\Gamma = 0.0257\,\tau_w + 1.68 \times 10^{-6}\,\tau_w^2$$

where Γ is in s^{-1} and τ_w is in dyn/cm^2. This relation is valid over the experimental range $0 < \tau_w < 7 \times 10^3$ dyn/cm^2. Calculate the volumetric flow rate, in cm³/s, for this fluid through 200 m of 2-cm-i.d. pipe with a pressure drop of 100 bars.

Solution. Given that a bar is 10^6 dyn/cm² (no, it's not what Davy Crockett shot in the woods), application of (16.5) gives $\tau_w = 2.5 \times 10^3$ dyn/cm². From the $\Gamma - \tau_w$ relation above, $\Gamma = 74.75$ s⁻¹, and from (16.23), $Q = 58.7$ cm³/s.

16.7   THE COUETTE VISCOMETER

Another common device for measuring viscous properties is the cup-and-bob, or Couette viscometer, a diagram of which is shown in Fig. 16.6. The fluid is confined in the gap between two concentric cylinders, one of which rotates relative to the other at a known angular velocity while the torque on one is measured. This is another classic example of a viscometric flow. In cylindrical coordinates, we assume only a tangential velocity component, so the 1 coordinate is the tangential or θ direction and the 2 coordinate is the radial direction.

Example 5. a. Neglecting end effects, determine the shear stress as function of radius in terms of the measured torque on the stationary inner cylinder (bob), $M(R_i)$, and the geometry of the apparatus as the outer cylinder (cup) is rotated with an angular velocity ω (radians/s).

b. This is the classic example of a viscometric flow in which the shear rate is not equal to the velocity gradient dv_θ/dr (see Problem 16.2). As a result, determining the shear rate as a function of radius is not easy. Bird et al.[7] do it by *assuming* a Newtonian relation between shear stress and shear rate. In general, however, the nature of the flow curve is not known a priori, and must be

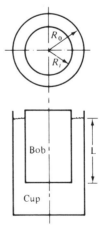

Figure 16.6   Schematic of Couette (cup-and-bob) viscometer.

determined by viscometry. One means of doing this is to make the gap between the cylinders $(R_o - R_i)$ very small compared to the radius of either cylinder. Letting $R_o - R_i = \delta$ and $R_o \approx R_i = R$, obtain the expression for shear rate in terms of ω and the geometry.

Solution. *a.* In a rotating system at equilibrium, Σ torques $= 0$, or there would be angular acceleration. Consider a ring of fluid with inner radius R_i and outer radius r: $M(r) = M(R_i)$

$$
\underbrace{2\pi r L \quad \tau(r)}_{\text{force}} \quad \begin{matrix} r \\ \text{Moment} \\ \text{arm} \end{matrix} \quad = \quad \begin{matrix} M(R_i) \\ \text{Measured} \\ \text{torque} \end{matrix}
$$

$$
\tau(r) = \frac{M(R_i)}{2\pi r^2 L} \tag{16.28}
$$

b. This situation approximates the case of two flat plates separated by a distance δ, one sliding past the other with a linear velocity equal to the tangential velocity $R\omega$:

$$
\dot{\gamma} = \frac{R\omega}{\delta} \quad \text{(for } \delta \ll R_i \text{ only)} \tag{16.29}
$$

Where the geometric approximations in Example 5b above are not applicable, Kreiger and Maron[8] have presented an analysis similar to the Rabinowitsch development for flow in tubes. It involves differentiation of the M vs. ω data, but unfortunately, is in the form of an infinite series. If $R_o/R_i < 1.2$, a closed-form approximation is given.

Obviously, the analysis above is not valid in the area beneath the bob at the bottom of the viscometer. This is best taken into account by making measurements with two fluid depths, the lower being well above the bottom of the bob, and using the *differences* between the torques and depths in (16.28), thereby subtracting out the effects of non-Couette flow. Another approach is illustrated in Example 7.

16.8 CONE-AND-PLATE VISCOMETER

Still another popular type of rotational viscometer is the cone-and-plate, in which the sample is sheared between a flat plate and a broad cone whose apex contacts the plate (Fig. 16.7). For small cone–plate angles α, this approximates a viscometric flow with (in spherical coordinates) flow in the tangential θ or 1 direction and the gradient in the azimuthal ϕ or 2 direction. Here, the radial

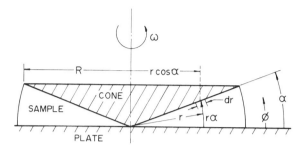

Figure 16.7   Schematic of cone-and-plate viscometer.

direction is the neutral or 3 coordinate.* It turns out that true viscometric flow of this type is inconsistent with the equations of motion if the inertial terms are included. (Formal flow solutions for rotational viscometers normally involve neglect of the inertial terms in the equations of motion, as we tacitly do here.) There must, therefore, be radial and azimuthal velocity components. These are minimized in practice by keeping α quite small—often less than $1°$. The great advantage of this type of device is that (for small α) the shear rate, and hence the shear stress, is uniform throughout the material.

Example 6. Obtain the expressions for shear rate and shear stress in a cone-and-plate viscometer in terms of the rate of cone rotation ω, the measured torque M, and the geometry.

Solution (see Fig. 16.7). The tangential velocity v_θ of a point on the cone relative to the plate is $v_\theta = \omega r \cos \alpha$. Fluid is sheared between that point and the plate over a distance $\delta = \alpha r$:

$$\dot{\gamma} = \frac{v_\theta}{\delta} = \frac{\omega r \cos \alpha}{\alpha r} \xrightarrow{\text{small } \alpha} \frac{\omega}{\alpha} \quad \text{(independent of } r\text{)} \qquad (16.30)$$

Torque = (Shear stress) (Area) (Moment arm)

$$dM = (\tau)(2\pi r \cos \alpha \, dr)(r \cos \alpha)$$

Integrating over the cone face gives

$$M = 2\pi\tau \cos^2\alpha \int_0^{R/\cos\alpha} r^2 \, dr = \frac{2\pi\tau R^3}{3\cos\alpha} \xrightarrow{\text{small } \alpha} \frac{2\pi\tau R^3}{3}$$

---

* At first glance, it might seem that there is a gradient component in the r direction because the tangential velocity increases with r. However, material points in a cone at constant ϕ do not move relative to one another; that is, they undergo *rigid-body rotation*, so there is no shearing in the r direction.

$$\tau = \frac{3M}{2\pi R^3} \qquad (16.31)$$

Note: Here, τ is constant, and can be taken outside the integral, because $\tau = \tau(\dot{\gamma})$, and $\dot{\gamma}$ is constant, as shown by (16.30).

Cone-and-plate viscometers of the type shown here are usually limited to fairly low shear rates. At higher shear rates, solutions tend to be flung from the gap by centrifugal force, and melts tend to "ball up" (like rubbing a finger over dry rubber cement). These problems can be overcome by enclosing the fluid around a biconical rotor, giving, in effect, two cone-and-plate viscometers back-to-back.[9] The flow curves in Chapter XV were obtained with such a device.

Example 7. One means of minimizing end effects in a Couette viscometer is to make the bottom of the bob a cone, the apex of which contacts the base of the cup, so that the area beneath the bob is a cone-and-plate viscometer. For a Couette geometry in which the gap δ is much smaller than the bob radius R_i, what must the cone angle α be to match the shear rates in the Couette and cone-and-plate regions?

Solution. By equating the expressions for shear rate from Examples 5(*b*) and 6, $\alpha = \delta/R$.

16.9   DISK–PLATE VISCOMETER

Another type of viscometer which finds occasional use is the disk–plate viscometer (Fig. 16.8). A disk of radius R rotates with an angular velocity of ω relative to a parallel plate. The disk and plate are separated by a distance $d\,(d \ll R)$, with the test fluid in between. The torque M on either the disk or plate is measured. This is known as torsional (twisting) flow, and is another example of a viscometric flow.

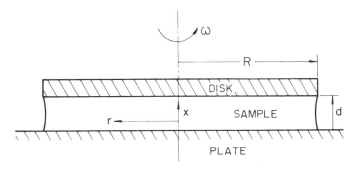

Figure 16.8   Schematic of disk–plate viscometer.

Example 8. a. Identify the 1 (flow), 2 (gradient), and 3 (neutral) directions with respect to the cylindrical coordinates in Fig. 16.8.*

b. Obtain an expression for $\dot{\gamma}$. Why is the cone-and-plate geometry usually preferred?

c. Obtain an expression for the torque M needed to maintain a steady rate of rotation ω when a power-law fluid is in the gap.

Solution. *a.* 1 (flow) $= \theta$, 2 (gradient) $= x$, 3 (neutral) $= r$.

b.

$$\dot{\gamma} = \frac{\text{Relative velocity, disk to plate}}{\text{Separation distance}} = \frac{r\omega}{d} \tag{16.32}$$

In this geometry, unlike the cone and plate, the shear rate increases linearly with radius, and so is not uniform throughout the fluid.

c. Consider a differential ring of the disk (or plate) surface, at radius r with thickness dr:

$$dA = 2\pi r\, dr$$

$$dM = r\tau\, dA = 2\pi r^2 \tau\, dr$$

For a power-law fluid, $\tau = K\dot{\gamma}^n = K(r\omega/d)^n$,

$$M = 2\pi K \left(\frac{\omega}{d}\right)^n \int_0^R r^{2+n}\, dr = \frac{2\pi K(\omega/d)^n}{3+n} R^{3+n}$$

or

$$M = \frac{2\pi K R^3}{3+n}(\dot{\gamma}_R)^n \quad \text{(power-law fluid)} \tag{16.33}$$

where

$$\dot{\gamma}_R = \frac{R\omega}{d} \tag{16.34}$$

When the form of the flow curve is not known a priori, it may be obtained through a Rabinowitsch-like development which relates shear stress and shear rate at $r = R$ (the derivation of which will be left as an end-of-chapter problem):

$$\tau_R = \frac{3M}{2\pi R^3}\left(1 + \frac{1}{3}\frac{d \ln M}{d \ln \dot{\gamma}_R}\right) \tag{16.35}$$

*See the footnote to Section 16.8. Here, points at constant x undergo rigid-body rotation like a phonograph record.

16.10 TURBULENT FLOW

On occasion, turbulent flow is encountered with dilute polymer solutions. As with the turbulent flow of Newtonian fluids, pressure drops are conveniently handled in terms of the Fanning friction factor

$$f \equiv \frac{Rg_c}{\rho V^2} \left(\frac{dP}{dx} \right) \tag{16.36}$$

(The dimensional constant $g_c = 32.2$ ft·lb_m/lb_f·s^2 is included here because these equations are still sometimes used with the English engineering system of units.) Since $\tau_w = (R/2)(dP/dx)$ (the minus sign has been dropped, it being understood that flow is in the direction of decreasing pressure), then

$$f = \frac{2\tau_w g_c}{\rho V^2} \tag{16.37}$$

Equations 16.36 and 16.37 apply to all fluids in both laminar and turbulent flow.

The next question is, How do you define a Reynolds number for a fluid that has a variable viscosity? Metzner and Reed[10] proposed that the known relation between friction factor and Reynolds number for the laminar flow of Newtonian fluids be applied to the laminar flow of non-Newtonians as well:

$$f \equiv \frac{16}{Re} \qquad \text{(laminar flow of all fluids)} \tag{16.38}$$

Since (16.38) has been defined to apply to all fluids in laminar flow, it may now be used to obtain a *generalized Reynolds number*—applicable to all fluids in both laminar and turbulent flow—by combining it with (16.37):

$$Re = \frac{8\rho V^2}{\tau_w g_c} \tag{16.39}$$

For a power-law fluid, $\tau g_c = K(\dot{\gamma})^n$, (16.5), (16.13), and (16.14) give

$$Re = \frac{8D^n V^{2-n} \rho}{K[6 + (2/n)]^n} \tag{16.40}$$

Actually, Metzner and Reed used an alternative formulation. Since the shear stress at the tube wall is a unique function of the apparent shear rate for laminar flow in cylindrical tubes, the power law may be written at the tube wall with the aid of (16.24) as

$$\tau_w g_c = K(\dot{\gamma}_w)^n = K\left(\Gamma \frac{3n+1}{4n} \right)^n = K'(\Gamma)^n \tag{16.41}$$

where

$$K' = K\left(\frac{3n + 1}{4n}\right)^n \tag{16.42}$$

Using these variables, the generalized Reynolds number becomes

$$Re = \frac{D^n V^{2-n} \rho}{K' 8^{n-1}} \tag{16.43}$$

Both (16.40) and (16.43) reduce to the familiar Newtonian relation, $Re = DV\rho/\eta$, with $\eta = K = K'$, when $n = 1$.

Equations 16.36–16.39 by definition should describe the behavior of all fluids in equilibrium laminar viscous flow. They do, provided that the true equilibrium pressure gradient is used in (16.36).[11] Approximation of the equilibrium gradient by $\Delta P/L$ can sometimes cause considerable error.

Example 9. Assuming that $\Delta P/L = dP/dx$, obtain an expression for the pressure drop in the laminar flow of a power-law fluid through a cylindrical tube of length L and inside diameter D.

Solution. Combination of (16.36), (16.38), and (16.43) gives

$$\Delta P = \frac{32K' 8^{n-1} L V^n}{g_c D^{n+1}} = \frac{4K'}{g_c}\left(\frac{8V}{D}\right)^n\left(\frac{L}{D}\right)$$

The same result can be obtained from (16.13), (16.14), and (16.42).

16.11   DRAG REDUCTION[11-16]

Having defined the Reynolds number for non-Newtonian fluids, it would be nice to be able to say that one could then use the standard friction factor–Reynolds number relations to make the usual turbulent-flow calculations. Such is not always the case, however, because of a startling phenomenon known as *drag reduction* which is often observed in the turbulent flow of non-Newtonian fluids (Fig. 16.9).[14] Small quantities of a polymeric solute can cut the friction factor significantly at higher Reynolds numbers, despite the fact that the solution viscosity is only slightly greater than that of the solvent. For example, 200 ppm of guar gum (a natural, water-soluble polymer) reduces the friction factor in Fig. 16.9 by approximately a factor of five at $Re = 10^5$, while the viscosity of the solution is but 24% greater than that of pure water.

This phenomenon has important practical applications. Pressure drops, and therefore pumping costs, can be reduced for a given flow rate, or the capacity of a pumping system (e.g., the Trans-Alaska pipeline) can be increased by the

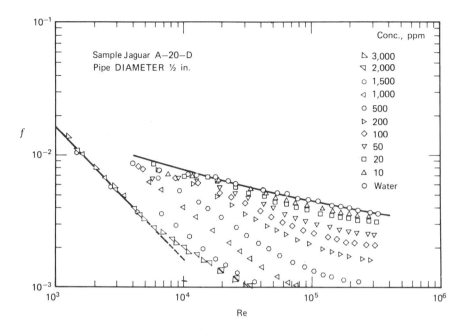

Figure 16.9 Turbulent drag reduction.[16] Copyright 1972 by the American Chemical Society. Reprinted with permission of the copyright owner.

addition of a drag-reducing solute. Fire departments add poly(ethylene oxide) to their pumpers to increase capacity. Experiments have been conducted with the aim of increasing ship speed by squirting out a drag-reducing additive at the bow.

The causes of drag reduction are not yet clear. It has been suggested that it may be due to a thickening of the laminar sublayer caused by elasticity of the polymer molecules (those polymer solutions that seem relatively inelastic are more likely to be amenable to the usual treatment). It has also been pointed out that most of the energy dissipation in turbulent flow occurs in small, high-frequency eddies. At these frequencies (the reciprocals of which are smaller than the material's relaxation time, giving high Deborah numbers, Chapter XVIII), the material responds elastically (like bouncing the Silly Putty rather than letting it flow down the wall), passing the stored elastic energy from eddy to eddy rather than dissipating it. Very high elongational viscosities (Chapter XVII) of polymer solutions have also been implicated, even though the regular (shear) viscosities of the solutions are not much higher than those of the solvent.

Drag reduction is known to be most pronounced for solutions of high molecular weight, flexible-chain polymers, possibly because these form highly elastic solutions. Its magnitude is reduced markedly as the polymers are degraded by continued flow. The effects of polymer molecular variables have been discussed.[17] On the other hand, drag reduction has also been observed with slurries of rigid particles or fibers in Newtonian fluids. Perhaps there's more than one cause.

As with Newtonians, it's a pretty safe bet that flow is laminar for Re < 2100, but drag-reducing additives often seem to delay the laminar–turbulent transition to higher Re's. The friction factor–Reynolds Number relation remains a function of pipe roughness. Unlike the Newtonian case, however, the f-Re curve seems to depend on pipe diameter when drag reduction is observed.

Unfortunately, the bottom line is that at present, in the absence of experimental data, there are no quantitative methods for predicting which polymeric solutes will produce drag reduction in a particular solvent, or over what range of concentrations drag reduction will be observed, or what its magnitude will be. Fortunately, neglecting drag reduction gives conservative designs; any drag reduction that does occur will be a bonus in terms of increased flow rate and/or reduced pressure drop.

REFERENCES

1. Collyer, A. A., and D. W. Clegg (eds.), *Rheological Measurement*, Elsevier, New York, 1988.

2. Dealy, J. M., *Rheometers for Molten Plastics*, Van Nostrand Reinhold, New York, 1982.

3. Pennwell, R. C., R. S. Porter, and S. Middleman, *J. Polymer Sci.* **A-2**(9), 731 (1971).

4. Bagley, E. B., *J. Appl. Phys.* **28**, 624 (1957).

5. Carley, J. F., *Soc. Plast. Eng. J.* **19**(12), 1263 (1963).

6. Van Wazer, et al., *Viscosity and Flow Measurement: A Laboratory Handbook of Rheology*, Interscience, New York, 1963.

7. Bird, R. B., W. E. Stewart, and E. N. Lightfoot, *Transport Phenomena*, Wiley, New York, 1960, p. 94, Example 3.5-1.

8. Kreiger, I. M., and S. H. Maron, *J. Appl. Phys.* **25**, 72 (1954).

9. Best, D. M., and S. L. Rosen, *Polym. Eng. Sci.* **8**(2), 116 (1968).

10. Metzner, A. B., and J. C. Reed, *Am. Inst. Chem. Eng. J.* **1**, 434 (1955).

11. Patterson, G. K., J. L. Zakin, and J. M. Rodriguez, *Ind. Eng. Chem.* **61**(1), 22 (1969).

12. Hoyt, J. W., *Trans. ASME J. Basic Eng.* **94D**, 258 (1972).

13. Lumley, J. L., *Macromol. Rev.* **7**, 263 (1973).

14. Wang, C. B., *Ind. Eng. Chem. Fundam.* **11**, 546 (1972).

15. Virk, P. S., *Am. Inst. Chem. Eng. J.* **21**, 625 (1975).

16. Sellin, R. H. J., and R. T. Moses, *Drag Reduction in Fluid Flows. Techniques for Friction Control*, Ellis Horwood, Chichester (distributed by Wiley, NY), 1989.

17. Zakin, J. L., and D. L. Hunston, *J. Macromol. Sci.-Phys.*, **B18**(4), 795 (1980).

PROBLEMS

1. The following data were obtained by Dr. N. D. Sylvester (Ph.D. thesis, Carnegie Institute of Technology, 1968) in the "world's largest capillary viscometer," an 8-ft long, $\frac{1}{2}$-in.-i.d. tube equipped with pressure taps to measure the pressure gradient down the tube. The fluid was a 1.34% solution of poly(ethylene oxide) in water.

Q (in.^{3}/s)	ΔP_{total} (psi)	ΔP_{ent} (psi)
2.10	1.898	0.036
3.11	2.510	.039
3.49	2.732	.081
4.48	2.947	.082
5.04	3.330	.123
6.16	3.599	.132
7.87	4.238	.211
10.74	4.965	.341
14.31	6.046	.620
15.49	6.249	.675
15.85	6.451	.726
20.13	7.597	1.183
23.35	8.670	1.607
29.09	10.130	2.335

Here, ΔP_{total} is the total pressure drop across the tube, and ΔP_{ent} is the above-equilibrium pressure loss caused by entrance effects.

a. Obtain the flow equation for this fluid, with constants in the cgs system. Note that at the higher flow rates, entrance losses are a significant portion of the total pressure drop.

b. Check the Reynolds number at the highest flow rate to make sure the data are in laminar flow.

2. Show that the shear rate in Couette flow between two coaxial cylinders is given by

$$\dot{\gamma} = r \frac{d}{dr}\left(\frac{v_\theta}{r}\right)$$

where $v_\theta = \omega r$ is the tangential velocity.

3. Consider Couette flow between two coaxial cylinders, the outer one at R_o fixed and the inner one at R_i rotating with a tangential velocity $v_\theta(R_i)$. Starting with (16.28) and the equation in Problem 2, obtain an expression for the dimensionless tangential velocity profile, $v_\theta(r)/v_\theta(R_i)$, for a power-law fluid. Write your expression in terms of the dimensionless quantities r/R_i, R_o/R_i, and the flow index n. Plot $v_\theta(r)/v_\theta(R_i)$ vs. r/R_i for $R_o/R_i = 2$ and $n = 1, \frac{1}{2}$, and $\frac{1}{4}$.

4. Derive Eq. 16.35. *Hints*:

    a. Start with the equation for dM in the solution to Example 8c and integrate it to give an expression for M. No integration by parts is required here.

    b. Transform the variable of integration from r to $\dot{\gamma}$ using (16.32) and the limits of integration with (16.34).

    c. The necessary Leibnitz differentiation is analogous to that between (16.17) and (16.18).

5. A cylindrical shaft of radius R_i and weight W slides vertically downward through a fixed bearing of inner radius R_o and length L. Assume that the shaft remains centered in the bearing and that annulus between them is filled with a power-law fluid. Neglect entrance and exit effects.

    a. Is flow in the annulus viscometric? If so, clearly identify the 1, 2, and 3 coordinate directions.

    b. Obtain an expression for the shear–stress distribution in the annulus $\tau(r)$.

    c. Obtain an expression for the velocity profile in the annulus.

    d. Obtain an expression for V, the steady-state velocity with which the shaft falls through the bearing.

6. Consider the experiment in which the pressure drop ΔP is measured as a function of volumetric flow rate Q through a cylindrical tube of length L and inner radius R. Entrance and exit effects are negligible. The data may be summarized over a limited range of Q by

$$\Delta P = AQ - BQ^2$$

where A and B are positive constants.

    a. Obtain an expression for the zero-shear viscosity of the fluid in terms of geometry and quantities in the above equation.

    b. Show how you would obtain the true shear stress–shear rate relation (flow curve) from the above data.

7. A power-law fluid is confined between two parallel, flat plates in simple shearing flow. The lower plate is fixed and the upper plate moves with a velocity V. The plates are separated by a distance δ. Calculate and sketch the velocity profiles for $n = 0.5$, $n = 1$, $n = 1.5$.

8. Injection molding consists of forcing a heated thermoplastic into a cooled mold. It has been noticed that short shots (material freezing before the mold is filled) can sometimes be cured by forcing the material through a smaller gate (narrower diameter entrance to the mold). In fact, so-called "pin gates" (very small orifices at the mold entrance) seem very effective at this. Why?

9. Aluminum foil to be made into capacitors is being coated with an insulating varnish by drawing it from a tank containing the varnish through a flat die.

The foil strip has a width W. The die length is L. It may be assumed that the foil remains centered in the die with a clearance of δ on each side. L and $W \gg \delta$, so entrance, edge and exit effects are negligible. The velocity of the foil through the die is V and it is pulled through with a force F. The viscous properties of the varnish are described by the power law. The pressure differential across the die is negligible. The y coordinate has its origin at the surface of the foil and is directed normal to the foil surface.

a. Obtain an expression for the shear stress in the varnish in the die in terms of F and geometry.

b. Obtain an expression for the dimensionless velocity profile in the die, $u(y)/V$.

c. Obtain an expression that relates F to V.

d. Obtain an expression for the thickness of the varnish layer on each side of the foil after it emerges from the die.

10. *a.* Obtain an expression for the dimensionless velocity profile $u(r)/V$ for a power-law fluid in a wire-coating die. The wire has an outer radius R_i and the die an inner radius R_o. The wire is pulled through the die with a velocity V. Assume that the wire remains centered in the die and that there is no pressure drop across the die.

b. Obtain an expression for the axial force F_x needed to pull the wire through a die of length L. Neglect entrance and exit effects and assume that all resistance is in the die.

11. *a.* Consider the laminar flow of a power-law fluid through an infinite slit under the influence of a pressure gradient (dP/dx). The walls of the slit are located at $y = \pm \delta/2$. Obtain the analogs to (16.5), (16.6), (16.10), (16.13), (16.14), and Fig. 16.3.

b. If you're really feeling masochistic, try for the analog of (16.21) or (16.22) for flow in the slit.

12. The entrance pressure loss, ΔP_{ent}, is the difference between $P_{\text{reservoir}}$ and the dotted (equilibrium-flow) line at $x = 0$ in Fig. 16.4. It and the Bagley entrance correction e are two different ways of expressing the same thing. Obtain the equation that relates ΔP_{ent} to e, ΔP, and geometry, and plot e vs. τ_w for the data in Problem 1.

13. The falling-level viscometer is a simple device that may be used to determine flow curves for relatively dilute polymer solutions in a single experiment. It consists of a vertical buret of diameter D connected to a horizontal capillary tube of length L and radius R. The buret is filled with the solution, which is then allowed to drain through the capillary. Data are h vs. t, where h is the height of the fluid in the buret above the capillary. It may be assumed that flow in the capillary is laminar, that all resistance to flow is in the capillary, and that entrance and exit effects are negligible. The fluid has a known density ρ. Describe clearly how to proceed from the data, h vs. t, to the flow curve, τ vs. $\dot{\gamma}$. Show all necessary equations.

14. A general scale-up equation for the laminar flow of a power-law fluid in cylindrical tubes may be written as

$$\left(\frac{\Delta P_2}{\Delta P_1}\right)^a \left(\frac{L_2}{L_1}\right)^b \left(\frac{D_2}{D_1}\right)^c \left(\frac{Q_2}{Q_1}\right)^d = 1$$

if we neglect entrance and exit effects. The subscripts 1 and 2 refer to two different flow situations with the same fluid. Given that exponent $a = 1$, determine exponents b, c, and d.

15. In their new plant, Crud Chemicals must transport 10 gal/min of a room-temperature polygunk solution 1000 ft with a maximum pressure drop of 500 psi. Laboratory data on the solution, obtained in a 0.10-in.-diameter, 10-in.-long tube, are summarized by

$$Q = 0.428\,\Delta P + 0.004\,81\,\Delta P^2$$

where ΔP is in psi and Q is in cm^3/min. This equation holds up to the maximum experimental ΔP of 40 psi. (*Note:* The fluid is pseudoplastic, but not power law.)

   Calculate the minimum pipe diameter needed to do the job, and check to see if flow will indeed be laminar.

16. It was pointed out in Section 16.7 that making the gap small can minimize the calculational difficulties associated with Couette viscometers. Unfortunately, a small gap poses some practical problems in the manufacture and operation of the device. What might these be?

17. Collins et al., (*Polymer Handbook*, 3d ed., J. Brandrup and E. H. Immergut (eds.), Wiley, New York, 1989, p. V/67–68) give apparent viscosities $\eta_a = \tau_w/\Gamma$ vs. Γ for a variety of PVCs at different temperatures. For a particular LVN = $[\eta]$ polymer, apply the Rabinowitsch correction at each temperature to get the true viscosity vs. shear rate relation.

18. A fluid's equilibrium viscous properties can be represented over a limited range of τ by

$$\dot{\gamma}(\tau) = A\tau + B\tau^2$$

where A and B are constants. Obtain expressions for $u(r)$ and Q for the laminar flow of this fluid in a tube of radius R under the influence of a pressure gradient $-dP/dx$.

CHAPTER XVII

Introduction to Continuum Mechanics

17.1 THREE-DIMENSIONAL STRESS AND STRAIN [1-7]

The stress existing at any point in a material may always be resolved into components acting on the faces of a differential element in three arbitrary coordinate directions (Fig. 17.1). The stress components acting on the faces of the element are of two types: *normal stresses* (forces acting normal to the surface per unit surface area) and *shear stresses* (forces acting parallel to the surface per unit surface area). The first subscript conventionally identifies the direction perpendicular to the surface in question, and the second identifies the direction of the force itself. In general, there are three normal stresses, τ_{11}, τ_{22}, and τ_{33}, and six shear stresses, τ_{12}, τ_{13}, τ_{21}, τ_{23}, τ_{31}, and τ_{32}. These nine quantities, necessary to specify completely the state of stress at a point, are the components of the *stress tensor* τ. They are conveniently written in matrix form:

$$\tau = \begin{vmatrix} \tau_{11} & \tau_{12} & \tau_{13} \\ \tau_{21} & \tau_{22} & \tau_{23} \\ \tau_{31} & \tau_{32} & \tau_{33} \end{vmatrix} \tag{17.1}$$

The stress tensor is usually broken into an isotropic or hydrostatic pressure and a *deviatoric stress tensor* τ':

$$\tau = - \begin{vmatrix} P & 0 & 0 \\ 0 & P & 0 \\ 0 & 0 & P \end{vmatrix} + \underbrace{\begin{vmatrix} \sigma_{11} & \tau_{12} & \tau_{13} \\ \tau_{21} & \sigma_{22} & \tau_{23} \\ \tau_{31} & \tau_{32} & \sigma_{33} \end{vmatrix}}_{\tau'} \tag{17.2}$$

289

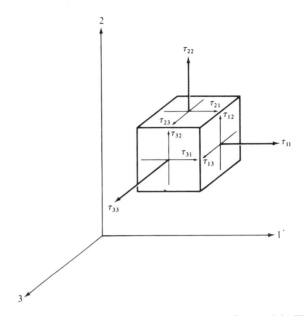

Figure 17.1 Stress components acting on an element of material. The equal-and-opposite stresses (necessary for equilibrium) on the hidden faces are not shown.

where the hydrostatic pressure P is rather arbitrarily defined as

$$P = -\frac{\tau_{11} + \tau_{22} + \tau_{33}}{3} = -\frac{1}{3}\,\text{tr }\tau \tag{17.3}$$

and the $\sigma_{ii} = \tau_{ii} + P$ are the *deviatoric normal stresses*. For static fluids at equilibrium the $\sigma_{ii} = 0$.

Caution 1: The negative signs in (17.2) and (17.3) arise from the historical conventions of treating both hydrostatic pressure (an inward force on the element) and *tensile* stress (an outward force on the element) as positive. These conventions are not adhered to universally, however.

Caution 2: There are about as many different forms of notation as there have been books and papers written in this area. Be sure to understand the notation before plowing through any of the literature—That's half the battle.

17.2   GENERAL CONSTITUTIVE RELATIONS

The three-dimensional rate of strain at the point is expressed by a rate-of-strain tensor analogous to (17.1). Given a general *constitutive relation* between the deviatoric stress and rate-of-strain tensors for a particular material, the mechanical response of that material can be described completely (in principle) under any circumstances. The goal of continuum mechanics is to develop such general constitutive relations and use them for predicting material response in the widest variety of situations. Ultimately, it is the goal of molecular mechanics to predict these constitutive relations from the molecular structure of the material.[8] Progress in both areas has been comprehensively reviewed.[6] Because of the unavoidable complexity of general, three-dimensional, time-dependent deformations, engineering application of some of these concepts has been slow.

17.3   EQUILIBRIUM VISCOMETRIC FLOWS[1]

For *equilibrium viscometric flows* of *incompressible* fluids (the latter is a reasonable assumption for polymer melts and solutions in most engineering circumstances), the situation is brighter. With the coordinate directions assigned as described for viscometric flows in the last chapter, the shear stresses τ_{12} and τ_{21} are equal (or there would be a rotational flow component), and $\tau_{13} = \tau_{31} = \tau_{32} = \tau_{23} = 0$. Because of the arbitrary nature of the static pressure definition, two independent *differences* of the deviatoric normal stresses are commonly defined:

$$N_1 = \sigma_{11} - \sigma_{22} \tag{17.4a}$$

$$N_2 = \sigma_{22} - \sigma_{33} \tag{17.4b}$$

where N_1 and N_2 are the so-called first and second normal stress differences. Furthermore, there is only one nonzero component of the rate-of-strain tensor, $\dot{\gamma}_{12} = \dot{\gamma}$. Thus, a complete description of an incompressible material in an equilibrium viscometric flow requires a knowledge of three functions:

$$\tau = \tau(\dot{\gamma}) \qquad \text{(flow curve)} \tag{17.5a}$$

$$N_1 = N_1(\dot{\gamma}) \qquad \text{(first normal-stress difference)} \tag{17.6a}$$

$$N_2 = N_2(\dot{\gamma}) \qquad \text{(second normal-stress difference)} \tag{17.7a}$$

or, alternatively,

$$\eta(\dot{\gamma}) \equiv \frac{\tau}{\dot{\gamma}} \qquad \text{(viscosity)} \tag{17.5b}$$

$$\mu_1(\dot{\gamma}) \equiv \frac{N_1}{\dot{\gamma}^2} \qquad \text{(first normal-stress coefficient)} \tag{17.6b}$$

$$\mu_2(\dot{\gamma}) \equiv \frac{N_2}{\dot{\gamma}^2} \qquad \text{(second normal-stress coefficient)} \tag{17.7b}$$

The first and second normal-stress coefficients μ_1 and μ_2 are defined in terms of the square of $\dot{\gamma}$ because the normal stresses are even functions of shear rate;

that is, their signs do not change when the direction of shearing (sign of $\dot{\gamma}$) changes. On the other hand, τ is an odd function of $\dot{\gamma}$; its sign changes with the sign of $\dot{\gamma}$.

The determination of (17.5) is discussed in Chapter 16. Techniques for measuring functions (17.6) and (17.7) have been comprehensively reviewed.[1,5,7] In general, N_1 is positive and, like τ, increases with $\dot{\gamma}$. The two are roughly comparable in magnitude for many polymer melts and solutions. This means that there can be significant tension in the flow direction. N_2 is considerably smaller (and therefore much more difficult to measure) and is often negative.

In nonequilibrium deformations, the response of the material depends not only on its current rate of strain, but on its complete strain history. Nonequilibrium deformations are treated in the next chapter.

Example 1. Consider a wire-coating die (Problem 16.10) in which a wire of radius R_i is drawn with a velocity V through a concentric die of inner radius R_o and length L by a force F_x. The annulus between wire and die is filled with a fluid. Here, however, the die is rotating about its (and the wire's) axis with an angular velocity ω. There is no pressure drop across the die, and we can neglect entrance and exit effects. This is known as *helical flow*, from the path followed by a fluid element.

a. Is this a viscometric flow? If so, define the 1, 2, and 3 coordinate directions.

b. Write the components of the rate-of-strain tensor for this flow using conventional cylindrical (x, r, θ) coordinates.

c. Make the flat-plate approximation, $\delta = R_o - R_i \ll R_i \approx R$, and apply a simple constitutive equation, the *generalized power law* (a generalization of (15.6b)),

$$\tau_{ij} = K|\dot{\gamma}|^{n-1}\dot{\gamma}_{ij} \qquad \text{(generalized power law)} \qquad (17.8)$$

to obtain an expression for F_x. How does F_x depend on ω? (Note that this constitutive equation says nothing about normal stresses.)

Solution. a. This *is* a viscometric flow, but it is unlike any we have seen so far. It is impossible to assign a unique coordinate direction to the velocity vector using a conventional, laboratory-fixed coordinate system. With conventional cylindrical coordinates, the velocity vector at the surface of the wire $(r = R_i)$ is in the axial (x) direction. At the die surface $(r = R_o)$ it is in the tangential (θ) direction. It varies from one extreme to the other between the bounding surfaces. Nevertheless, a velocity vector always lies in a surface at constant r and the gradient vector is always in the r direction, so flow is viscometric.

b. The general rate-of-strain tensor is analogous to (17.1):

$$\dot{\gamma} = \begin{vmatrix} \dot{\varepsilon}_{xx} & \dot{\gamma}_{x\theta} & \dot{\gamma}_{xr} \\ \dot{\gamma}_{\theta x} & \dot{\varepsilon}_{\theta\theta} & \dot{\gamma}_{\theta r} \\ \dot{\gamma}_{rx} & \dot{\gamma}_{r\theta} & \dot{\varepsilon}_{rr} \end{vmatrix}$$

In this particular case, there are no normal (tensile or compressive) strains (ε_{ii}), and the only shear-strain components are those with the gradient in the r direction, $\dot{\gamma}_{rx}$ and $\dot{\gamma}_{r\theta}$. The tensor thus reduces to

$$\dot{\gamma} = \begin{vmatrix} 0 & 0 & 0 \\ 0 & 0 & 0 \\ \dot{\gamma}_{rx} & \dot{\gamma}_{r\theta} & 0 \end{vmatrix}$$

c. In the flat-plate approximation, we have two parallel plates of length L (in the x direction) and width $2\pi R$ (in the θ direction) separated by a distance δ. One moves with a velocity V in the x direction and the other moves with a velocity $R\omega$ in the θ direction. The shear-rate components are therefore

$$\dot{\gamma}_{rx} = \frac{V}{\delta} \quad \text{and} \quad \dot{\gamma}_{r\theta} = \frac{R\omega}{\delta}$$

so

$$|\dot{\gamma}| = [(\dot{\gamma}_{rx})^2 + (\dot{\gamma}_{r\theta})^2]^{1/2} = \left[\left(\frac{V}{\delta}\right)^2 + \left(\frac{R\omega}{\delta}\right)^2\right]^{1/2}$$

Note that in this case, both the $\dot{\gamma}_{ij}$ and $|\dot{\gamma}|$ are uniform throughout the fluid. Applying the constitutive equation, we see that

$$\tau_{ij} = \underbrace{K[(V/\delta)^2 + (R\omega/\delta)^2]^{(n-1)/2}}_{\eta(|\dot{\gamma}|)} \dot{\gamma}_{ij}$$

and

$$F_x = 2\pi RL\tau_{rx} = 2\pi RLK\left[\left(\frac{V}{\delta}\right)^2 + \left(\frac{R\omega}{\delta}\right)^2\right]^{(n-1)/2}\left(\frac{V}{\delta}\right)$$

To examine the effect of rotation on F_x, we can look at the ratio

$$\frac{F_x}{F_x|_{\omega=0}} = \left[1 + \left(\frac{R\omega}{V}\right)^2\right]^{(n-1)/2}$$

where $R\omega/V$ is the ratio of rotational to axial velocities. (Note the similarity of this to the Carreau equation (15.8) with $\eta_\infty = 0$.) With a pseudoplastic fluid, $n < 1$, we see that the ratio drops with increasing rotational speed. Thus, the rotational component of shear rate contributes to lowering the viscosity, which in turn reduces the force necessary to pull the wire in the axial direction.

It must be emphasized that we were able to obtain a simple analytical solution here only because the viscosity is uniform throughout the fluid. Suppose that instead of a moving wire, we had a fixed core and an axial pressure gradient (as in an extruder die for making pipe). Now, $\dot{\gamma}_{rx}$ and therefore $\eta(|\dot{\gamma}|)$

vary with r. This influences the velocity profiles, which in turn determine the shear rates. The solution is a nasty numerical problem, even with this simple constitutive equation.

Example 2. Consider the classical example of a nonviscometric flow, *simple elongation*. This situation arises, for example, when a weight is suspended from the end of a vertical rod of material. Note that here the velocity gradient is not perpendicular to the direction of fluid motion. Discuss the nature of the stress and rate of strain tensors.

Solution. For this unconfined flow, we can unambiguously let $P = 1$ atm. We neglect such things as surface-tension forces on the sides of the rod and find that the deviatoric stress tensor has only one component, σ_{11}, where the subscript 1 represents the axial direction. This is the familiar tensile stress. It will obviously produce a tensile elongation, so there will be a rate of tensile elongation component of the rate-of-strain tensor $\dot{\varepsilon}_{11}$. For an incompressible material (Poisson's ratio = 0.5), the lateral dimensions of the rod will contract to maintain the volume constant as the rod is extended, and the lateral *contractile* strains will be one-half the axial extension strain. Thus, $\dot{\varepsilon}_{11} = -2\dot{\varepsilon}_{22} = -2\dot{\varepsilon}_{33}$. There is no shearing, so all the shear strains are zero. Therefore, the deviatoric stress and rate-of-strain tensors are

$$\tau' = \begin{vmatrix} \sigma_{11} & 0 & 0 \\ 0 & 0 & 0 \\ 0 & 0 & 0 \end{vmatrix} \qquad \dot{\gamma} = \begin{vmatrix} \dot{\varepsilon}_{11} & 0 & 0 \\ 0 & -\tfrac{1}{2}\dot{\varepsilon}_{11} & 0 \\ 0 & 0 & -\tfrac{1}{2}\dot{\varepsilon}_{11} \end{vmatrix}$$

To describe this nonviscometric deformation, a new material function η_e, the elongational or Trouton viscosity, is needed to relate tensile stress to the rate of tensile strain, $\sigma_{11} = \eta_e(\dot{\varepsilon}_{11})\dot{\varepsilon}_{11}$, or equivalently the function $\sigma_{11} = \sigma_{11}(\dot{\varepsilon}_{11})$. For incompressible Newtonian fluids, it may be shown that $\eta_e = 3\eta$.

Fiber drawing (stretching) is an important practical example of uniaxial extension. Converging and diverging flows (in which the fluid passes from a large channel into a smaller one and vice versa) are examples of nonviscometric flows that combine shearing and elongational deformations. Such flows are frequently encountered in polymer processing operations. Nevertheless, even for Newtonian fluids, analytical solutions have not been obtained. However, considerable progress has been made using numerical solutions with appropriate constitutive equations.[9]

17.4 INTERPRETATION OF THE NORMAL STRESSES

The presence of deviatoric normal stress components in steady viscometric flow implies fluid elasticity, and they are physically real quantities. The well-known *Weissenberg effect* is a good example. When a vertical rotating rod is immersed in a container of inelastic fluid, the fluid is flung outward by centrifugal force, and its surface assumes a profile with the lowest point at the rod. A viscoelastic fluid, on the other hand, will actually climb up the rod. This is a viscometric Couette flow, as described in the previous chapter. The behavior of the viscoelastic fluid can be rationalized by imagining it to consist of rubber bands, stretched by the rotation along a fluid streamline in the tangential (θ or 1) direction, at constant radius. The rubber bands are obviously in tension, hence $\sigma_{11} = \sigma_{\theta\theta}$ is positive (a "hoop" stress, as in the bands of a barrel). This positive hoop stress places the cylinder of material within under a compressive stress in the radial direction (think of what wrapping a rubber band around your finger does), a negative $\sigma_{22} = \sigma_{rr}$. Since the inwardly compressed material is prevented from moving inward by the presence of the rod, it has nowhere to go but up the rod under the influence of a compressive axial stress, a negative $\sigma_{33} = \sigma_{zz}$. This does not mean to imply that centrifugal force is no longer acting; merely that it is overwhelmed by the elastic normal stresses.

Similar reasoning leads to the conclusion that in cone-and-plate or disk-plate viscometers (Sections 16.7 and 16.8) there will be forces tending to push the cone and plate or the disk and plate apart. Indeed, there are, and the measurement of these forces provides a means of determining the functions $N_1(\dot{\gamma})$ and $N_2(\dot{\gamma})$.[1, 5, 10, 11] Also, for the case of torsional flow, Maxwell and Scalora[12] showed that if a hole is drilled through the plate along the axis of rotation, a screwless extruder is obtained, as the axial normal stress pushes a polymer melt through the hole. Unlike a screw extruder, however, this device works only with viscoelastic fluids.

REFERENCES

1. Coleman, B. D., H. Markovitz, and W. Noll, *Viscometric Flows of Non-Newtonian Fluids*, Springer, New York, 1966.

2. Middleman, S., *The Flow of High Polymers*, Wiley, New York, 1968.

3. Christensen, R. M., *Theory of Viscoelasticity: An Introduction*, Academic, New York, 1971.

4. Darby, R., *Viscoelastic Fluids: An Introduction to their Properties and Behavior*, Dekker, New York, 1976.

5. Bird, R. B., R. C. Armstrong, and O. A. Hassager, *Dynamics of Polymeric Liquids*, Vol. 1, *Fluid Mechanics*, Wiley, New York, 1977.

6. Larson, R. G., *Constitutive Equations for Polymer Melts and Solutions*, Butterworths, Boston, 1988.

7. Tanner, R. I., *Engineering Rheology*, Oxford UP, New York, 1985.

8. Bird, R. B., O. A. Hassager, R. C. Armstrong, and C. F. Curtiss, *Dynamics of Polymeric Liquids*, Vol. 2, *Kinetic Theory*, Wiley, New York, 1977.

9. Pearson, J. R. A., and S. M. Richardson, *Computational Analysis of Polymer Processing*, Elsevier, New York, 1983.

10. Collyer, A. A., and D. W. Clegg, *Rheological Measurement*, Elsevier, New York, 1988.

11. Walters, K., *Rheometry*, Halsted, New York, 1975.

12. Maxwell, B., and A. J. Scalora, *Modern Plastics* **37**, 107 (1959).

PROBLEMS

1. A vertical, cylindrical rod of a (very) viscous fluid with a constant extensional viscosity η_e has an initial length l_0. An axial tensile stress σ_{11} is applied to one end of the rod, while the other end is held fixed. Obtain an expression for the extension ratio $\alpha = l/l_0$ as a function of time. Neglect the effects of gravity on the rod material. *Hint:* The local tensile strain in the material is $d\varepsilon_{11} = dl/l$. Consider two cases:

   a. The applied stress σ_{11} is constant.

   b. The axial *force* (e.g., a weight) applied to the end of the rod is constant, but because the material is incompressible, the cross-sectional area decreases and therefore the stress σ_{11} increases as the length increases. The initial stress is σ_{110}. According to this analysis, how long will the rod support the weight?

2. Repeat Problem 1 for a material that follows a power-law type of constitutive equation in tension: $\sigma_{11} = K' (\dot{\varepsilon}_{11})^m$.

3. Show that $\dot{\varepsilon}_{11} = -2\dot{\varepsilon}_{22} = -2\dot{\varepsilon}_{33}$ for an incompressible material.

4. It is tempting to say that the velocity vectors in uniaxial extension are all in the axial direction. Consider a homogeneous, incompressible cylindrical rod of initial radius R_0 and length l_0 that is suspended vertically with one end fixed at $x = 0$. The other end of the rod is extended in the axial (x) direction until $l = 2l_0$. Using conventional cylindrical coordinates, determine the final position of a point that was located at $r = R_0$, $x = l_0/2$ prior to extension.

5. *Biaxial extension* is another example of a nonviscometric deformation. It can be achieved, for example, by simultaneously stretching a sheet of material in its length and width directions or by blowing up a balloon. It plays a prominent role in the blow-molding process (Section 19.4). Show the components of the deviatoric stress and the rate-of-strain tensors for the biaxial extension of an incompressible material. What material function(s) would be needed to describe the equilibrium biaxial extension of a homogeneous fluid?

6. Fluid is confined to the space between two concentric hemispheres, one of which rotates relative to the other about an axis through their common center. Describe the velocity and gradient directions. Is this a viscometric flow?

7. Speculate about the signs of $\sigma_{\theta\theta}$, $\sigma_{\phi\phi}$, and σ_{rr} in cone-and-plate flow (Fig. 16.7) and $\sigma_{\theta\theta}$, σ_{xx}, and σ_{rr} in torsional flow (Fig. 16.8).

8. Consider a Couette viscometer (Fig. 16.6) in which pressure taps are drilled radially in the cup and bob at the same height. These taps are connected to vertical manometer tubes, and the difference in fluid height in these tubes

$h_{bob} - h_{cup}$ is measured as a function of rotation rate ω. Sketch qualitatively $h_{bob} - h_{cup}$ vs. ω for two fluids of the same density, one Newtonian and the other highly elastic.

9. It has been pointed out that the use of manometers or recessed transducers to measure pressures in the steady, laminar flow of viscoelastic fluids can give erroneous results ("hole-pressure errors") because the tap hole distorts the streamlines in its immediate vicinity. The streamlines sort of "dip into" the hole as they go by. With a polymer fluid, do you think that the gauge pressure would be greater than, less than, or the same as that measured with a flush transducer (with no distortion of the streamlines)? Why? (A sketch is worth at least a dozen words here.) This situation is analyzed in Refs. 7 (p. 150), 10 (Chapter 11), and 11 (p. 75), but don't cheat and look there first.

10. Consider a piping system with a 90° elbow in the horizontal plane. The elbow is fitted with two flush pressure transducers, one at the inside and the other at the outside on the 45° line. Sketch qualitatively the measured pressure difference $P_{in} - P_{out}$ vs. Q for the laminar flow of two fluids with the same density, one Newtonian and the other highly elastic.

11. Consider two flat, parallel disks of radius R, separated by a fixed distance δ ($\delta \ll R$). Fluid is injected to the space between the disks through a small hole at the center of one of the disks. Flow is steady and laminar. Neglect the area immediately surrounding the injection point.

    Is this a viscometric flow? If so, define a coordinate system with a sketch and identify the 1, 2, and 3 coordinate directions. If not, explain.

CHAPTER XVIII

Linear Viscoelasticity

18.1 INTRODUCTION

Engineers have traditionally dealt with two separate and distinct classes of materials: the viscous fluid and the elastic solid. Design procedures based on these concepts have worked pretty well because most traditional materials (water, air, steel, concrete), at least to a good approximation, fit into one of these categories. The realization has grown, however, that these categories represent only the extremes of a broad spectrum of material response. Polymer systems fall somewhere in between, giving rise to some of the unusual properties of melts and solutions described previously. Other examples are important in the structural applications of polymers. In a common engineering stress–strain test, a sample is strained at an approximately constant rate, and the stress is measured as a function of strain. With traditional solids, the stress–strain curve is pretty much independent of the rate at which the material is strained. The stress–strain properties of many polymers are markedly *rate dependent*, however. Similarly, polymers often exhibit pronounced creep and stress relaxation (to be defined shortly). While such behavior is exhibited by other materials (metals near their melting points, for example), at normal temperatures it is negligible, and is not usually included in design calculations. If the time-dependent behavior of polymers is ignored, the results can sometimes be disastrous.

18.2 MECHANICAL MODELS FOR LINEAR VISCOELASTIC RESPONSE

As an aid in visualizing viscoelastic response, we introduce two linear mechanical models to represent the extremes of the mechanical response spectrum. The spring in Fig. 18.1a represents a *linear elastic* or Hookean *solid* whose constitutive equation (relation of stress to strain and time) is simply $\tau = G\gamma$, where G is

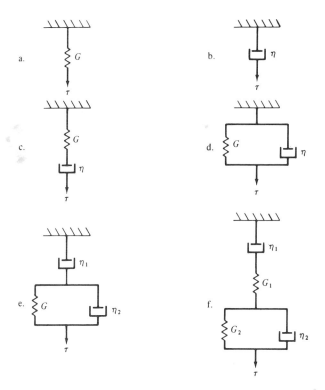

Figure 18.1 Linear viscoelastic models: (*a*) linear elastic; (*b*) linear viscous; (*c*) Maxwell element; (*d*) Voigt–Kelvin element; (*e*) three-parameter; (*f*) four-parameter.

a (constant) shear modulus. Similarly, a *linear viscous* or Newtonian *fluid* is represented by a dashpot (some sort of piston moving in a cylinder of Newtonian fluid) whose constitutive equation is $\tau = \eta\dot{\gamma}$, where η is a (constant) viscosity. The strain is represented by the extension (stretching) of the model.

Although the models developed here are to be visualized in tension, the notation used is for pure shear (viscometric) deformation. The equations are equally applicable to tensile deformation by replacing the shear stress τ with the tensile stress σ, the shear strain γ with the tensile strain ε, Hooke's modulus G with Young's (tensile) modulus E, and the Newtonian (shear) viscosity η with the elongational (Trouton) viscosity η_e.

Some authorities object strongly to the use of mechanical models to represent materials. They point out that real materials are not made of springs and dashpots. True, but they're not made of equations either, and it's a lot easier for most people to visualize the deformation of springs and dashpots than the solutions to equations.

A word is needed about the meaning of the term *linear*. For the present, a linear response will be defined as one in which the *ratio* of overall stress to overall strain, the overall modulus $G(t)$, is a function of *time only*, not of the magnitudes of stress or strain:

$$G(t) \equiv \frac{\tau}{\gamma} = \text{function of } t \text{ only for linear response} \qquad (18.1)$$

The Hookean spring responds instantaneously to reach an equilibrium strain γ upon application of a constant stress τ_0, and the strain remains constant as long as the stress is maintained constant. Sudden removal of the stress results in instantaneous recovery of the strain (Fig. 18.2). Doubling the stress on the spring simply doubles the resulting strain, so the spring is linear, with $G(t) = G$, a constant, according to (18.1). (In assuming that the spring instantaneously reaches an equilibrium strain under the action of a suddenly applied constant stress, we have neglected inertial effects. Although it is not necessary to do so, including them would contribute little to the present discussion.)

If a constant stress τ_0 is suddenly applied to the dashpot, the strain increases with time according to $\gamma = (\tau_0/\eta)t$ (considering the strain to be zero when the stress is applied) (Fig. 18.3). Doubling the stress doubles the slope of the strain–time line, and at any time, the modulus $G(t) \equiv \tau/\gamma = \eta/t = G(t$ only). So the dashpot is also linear.

It may be shown that *any combination of linear elements must be linear*, so any models based on these linear elements, no matter how complex, can represent only linear response. Just how realistic is linear response? Its most conspicuous shortcoming is that it permits only Newtonian behavior (constant viscosity) in equilibrium viscous flow. For most polymers at strains greater than a few percent or so (or rates of strain greater than $0.1\,\text{s}^{-1}$), linear response is not a good quantitative description. Moreover, even within the limit of linear viscoelasticity, a fairly large number of linear elements (springs and dashpots) are usually

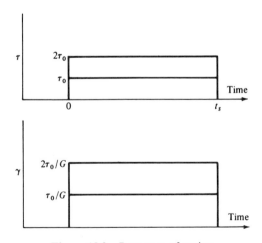

Figure 18.2   Response of spring.

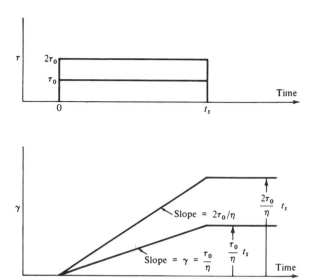

Figure 18.3   Response of dashpot.

needed to provide an accurate quantitative description of response. Hence, the quantitative applicability of simple linear models (those with a few springs and dashpots) is limited, but the models are extremely valuable in visualizing viscoelastic response, and in understanding how and why variations in molecular structure influence that response.

A   The Maxwell Element

James Clerk Maxwell realized that neither a linear viscous element (dashpot) nor a linear elastic element (spring) was sufficient to describe his experiments on the deformation of asphalts, so he proposed a simple series combination of the two, the *Maxwell element* (Fig. 18.1c). In a Maxwell element, the spring and dashpot support the same stress, so

$$\tau = \tau_{\text{spring}} = \tau_{\text{dashpot}} \tag{18.2}$$

Furthermore, the total strain (extension) of the element is the sum of the strains in the spring and dashpot:

$$\gamma = \gamma_{\text{spring}} + \gamma_{\text{dashpot}} \tag{18.3}$$

Differentiating (18.3) with respect to time gives

$$\dot{\gamma} = \dot{\gamma}_{\text{spring}} + \dot{\gamma}_{\text{dashpot}} \tag{18.4}$$

Realizing that $\dot{\gamma}_{\text{dashpot}} = \tau/\eta$ and $\dot{\gamma}_{\text{spring}} = \dot{\tau}/G$, plugging in and rearranging gives the differential equation for the Maxwell element:

$$\tau = \eta\dot{\gamma} - \frac{\eta}{G}\dot{\tau} = \eta\dot{\gamma} - \lambda\dot{\tau} \tag{18.5}$$

The quantity $\lambda = \eta/G$ has the dimension of time and is known as a *relaxation time*. Its physical significance will be apparent shortly.

Creep Testing

Let's examine the response of the Maxwell element in two mechanical tests commonly applied to polymers. First consider a *creep* test, in which a constant stress is instantaneously (or at least very rapidly) applied to the material, and the resulting strain is followed as a function of time. Deformation after removal of the stress is known as *creep recovery*.

As shown in Fig. 18.4, the sudden application of stress to a Maxwell element causes an instantaneous stretching of the spring to an equilibrium value of τ_0/G, where τ_0 is the constant applied stress. The dashpot extents linearly with time with a slope of τ_0/η, and will continue to do so as long as the stress is maintained. Thus, *the Maxwell element is a fluid*, because it will continue to deform as long as it is stressed. The creep response of a Maxwell element is therefore

$$\gamma(t) = \frac{\tau_0}{G} + \frac{\tau_0}{\eta}t \tag{18.6a}$$

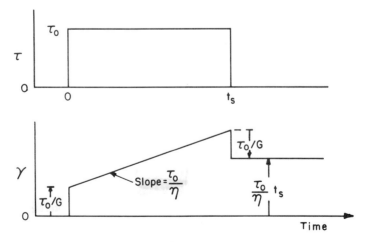

Figure 18.4 Creep response of Maxwell element.

or, in terms of a *creep compliance*, $J_c(t) \equiv \gamma(t)/\tau_0$,

$$J_c(t) \equiv \frac{\gamma(t)}{\tau_0} = \frac{1}{G} + \frac{t}{\eta} \qquad (18.6b)$$

The creep compliance $J_c(t)$, being independent of the applied stress τ_0 (for a linear material), is a more general way to represent the creep response. When the stress is removed at time t_s, the spring immediately contracts by an amount equal to its original extension, a process known as *elastic recovery*. The dashpot, of course, does not recover, leaving a *permanent set* of $(\tau_0/\eta)t_s$, the amount the dashpot has extended during the application of stress. Although real materials never show sharp breaks in a creep test as does the Maxwell element, the Maxwell element does exhibit the phenomena of elastic strain, creep, recovery, and permanent set, which are often observed with real materials.

Stress Relaxation

Another important test used to study viscoelastic response is *stress relaxation*. A stress-relaxation test consists of suddenly applying a strain to the sample, and following the stress as a function of time as the strain is held constant. When the Maxwell element is strained instantaneously, only the spring can respond initially (for an infinite rate of strain, the resisting force in the dashpot is infinite) to a stress of $G\gamma_0$, where γ_0 is the constant applied strain. The extended spring then begins to contract, but the contraction is resisted by the dashpot. The more the spring retracts, the smaller is its restoring force, and the rate of retraction drops correspondingly. Solution of the differential equation with $\dot{\gamma} = 0$ and the initial condition $\tau = G\gamma_0$ at $t = 0$ shows that the stress undergoes a first-order exponential decay:

$$\tau(t) = G\gamma_0 \, e^{-t/\lambda} \qquad (18.7a)$$

or, in terms of a *relaxation modulus*, $G_r(t) \equiv \tau(t)/\gamma_0$,

$$G_r(t) \equiv \frac{\tau(t)}{\gamma_0} = Ge^{-t/\lambda} \qquad (18.7b)$$

Again, the relaxation modulus $G_r(t)$ is a more general means of representing stress-relaxation response because it is independent of the applied strain for linear materials.

From (18.7), we see that the relaxation time λ is the time constant for the exponential decay, that is, the time required for the stress to decay to a factor of $1/e$ or 37% of its initial value. The stress asymptotically drops to zero as the spring approaches complete retraction (Fig. 18.5).

Stress-relaxation data for linear polymers actually look like the curve for the Maxwell element. Unfortunately, they can't often be fitted quantitatively with

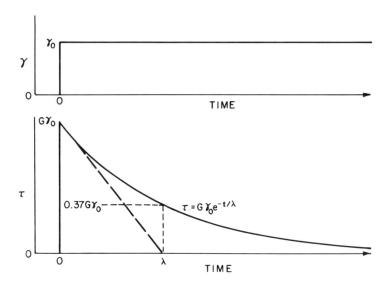

Figure 18.5   Stress relaxation of Maxwell element.

a single value of G and a single value of λ; that is, the decay is not really first order.

Example 1. Examine the response of a Maxwell element in an engineering stress–strain test, a test in which the rate of tensile strain is maintained (approximately) constant at $\dot{\varepsilon}_0$.

Solution. Rewriting (18.5) in tensile notation gives

$$\sigma = \eta_e \dot{\varepsilon}_0 - \frac{\eta_e}{E}\dot{\sigma} \quad \text{or} \quad \sigma + \left(\frac{\eta_e}{E}\right)\left(\frac{d\sigma}{dt}\right) = \eta_e \dot{\varepsilon}_0 = \text{constant}$$

The solution to this differential equation with the initial condition $\sigma = 0$ at $t = 0$ is

$$\sigma = \eta_e \dot{\varepsilon}_0 \left[1 - e^{-(E/\eta_e)t}\right]$$

Since $d\varepsilon/dt = \dot{\varepsilon}_0$ is constant and we can assume that $\varepsilon = 0$ when $t = 0$

$$\varepsilon = \dot{\varepsilon}_0 t$$

The stress–strain curve is then

$$\sigma = \eta_e \dot{\varepsilon}_0 \left[1 - e^{-(E/\eta_e \dot{\varepsilon}_0)\varepsilon}\right]$$

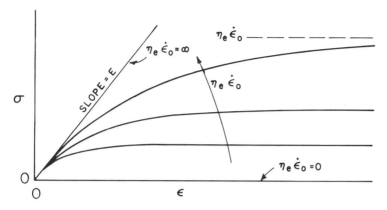

Figure 18.6 Response of a Maxwell element under constant rate of tensile strain (Example 1).

This response is sketched in Fig. 18.6. Note that at any given strain, σ increases with the rate of strain $\dot{\varepsilon}_0$; that is, the material appears "stiffer" (has a higher modulus). As crude as this model might prove to be in fitting experimental data, it does account, at least qualitatively, for some of the observed properties of linear polymers in engineering stress–strain tests.

B   The Voigt–Kelvin Element

If a series combination of a spring and dashpot has its drawbacks, the next logical thing to try is a parallel combination, a Voigt or Voigt–Kelvin element (Fig. 18.1d). Here, it is assumed that the crossbars supporting the spring and dashpot always remain parallel, so that the strain is the same in the spring and dashpot:

$$\gamma = \gamma_{\text{spring}} = \gamma_{\text{dashpot}} \tag{18.8}$$

The stress supported by the element is the sum of the stresses in the spring and the dashpot:

$$\tau = \tau_{\text{spring}} + \tau_{\text{dashpot}} \tag{18.9}$$

Combination of (18.8) and (18.9) with the equations for the deformation of the spring and dashpot gives the differential equation for the Voigt–Kelvin element:

$$\tau = \eta\dot{\gamma} + G\gamma \tag{18.10}$$

When the stress is suddenly applied in a creep test, only the dashpot offers an initial resistance to deformation, so the initial slope of the strain vs. time curve is

τ_0/η. As the element is extended, the spring provides an increasingly greater resistance to further extension, so the rate of creep decreases. Eventually, the system comes to equilibrium with the spring alone supporting the stress (with the rate of strain zero, the resistance of the dashpot is zero). The equilibrium strain is simply τ_0/G. Quantitatively, the response is an exponential rise,

$$\gamma(t) = \frac{\tau_0}{G}(1 - e^{-t/\lambda}) \tag{18.11a}$$

or, in terms of the creep compliance

$$J_c(t) = \frac{1}{G}(1 - e^{-t/\lambda}) \tag{18.11b}$$

If the stress is removed *after equilibrium has been reached*, the strain decays exponentially,

$$\gamma(t) = \frac{\tau_0}{G}e^{-t/\lambda} \tag{18.12}$$

Note that the Voigt–Kelvin element does not continue to deform as long as stress is applied, and it does not exhibit any permanent set (Fig. 18.7). It

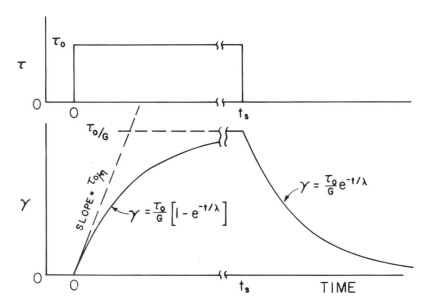

Figure 18.7   Creep response of a Voigt–Kelvin element.

therefore represents a viscoelastic *solid*, and gives a fair qualitative picture of the creep response of some cross-linked polymers.

The Voigt–Kelvin element is not suited for representing stress relaxation. The instantaneous application of strain would be met by an infinite resistance in the dashpot, and so would require the application of an infinite stress, which is obviously unrealistic.

C   The Three-Parameter Model

The next step in the development of linear viscoelastic models is the so-called three-parameter model (Fig. 18.1e). By adding a dashpot in series with the Voigt–Kelvin element, we get a liquid. The differential equation for this model may be written in operator form as

$$\left(1 + \lambda_1 \frac{d}{dt}\right)\tau = \eta_1\left(1 + \lambda_2 \frac{d}{dt}\right)\dot{\gamma} \tag{18.13}$$

where $\lambda_1 = (\eta_1 + \eta_2)/G$ and $\lambda_2 = \eta_2/G$. Further, the form of (18.13) suggests modification by adding higher order derivatives and more constants:

$$\left(1 + \lambda_1 \frac{d}{dt} + \xi_1 \frac{d^2}{dt^2} + \cdots\right)\tau = \eta_1\left(1 + \lambda_2 \frac{d}{dt} + \xi_2 \frac{d^2}{dt^2} + \cdots\right)\dot{\gamma} \tag{18.14}$$

This, of course, will fit data to any desired degree of accuracy if enough terms are used. With τ and $\dot{\gamma}$ replaced with the corresponding tensors and applied in convected material coordinates, (18.14) has had limited application as a constitutive equation.

18.3   THE FOUR-PARAMETER MODEL AND MOLECULAR RESPONSE

The four-parameter model (Fig. 18.1f) is a series combination of a Maxwell element with a Voigt–Kelvin element. Its differential equation is

$$\ddot{\tau} + \left(\frac{G_1}{\eta_2} + \frac{G_1}{\eta_1} + \frac{G_2}{\eta_2}\right)\dot{\tau} + \frac{G_1 G_2}{\eta_1 \eta_2}\tau = G_1\ddot{\gamma} + \frac{G_1 G_2}{\eta_2}\dot{\gamma} \tag{18.15}$$

Its creep response is the sum of the creep responses of the Maxwell and Voigt–Kelvin elements,

$$\gamma(t) = \frac{\tau_0}{G_1} + \frac{\tau_0}{\eta_1}t + \frac{\tau_0}{G_2}[1 - e^{-(G_2/\eta_2)t}] \tag{18.16a}$$

or, in terms of creep compliance

$$J_c(t) = \frac{1}{G_1} + \frac{1}{\eta_1}t + \frac{1}{G_2}[1 - e^{-(G_2/\eta_2)t}] \tag{18.16b}$$

This is summarized in Fig. 18.8.

The four-parameter model provides at least a qualitative representation of all the phenomena generally observed in the creep of viscoelastic materials: instantaneous elastic strain, retarted elastic strain, equilibrium viscous flow, instantaneous elastic recovery, retarded elastic recovery, and permanent set. It also describes at least qualitiatively the behavior of viscoelastic materials in other types of deformation. Of equal importance is the fact that the model parameters can be identified with the various molecular response mechanisms in polymers, and can therefore be used to predict the influences that changes in molecular

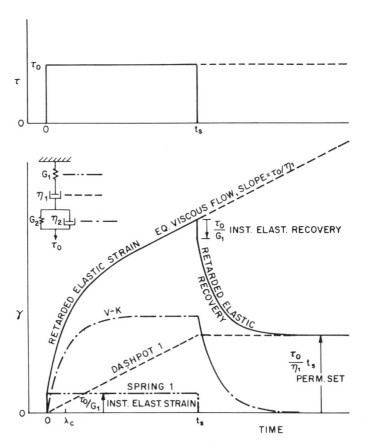

Figure 18.8 Creep response of a four-parameter model.

structure will have on mechanical response. The following analogies may be drawn:

1. *Dashpot 1* represents molecular slip—the translational motion of molecules. This slip of polymer molecules past one another is responsible for flow. The value of η_1 alone (molecular friction in slip) governs the equilibrium flow of the material.

2. *Spring 1* represents the elastic straining of bond angles and lengths. All bonds in polymer chains have equilibrium angles and lengths. The value of G_1 characterizes the resistance to deformation from these equilibrium values. Since these deformations involve interatomic bonding, they occur essentially instantaneously from a macroscopic point of view. This type of elasticity is known thermodynamically as *energy elasticity* (Chapter XIV).

3. *Dashpot 2* represents the resistance of the polymer chains to uncoiling and coiling, caused by temporary mechanical entanglements of the chains and molecular friction during these processes. Since coiling and uncoiling require cooperative motion of many chain segments, they cannot occur instantaneously, and hence account for retarded elasticity.

4. *Spring 2* represents the restoring force brought about by the thermal agitation of the chain segments, which tends to return chains oriented by a stress to their most random, or highest entropy, configuration. This is known, therefore, as *entropy elasticity* (Chapter XIV).

The magnitude of the time scale shown in Fig. 18.8 will of course depend on the values of the model parameters. The two viscosities, in particular, depend strongly on temperature. Well below T_g, for example, where η_1 and η_2 are very large, t_s might be on the order of days or weeks to observe appreciable retarded elasticity and flow. Well above T_g, t_s might be only seconds or less to permit the deformation shown. An important thing to keep in mind is that designs based on short-term property measurements will be inadequate if the object must support a stress for longer periods of time.

Example 2. Using the four-parameter model as a basis, sketch qualitatively the effects of (*a*) increasing molecular weight and (*b*) increasing degrees of crosslinking on the creep response of a linear, amorphous polymer.

Solution. *a.* As discussed in Chapter XV, the equilibrium zero-shear (linear) viscosity of polymers, represented by η_1 in the model, increases with the 3.4 power of $\bar{M}_w$. Thus, the slope in the equilibrium flow region τ_0/η_1 is greatly decreased as the molecular weight increases, and the permanent set $(\tau_0/\eta_1)t_s$ is reduced correspondingly (Fig. 18.9).

b. Light crosslinking represents the limit of case (*a*) above, when the molecular weight reaches infinity, since all the chains are hooked together by crosslinks. Under these conditions, they *can't* slip past one another, so η_1 becomes infinite. If the crosslinking is light (crosslinks few and far between), as in a rubber band,

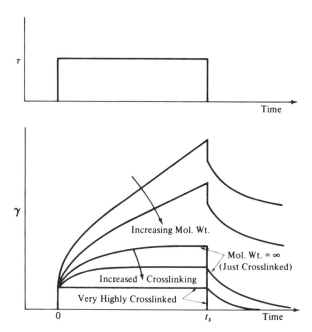

Figure 18.9 The effects of molecular weight and crosslinking on the creep response of an amorphous polymer.

coiling and uncoiling won't be hindered appreciably. Note that crosslinking converts the material from a fluid to a solid (it eventually reaches an equilibrium strain under the application of a constant stress) and it eliminates permanent set. The equilibrium modulus will be on the order of 10^6 to 10^7 dyn/cm^2 (10^5 to 10^6 N/m^2 = Pa), the characteristic "rubbery" modulus. Further crosslinking begins to hinder the ability of the chains to uncoil and raises the restoring force (increases η_2 and G_2). At high degrees of crosslinking, as in hard rubber (ebonite), the only response mechanism left is straining bond angles and lengths, giving rise to an almost perfectly elastic material with a modulus on the order of 10^{10} to 10^{11} dyn/cm^2 (10^9 to 10^{10} N/m^2), the characteristic "glassy" modulus.

The four-parameter model nicely accounts for the interesting examples of viscoelastic response mentioned earlier. For example, dashpot 1 allows viscous flow, while the elastic restoring forces of springs 1 and 2 provide the "rubber band" elasticity responsible for the Weissenberg effect. In engineering stress–strain tests, the moduli of polymers are observed to increase with the applied rate of strain. At high rates of strain, spring 1 provides the major response mechanism. As the rate of strain is lowered, dashpot 1 and the Voigt–Kelvin element contribute more and more to the overall deformation, giving a greater strain at any stress, that is, a lower modulus. When Silly Putty is bounced (stress applied rapidly for a short period of time), spring 1 again provides the major response mechanism. There isn't time for appreciable flow of the dashpots 1 and 2, so not much of the initial potential energy is converted to heat through the molecular friction involved in slippage and uncoiling, and the material behaves in an almost perfectly elastic fashion. When it's stuck on the wall, the stress (in this case due to its own weight) is applied for a long period of time, and it flows downward as a result of the molecular slip represented by dashpot 1.

18.4   VISCOUS OR ELASTIC RESPONSE? THE DEBORAH NUMBER[1]

Thus, whether a viscoelastic fluid behaves as an elastic solid or a viscous liquid depends on the relation between the time scale of the deformation to which it is subjected and the time required for the material's time-dependent mechanisms to respond. Strictly speaking, the concept of a single relaxation time applies only to first-order response, and thus is not applicable to real materials, in general. Nevertheless, a characteristic time λ_c for any material can always be defined as, for example, the time required for the material to reach $1 - 1/e$ or 63.2% of its ultimate retarded elastic response to a step change. A precise value is rarely necessary. The characteristic time is simply a means of characterizing the rate of a material's time-dependent elastic response, short λ_c's indicating rapid response and large λ_c's indicating sluggish response. The ratio of this characteristic material time to the time scale of the deformation is the *Deborah number*:

$$\text{De} \equiv \frac{\lambda_c}{t_s} \tag{18.17}$$

Response will appear elastic at high Deborah numbers ($\text{De} \gg 1$) and viscous at low Deborah numbers ($\text{De} \to 0$)*.

Consider, for example, the creep response of the four-parameter model (Fig. 18.8). For this model, a logical choice for λ_c would be the time constant for its Voigt–Kelvin component, η_2/G_2. For $\text{De} \gg 1$ ($t_s \ll \lambda_c$), the Voigt–Kelvin element and dashpot 1 will be essentially immobile, and the response will be due almost entirely to spring 1, that is, almost purely elastic. For $\text{De} \to 0$ ($t_s \gg \lambda_c$), the instantaneous and retarded elastic response mechanisms have long since reached equilibrium, so the only remaining response will be the purely viscous flow of dashpot 1, and the deformation due to viscous flow will completely overshadow that due to the elastic response mechanisms (imagine the creep

---

* "The mountains flowed before the Lord" [Song of Deborah, Judges 5:5]. For the Lord, t_s is exceedingly large—see Reiner.[1] (In some English versions of the Old Testament, one finds "melted" in place of "flowed." That makes no sense in the present context. I have been assured by people familiar with the original language that "flowed" is a more accurate translation, however.)

curve of Fig. 18.8 extended a meter or so beyond the page). *Under conditions where De → 0, materials can be treated by the techniques outlined in Chapter XVI for purely viscous fluids.*

Modifications of the devices described in Chapter XVI can also be used to obtain information on the material's elastic response. For example, if the stress is suddenly removed from a rotational viscometer, the creep recovery or elastic recoil of the material can be followed. This provides a value of λ_c for the material.

Example 3. Thermocouples and Pitot tubes inserted in a flowing stream of a viscoelastic fluid often give erroneous results. Explain.

Solution. When a viscoelastic fluid in equilibrium flow (De → 0) encounters a probe, it must make a sudden (De ≫ 1) jog to get arount it. The retarded elastic response mechanisms simply cannot respond fast enough in the immediate vicinity of the probe, which for all practical purposes behaves as if it were covered with a solid plug. What is measured, therefore, is not characteristic of the fluid in an unobstructed stream.

Example 4. (This is believed due to Professor A. B. Metzner.) A paper cup containing water is placed on a stump. A .22-cal bullet fired at the cup passes cleanly through, leaving the cup sitting on the stump. The water is replaced by a dilute polymer solution in a second cup. This time, the bullet knocks the cup 25 ft beyond the stump. Explain.

Solution. The characteristic time for a low molecular weight fluid such as water is extremely short, much shorter than the time it takes the bullet to pass through the cup (t_s). This, then, is a low-De situation. The water behaves as a viscous fluid. The bullet transfers a little momentum to it through viscous friction, but not enough to dislodge the cup. Adding a polymeric solute raises the characteristic time many orders of magnitude, to the point where this becomes a high-De experiment (the polymer chains can't respond fast enough to get out of the way of the bullet). The bullet, in effect, slams into a solid, and transfers much of its momentum to the fluid–cup system, carrying it beyond the stump.

18.5   QUANTITATIVE APPROACHES[2-5]

Although the four-parameter model is useful from a conceptual standpoint, it doesn't often provide an accurate fit of experimental data and therefore can't be used to make quantitative predictions of material response. To do so, and to infer some detailed information about molecular response, more general models have been developed. The *generalized Maxwell model* (Fig. 18.10) is used to describe stress-relaxation experiments. The stress relaxation of an individual

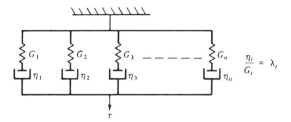

Figure 18.10 Generalized Maxwell model.

Maxwell element is given by

$$\tau_i(t) = \gamma_0 G_i e^{-t/\lambda_i} \tag{18.18}$$

where $\lambda_i = \eta_i/G_i$. The relaxation of the generalized model, in which the individual elements are all subjected to the same constant strain γ_0 is then

$$\tau(t) = \sum_{i=1}^{n} \tau_i(t) = \gamma_0 \sum_{i=1}^{n} G_i e^{-t/\lambda_i} \tag{18.19a}$$

Expressed in terms of the time-dependent relaxation modulus $G_r(t)$, the response is

$$G_r(t) \equiv \frac{\tau(t)}{\gamma_0} = \sum_{i=1}^{n} G_i e^{-t/\lambda_i} \tag{18.19b}$$

Now, if n is large, the summation in (18.19) may be approximated by the integral over a *continuous distribution of relaxation times* $G(\lambda)$:

$$G_r(t) = \int_0^{\infty} G(\lambda) e^{-t/\lambda} d\lambda \tag{18.20}$$

Note that while the G_i have modulus units (e.g., dyn/cm^2), $G(\lambda)$ is in modulus/time units. Note also that if the generalized Maxwell model is to represent a viscoelastic solid such as a crosslinked polymer, at least one of the viscosities has to be infinite.

For creep tests, a generalized Voigt–Kelvin model is used (Fig. 18.11). The creep response of an individual Voigt–Kelvin element is given by

$$\gamma_i(t) = \tau_0 J_i(1 - e^{-t/\lambda_i}) \tag{18.21}$$

where $J_i = 1/G_i$ is the individual spring compliance. The response of the array, in which each element is subjected to the same constant applied stress τ_0 is then

$$\gamma(t) = \tau_0 \sum_{i=1}^{n} J_i(1 - e^{-t/\lambda_i}) \tag{18.22a}$$

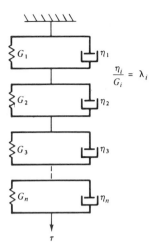

Figure 18.11   Generalized Voigt–Kelvin model.

or, in terms of the overall creep compliance $J_c(t)$,

$$J_c(t) \equiv \frac{\gamma(t)}{\tau_0} = \sum_{i=1}^{n} J_i(1 - e^{-t/\lambda_i}) \tag{18.22b}$$

Again, for large n, the discrete summation above may be approximated by

$$J_c(t) = \int_0^\infty J(\lambda)(1 - e^{-t/\lambda})\,d\lambda \tag{18.23}$$

where $J(\lambda)$ is the *continuous distribution of retardation times* (1/modulus·time).

If the generalized Voigt–Kelvin model is to represent a viscoelastic liquid such as a linear polymer, the modulus of one of the springs must be zero (infinite compliance), leaving a simple dashpot in series with all the other Voigt–Kelvin elements. Sometimes, the steady-flow response of this lone dashpot, $\gamma_{\text{dashpot}} = (\tau_0/\eta_0)t$, is subtracted from the overall response, leaving the compliances to represent only the *elastic* contributions to the overall response:

$$\gamma(t) = \frac{\tau_0}{\eta_0}t + \tau_0 \sum_{i=1}^{n} J_i^\dagger(1 - e^{-t/\lambda_i}) \tag{18.24a}$$

$$J_c(t) = \frac{t}{\eta_0} + \sum_{i=1}^{n} J_i^\dagger(1 - e^{-t/\lambda_i}) \tag{18.24b}$$

$$J_c^\dagger(t) = J_c(t) - \frac{t}{\eta_0} = \sum_{i=1}^{n} J_i^\dagger(1 - e^{-t/\lambda_i}) \tag{18.25}$$

$$J_c^\dagger(t) = J_c(t) - \frac{t}{\eta_0} = \int_0^\infty J^\dagger(\lambda)(1 - e^{-t/\lambda})\,d\lambda \qquad (18.26)$$

Here, η_0 is the equilibrium (Newtonian) viscosity. The daggers indicate that equilibrium viscous flow has been removed and is treated separately.

Application of the discrete equations (18.19) and (18.22) often involves a fairly large n to describe data accurately, thus requiring an impractically large number of parameters λ_i and $G_i = 1/J_i$. It has been suggested, however, that the individual parameters are related by[6]

$$\lambda_i = \frac{\lambda_0}{i^\alpha} \qquad (18.27)$$

$$G_i = \frac{1}{J_i} = \frac{\eta_0}{\sum\limits_{i=1}^n \lambda_i} \qquad (18.28)$$

Equations 18.27 and 18.28 require that the G_i all be the same and that $\eta_0 = \Sigma \eta_i$ (with $\eta_i = \lambda_i G_i$). They reduce the number of necessary parameters to three: η_0, the *steady-state* zero-shear viscosity; λ_0, a maximum relaxation time; and α, an empirical constant. The Rouse theory[7] for dilute polymer solutions predicts $\alpha = 2$, but for concentrated solutions and melts, better fits are obtained with α's between 2 and 4.[6]

Often, enough discrete parameters to provide reasonable response models can be extracted from experimental creep or stress-relaxation data using Tobolsky's "Procedure X."[4] This procedure will be illustrated for stress-relaxation data in the form of the relaxation modulus $G_r(t)$. According to (18.19b),

$$G_r(t) = G_1 e^{-t/\lambda_1} + G_2 e^{-t/\lambda_2} + G_3 e^{-t/\lambda_3} + \cdots \qquad (18.19b)$$

The procedure is based on two assumptions. First, the G_i do not differ radically in magnitude. This often turns out to be the case. As noted in (18.28), theory suggests that the G_i should be identical. Second, there are a few discrete λ_i, with $\lambda_1 > \lambda_2 > \lambda_3 > \cdots$, and they differ enough so that at long times, the second-, third-, and higher-order terms approach zero, leaving only the first term to determine $G_r(t)$. These assumptions are easily tested. If a plot of $\ln G_r(t)$ vs. t becomes linear at large t, the assumptions are valid. If that turns out to be the case, the slope of the linear region at long times is $-1/\lambda_1$ and its $t = 0$ intercept is $\ln G_1$. The known response of the first Maxwell element can then be subtracted from the overall response:

$$G_r(t) - G_1 e^{-t/\lambda_1} = G_2 e^{-t/\lambda_2} + G_3 e^{-t/\lambda_3} + \cdots \qquad (18.29)$$

and $\ln[G_r(t) - G_1 e^{-t/\lambda_1}]$ is plotted vs. t. Again, if a linear region is reached at long times, the slope of that region is $-1/\lambda_2$ and its intercept is $\ln G_2$, and so on.

In principle, this procedure can be repeated indefinitely. In practice, the precision and time scale of typical single-temperature experimental data rarely justify going beyond $i = 3$. Even so, the resulting three-element generalized Maxwell model can often give a good fit to the data used to establish it. More importantly, the model can then be used to predict material response in other types of deformation, at least over similar time scales. The time–temperature superposition principle, which is discussed in Section 18.8, can extend time scales to the point where the parameters may be established to $i = 6$ or 7 or so.

An analogous procedure based on (18.25) can be used to extract the λ_i and $J_i = 1/G_i$ in a generalized Voigt–Kelvin model from creep data, $J_c^\dagger(t)$. It requires a knowledge of the *equilibrium* elastic compliance, $J_c^\dagger(\infty)$. Its development will be left as an end-of-chapter exercise.

Procedures are also available for extracting the continuous distributions $G(\lambda)$ or $J(\lambda)$ from experimental data.[2,4] Once known, they can be used in (18.20) or (18.23) for predicting creep or stress relaxation response in the linear region. So what, if you first had to measure the response to determine these functions? Well, for one thing, $G(\lambda)$ can in principle be obtained from $J(\lambda)$ and vice versa, if one distribution is known over the range $0 < \lambda < \infty$. Although the functions are never obtainable over the complete range, the time–temperature superposition principle, Section 18.8, can often extend the range to the point where the necessary approximations are adequate. Thus, creep response can be predicted from stress-relaxation measurements, and vice versa. This interconvertability also applies to a variety of linear mechanical responses in addition to the two types discussed here, as is illustrated in the next section. The interconversion procedures have been discussed in detail by Schwarzl.[8,9] Furthermore, the shape of the distributions provides the polymer scientist with information on molecular response mechanisms within the polymer. For example, peaks in a certain region of λ might imply motion of side chains on the molecules. This type of information can lead to the "design" of polymers with the type of side chains needed to provide particular mechanical properties.

18.6   THE BOLTZMANN SUPERPOSITION PRINCIPLE

Suppose a material initially free of stress and strain is subjected to a test in which a strain $\gamma(t_0)$ is suddenly imposed at $t = 0$ and maintained constant for a while. This is classical stress relaxation, and the stress will decay according to the material's time-dependent relaxation modulus $G_r(t)$; that is, $\tau(t) = G_r(t)\gamma(t_0)$. Now, however, at time t_1, the strain is suddenly changed to a new level $\gamma(t_1)$, held there for a while, then at t_2 changed to $\gamma(t_2)$, and so on, as sketched in Fig. 18.12a. What happens to the stress as a result of this strain history? Well, way back in 1876, Boltzmann suggested that the stresses resulting from each individual strain *increment* should be linearly additive; that is,

$$\tau(t) = \sum_{i=0}^{n} \Delta\tau_i = \sum_{i=0}^{n} G_r(t - t_i)\Delta\gamma(t_i) \qquad \text{(for } t > t_i) \qquad (18.30)$$

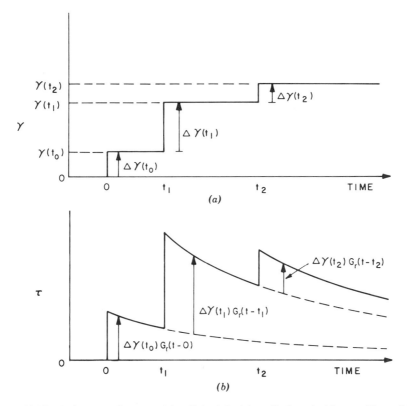

Figure 18.12   Boltzmann Superposition Principle: (a) applied strain history; (b) resulting stress history.

where

$$\Delta\gamma(t_i) = \gamma(t_i) - \gamma(t_{i-1}) \qquad (18.30a)$$

and

$$\Delta\tau_i = G_r(t - t_i)\Delta\gamma(t_i) \qquad (18.30b)$$

Here, $\Delta\tau_i$ is the stress increment which results from the strain increment $\Delta\gamma_i$. The argument $t - t_i$ is the time after the application of a particular strain increment $\Delta\gamma_i$. This behavior is sketcked in Fig. 18.12b.

According to Boltzmann, the stress in the material at any time t depends on its entire past strain history, although since $G_r(t)$ is a decreasing function of time, the further back a $\Delta\gamma(t_i)$ has occurred, the smaller will be its influence in the present. This leads to the anthropomorphic concept of viscoelastic materials having a fading memory (like an aging professor), with $G_r(t)$ sometimes known as the *memory function*. (The concept, of course, is valid even in the absence of linear additivity—it's just much more difficult to quantify.)

Example 5. A Maxwell element is initially free of stress and strain. At time $t = 0$, a strain of magnitude γ_0 is suddenly applied and maintained constant until $t = \lambda/2$, at which time the strain is suddenly reversed to a value of $-\gamma_0$ and maintained at that value (Fig. 18.13a). Obtain an expression for $\tau(t)$ and plot the result.

Solution. The relaxation modulus (memory function) $G_r(t)$ for a Maxwell element is given by (18.7b). For this particular strain history, $\Delta\gamma(t_0) = +\gamma_0$ and $\Delta\gamma(t_1) = -2\gamma_0$ (remember, we need the increment, not the absolute value).

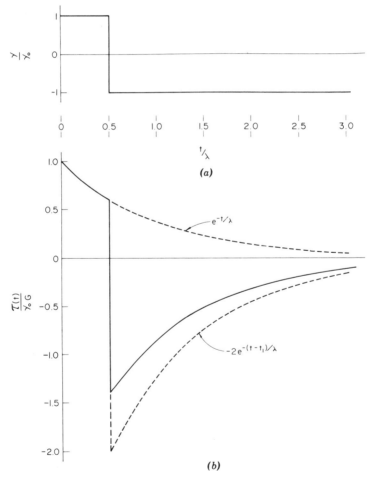

Figure 18.13   Response of a Maxwell element (Example 5).

Plugging these values into (18.30) gives

$$\tau(t) = G_r(t - 0)\Delta\gamma(t_0) + G_r(t - t_1)\Delta\gamma(t_1)$$
$$= \gamma_0 Ge^{-t/\lambda} - 2\gamma_0 Ge^{-(t-t_1)/\lambda}$$
$$= \gamma_0 G[e^{-t/\lambda} - 2e^{-(t-t_1)/\lambda}]$$

This result is plotted in dimensionless form in Fig. 18.13b. Keep in mind that the second term applies only at $t > t_1 = \lambda/2$.

Of course, not all strain histories consist of a nice series of finite step changes. No matter how an applied strain varies with time, however, it can always be approximated by a series of differential step changes, for which (18.30) becomes

$$\tau(t) = \int_{\gamma(-\infty)}^{\gamma(t)} G_r(t - t')d\gamma(t') = \int_{-\infty}^{t} G_r(t - t')\frac{d\gamma(t')}{dt'}dt' = \int_{-\infty}^{t} G_r(t - t')\dot{\gamma}(t')dt'$$

$$(18.31)$$

where $t = present$ time and $t' = past$ time. A word is needed about the somewhat bizarre (but fairly standard) notation in (18.31). The implication is that to evaluate the stress at the present time t, we must integrate over the entire past strain history of the sample; hence, the lower limit of $t' = -\infty$. In some (but not all) cases, it is convenient to assume that $\tau = \gamma = 0$ for $t' < 0$, in which case the lower limit on the integrals becomes zero.

Equation 18.31 allows calculation of $\tau(t)$ from stress-relaxation data $G_r(t)$ for any applied strain history as long as response is linear. Furthermore, by inversion of (18.31), it is possible (at least in principle) to obtain $G_r(t)$ from any test in which both $\tau(t)$ and $\gamma(t)$ are measured. When the independent variable is $\tau(t)$ and you wish to calculate $\gamma(t)$, analogs of (18.30) and (18.31) may be written in terms of the creep compliance $J_c(t)$:

$$\gamma(t) = \sum_{i=0}^{n} \Delta\gamma_i = \sum_{i=0}^{n} J_c(t - t_i)\Delta\tau(t_i) \qquad \text{(for } t > t_i) \qquad (18.32)$$

$$\gamma(t) = \int_{\tau(-\infty)}^{\tau(t)} J_c(t - t')d\tau(t') = \int_{-\infty}^{t} J_c(t - t')\frac{d\tau(t')}{dt'}dt' = \int_{-\infty}^{t} J_c(t - t')\dot{\tau}(t')dt'$$

$$(18.33)$$

When all is said and done, probably the best definition of a linear material is simply one that follows Boltzmann's principle. Thus, spring–dashpot models, which are linear, automatically follow Boltzmann's principle. However, it is important not to infer a dependence of the Boltzmann principle on spring–dashpot models. The Boltzmann principle applies to linear response regardless of whether it can be described with a spring–dashpot model. All that's needed are experimental $G_r(t)$ or $J_c(t)$ data. Models are used here simply as a matter of convenience to illustrate application of the principle.

Example 6. Solve Example 1 by applying the Boltzmann Superposition Principle, thereby demonstrating how stress–time response in an engineering stress–strain test may be predicted from stress-relaxation data.

Solution. The *tensile* stress-relaxation modulus for a Maxwell element is

$$E_r(t) \equiv \frac{\sigma(t)}{\varepsilon_0} = E e^{-t/\lambda}$$

Equation 18.31 becomes, in tensile notation, with the assumption that $\sigma = \varepsilon = 0$ for $t' < 0$,

$$\sigma(t) = \int_0^t E_r(t - t') \frac{d\varepsilon(t')}{dt'} dt'$$

But, for an engineering stress–strain test

$$\frac{d\varepsilon(t')}{dt'} = \dot{\varepsilon} \approx \text{constant} = \dot{\varepsilon}_0$$

Thus,

$$\sigma(t) = E\dot{\varepsilon}_0 \int_0^t e^{-(t-t')/\lambda} dt'$$

The integration is performed in the *present* time (i.e., t is a constant) over the material's past history, from $t' = 0$ to $t' = t$, with the result

$$\sigma(t) = E\dot{\varepsilon}_0\lambda[1 - e^{-t/\lambda}] = \eta_e\dot{\varepsilon}_0[1 - e^{-(E/\eta_e)t}]$$

This was obtained by direct integration of the differential equation for the Maxwell element in Example 1.

Example 7. To demonstrate the fact that the stress in a viscoelastic material depends on its past strain history, calculate the stress $\tau(t_s)$ in a Maxwell element initially free of stress and strain that is brought to a strain γ_0 at time t_s by three different paths:

a. For $t' < 0$, $\gamma = 0$; for $0 \leqslant t' \leqslant t_s$, $\gamma = \gamma_0$.
b. For $t' < t_s$, $\gamma = 0$; for $t' \geqslant t_s$, $\gamma = \gamma_0$.
c. $\gamma(t') = (\gamma_0/t_s)t'$.

Solution

a. This is good old stress relaxation. From (18.7a)

$$\tau(t_s) = G\gamma_0 e^{-t_s/\lambda}$$

b. This corresponds to the initial extension in stress relaxation:

$$\tau(t_s) = G\gamma_0$$

c. This is the shear analog of an engineering stress–strain test with the constant shear rate $\dot\gamma_0 = \gamma_0/t_s$. By analogy to the solution of Example 6 above

$$\tau(t_s) = G\dot\gamma_0\lambda(1 - e^{-t_s/\lambda}) = G(\gamma_0/t_s)\lambda(1 - e^{-t_s/\lambda})$$

18.7 DYNAMIC MECHANICAL TESTING

Creep and stress-relaxation measurements correspond to the use of step-response techniques to analyze the dynamics of electrical and process systems. Those familiar with these areas know that frequency-response analysis is perhaps a more versatile tool for investigating system dynamics. An analogous

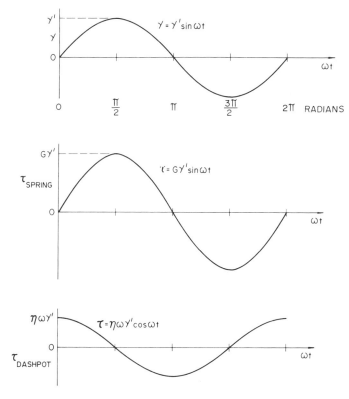

Figure 18.14 Stress in a linear spring and a linear dashpot in response to a sinusoidal applied strain.

procedure, *dynamic mechanical testing*, is applied to the mechanical behavior of viscoelastic materials. It is based on the fundamentally different response of viscous and elastic elements to a sinusoidally varying stress or strain.

If a sinusoidal strain, $\gamma = \gamma' \sin \omega t$ (where ω is the angular frequency, radian/s) is applied to a linear spring, since $\tau = G\gamma$, the resulting stress $\tau = G\gamma' \sin \omega t$ is in phase with the strain. For a linear dashpot, however, because the stress is proportional to the rate of strain rather than the strain, $\tau = \eta\dot{\gamma} = \eta\omega\gamma' \cos \omega t$, the stress is 90° out of phase with the strain. These relations are sketched in Fig. 18.14.

As might be expected, viscoelastic materials exhibit some sort of intermediate response, which might look like Fig. 18.15b. This can be thought of as being a projection of two vectors, τ^* and γ^*, rotating in the complex plane (Fig. 18.15a). The angle between these vectors is the *phase angle* δ ($\delta = 0$ for a purely elastic material and 90° for a purely viscous material). It is customary to resolve the vector representing the dependent variable into components in phase (designated by a prime) and 90° out of phase (designated by a double prime) with the independent variable. In this example, the applied strain is the independent variable, so the stress vector (τ^*) is resolved into its in-phase (τ') and out-

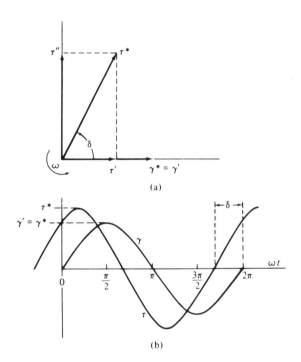

(a)

(b)

Figure 18.15   Quantities in dynamic testing: (a) rotating vector diagram; (b) stress and strain.

of-phase (τ'') components, $|\gamma^*| = \gamma'$ and $\gamma'' = 0$. In complex notation,

$$\tau^* = \tau' + i\tau'' \tag{18.34}$$

where i is the out-of-phase unit vector.

An *in-phase* or *storage* modulus is defined by

$$G' \equiv \frac{\tau'}{\gamma'} \qquad \text{storage modulus (in-phase component)} \tag{18.35}$$

and an *out-of-phase* or *loss* modulus is defined by

$$G'' \equiv \frac{\tau''}{\gamma'} \qquad \text{loss modulus (out-of-phase component)} \tag{18.36}$$

The *complex modulus* G^* is the *vector sum* of the in-phase and out-of-phase moduli:

$$G^* \equiv G' + iG'' = \frac{\tau' + i\tau''}{\gamma'} = \frac{\tau^*}{\gamma^*} \qquad \text{Complex modulus} \tag{18.37}$$

Additionally, a *complex viscosity* η^* may be defined:

$$\eta^* \equiv \eta' - i\eta'' = \frac{\tau^*}{\dot{\gamma}^*} \qquad \text{Complex viscosity} \tag{18.38}$$

Also,[10]

$$\dot{\gamma}^* = i\omega\gamma^* \tag{18.39}$$

By combining (18.37)–(18.39) (and recalling that $i^2 = -1$), we get

$$\eta''\omega + i\omega\eta' = G' + iG'' \tag{18.40}$$

Comparison of the real (in-phase) and imaginary (out-of-phase) parts of (18.40) gives

$$G' = \eta''\omega \tag{18.41}$$

and

$$G'' = \eta'\omega \tag{18.42}$$

Furthermore, combination of (18.38), (18.40), and (18.41) reveals that

$$G^* = G' + iG'' = i\omega\eta^* \tag{18.43}$$

From the geometry of Fig. 18.15 and the relations above, the *loss tangent*, tan δ, is

$$\tan \delta = \frac{\tau''}{\tau'} = \frac{G''}{G'} = \frac{\eta'}{\eta''} \qquad \text{Loss tangent} \qquad (18.44)$$

What is the physical significance of the quantities just defined? This can best be appreciated by considering what happens to the energy applied to a sample undergoing cyclic deformation. From (14.3), the work done on a unit volume of material undergoing a pure shear deformation is

$$W = \int \tau \, d\gamma \qquad (18.45)$$

From Fig. 18.15,

$$\gamma = \gamma' \sin \omega t \qquad (18.46)$$

and

$$\tau = |\tau^*| \sin (\omega t + \delta) \qquad (18.47)$$

Differentiating (18.46) with respect to (ωt) gives

$$d\gamma = \gamma' \cos \omega t \, d(\omega t) \qquad (18.48)$$

Inserting (18.47) and (18.48) into (18.45) gives

$$W = |\tau^*|\gamma' \int \sin (\omega t + \delta) \cos \omega t \, d(\omega t) \qquad (18.49)$$

Let's first consider the work done on the first quarter-cycle of applied strain, that is, integrate (18.49) between 0 and $\pi/2$. Using appropriate trigonometric identities and a good set of integral tables gives

$$W(\text{1st } \tfrac{1}{4} \text{ cycle}) = |\tau^*|\gamma' \left(\frac{\cos \delta}{2} + \frac{\pi}{4} \sin \delta \right) \qquad (18.50)$$

Putting it in terms of moduli or viscosities, using the trigonometry of Fig. 18.15 and (18.41) and (18.42), we find

$$W(\text{1st } \tfrac{1}{4} \text{ cycle}) = \frac{(\gamma')^2}{2} G' + \frac{\pi}{4}(\gamma')^2 G'' \qquad (18.51a)$$

$$W(\text{1st } \tfrac{1}{4} \text{ cycle}) = \frac{(\gamma')^2}{2} \omega\eta'' + \frac{\pi}{4}(\gamma')^2 \omega\eta' \qquad (18.51b)$$

The first term on the right side of (18.51a) is simply the work done in straining a linear spring of modulus G' an amount γ' (the area under the spring's stress–strain curve). It therefore represents the energy stored elastically in the material during its straining in the first quarter cycle. Hence, G' is the storage modulus. If the applied mechanical energy (work) is not stored elastically, it must be "lost"—converted to heat through molecular friction, that is, viscous dissipation, within the material. This is precisely what the second term on the right represents, so G'' is known as the loss modulus. Likewise, from (18.51b), stored energy is proportional to η'' and dissipated energy is proportional to η'.

Considering the second quarter of the cycle, integrating (18.49) from $\pi/2$ to π gives results identical to (18.51) except that the sign on the first (storage) term is negative. This simply means that the energy stored elastically in straining the material from 0 to γ' is recovered when it returns from γ' to 0. Thus, over a half cycle or a full cycle, there is no net work done or energy lost through the elastic component. The sign of the second term, however, is positive for any quarter cycle, so the net energy loss (converted to heat within the material) for a full cycle (also obtainable by integrating (18.49) between 0 and 2π) is simply

$$W(\text{complete cycle}) = \pi(\gamma')^2 G'' = \pi(\gamma')^2 \,\omega\eta' \tag{18.52}$$

The average power dissipated as heat within the material $\langle \dot{W} \rangle$ is obtained by dividing the energy dissipated per cycle by the period (time) of a cycle, $2\pi/\omega$:

$$\langle \dot{W} \rangle \ (\text{avg. power dissipation}) \ = \tfrac{1}{2}(\gamma')^2 \omega G'' = \tfrac{1}{2}(\gamma')^2 \omega^2 \eta' \tag{18.53}$$

These results are of direct importance in the design of polymeric objects that are subjected to cyclic deformation. In a tire, for example, high temperatures contribute to rapid degradation and wear. A rubber compound with a low G'' (or η') therefore helps to minimize energy dissipation and the resultant heat buildup. Moreover, dissipated energy wastes gasoline, so such a compound also contributes to better gas mileage. In the design of an engine mount, however, the goal is usually to prevent the transmission of vibration from the engine. Here, a material with a large G'' (or η') would dissipate considerable vibrational energy as heat rather than transmit it to the passengers.

Example 8. Obtain the expressions for the quantities G', G'', $|G^*|$, $\tan \delta$, η', η'', and $|\eta^*|$ for a Maxwell element.

Solution. This problem is solved in Reference 10 (p. 56ff) by direct integration of the differential equation for the Maxwell element. Here, we will apply Boltzmann's superposition principle to obtain the results, and in so doing, again illustrate how information from one type of linear test (stress relaxation) may be used to predict the response in another (dynamic testing).

The shear stress-relaxation modulus (memory function) for a Maxwell element is

$$G_r(t) = Ge^{-t/\lambda}$$

We will assume that the element has *always* been subjected to a shear strain

$$\gamma(t') = \gamma' \sin(\omega t')$$

which, when differentiated with respect to past time t', gives

$$\dot{\gamma}(t') = \gamma' \omega \cos(\omega t')$$

(Keep in mind that the prime here has two entirely different meanings: when applied to t, it designates past time; when applied to γ, it means the in-phase component of strain.) Plugging these results into (18.31), we get

$$\tau(t) = G\gamma' \omega \int_{-\infty}^{t} e^{-(t-t')/\lambda} \cos(\omega t') \, dt'$$

Evaluating the above integral is a nontrivial exercise, but again, with appropriate trigonometric identities and a good table of integrals, the result is

$$\tau(t) = \frac{G\gamma'(\omega\lambda)^2}{1+(\omega\lambda)^2} \sin(\omega t) + \frac{G\gamma'(\omega\lambda)}{1+(\omega\lambda)^2} \cos(\omega t)$$

(By assuming that the element has *always* been subjected to the periodic strain and integrating from $t' = -\infty$, we eliminate the transient part of the solution which would arise had we integrated from $t' = 0$.) It is clear that the first (sin) term in the above expression is in phase with the applied strain, while the second (cos) term is 90° out-of-phase with the applied strain. Therefore,

$$\tau' = \frac{G\gamma'(\omega\lambda)^2}{1+(\omega\lambda)^2} \quad \text{and} \quad \tau'' = \frac{G\gamma'\omega\lambda}{1+(\omega\lambda)^2}$$

From here on, it's definitions and algebra:

$$G' = \frac{\tau'}{\gamma'} = \frac{G(\omega\lambda)^2}{1+(\omega\lambda)^2} \quad \text{and} \quad G'' = \frac{\tau''}{\gamma'} = \frac{G\omega\lambda}{1+(\omega\lambda)^2}$$

$$|G^*| = [(G')^2 + (G'')^2]^{1/2} = \frac{G\omega\lambda}{[1+(\omega\lambda)^2]^{1/2}}$$

$$\tan \delta = \frac{G''}{G'} = \frac{1}{\omega\lambda}$$

$$\eta' = \frac{G''}{\omega} = \frac{G\lambda}{1+(\omega\lambda)^2} = \frac{\eta}{1+(\omega\lambda)^2}$$

$$\eta'' = \frac{G'}{\omega} = \frac{G\lambda^2\omega}{1+(\omega\lambda)^2} = \frac{\eta\omega\lambda}{1+(\omega\lambda)^2}$$

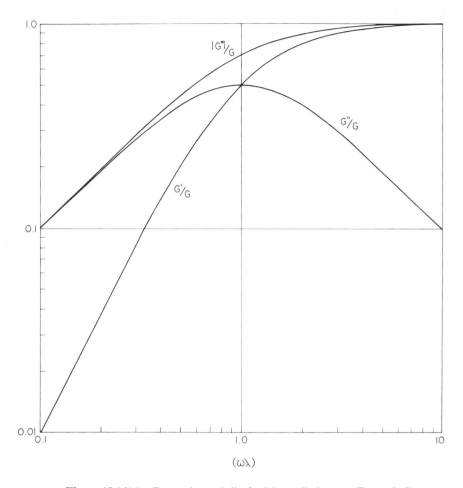

Figure 18.16(a)   Dynamic moduli of a Maxwell element (Example 8).

$$|\eta^*| = [(\eta')^2 + (\eta'')^2]^{1/2} = \frac{\eta}{[1 + (\omega\lambda)^2]^{1/2}}$$

The dynamic moduli are plotted in dimensionless form in Fig. 18.16a, and the dynamic viscosities in Fig. 18.16b.

As when applied to other types of mechanical tests, the Maxwell element won't win any prizes for quantitatively fitting dynamic data for real materials. Nevertheless, Example 8 does serve to illustrate the frequency dependence of dynamic mechanical properties, and Figs. 18.16a and b do in some ways resemble the variation in isothermal dynamic data with frequency for real materials. In particular, the apparent "stiffness" $|G^*|$ increases to a limiting

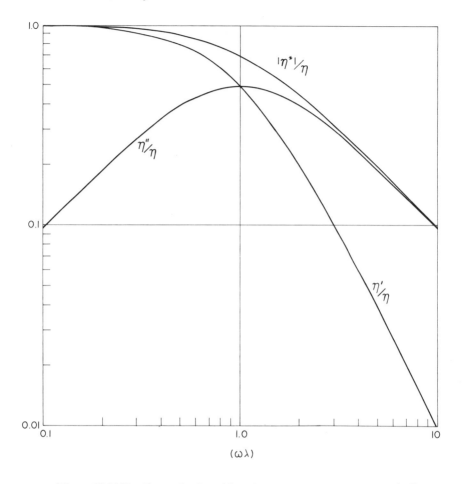

Figure 18.16(b)   Dynamic viscosities of a Maxwell element (Example 8).

value with frequency. In the model, the dashpot simply can't keep up with high frequencies, leaving only the spring to respond. Also, the maximum in G'' is usually observed at frequencies in the range where G' and $|G^*|$ are falling from their high-frequency limit. In the model, at low frequencies, the dashpot offers little resistance to motion, and so dissipates little energy. At high frequencies, its high resistance prevents its motion, limiting response to the spring, and energy dissipation again falls off.

With real materials, the high-frequency limit of $|G^*|$ and G' corresponds quantitatively to the moduli obtained in the limit of short times in stress-relaxation and creep measurements; that is,

$$|G^*|\,(\omega \to \infty) = G'(\omega \to \infty) = G_r(t \to 0) = \frac{1}{J_c(t \to 0)} \qquad (18.54)$$

Furthermore, the measured low-frequency limit of $|\eta^*|$ and η' agrees with the zero-shear-rate steady-flow viscosity η_0:

$$|\eta^*|(\omega \to 0) = \eta'(\omega \to 0) = \eta_0 = \eta(\dot{\gamma} \to 0) \tag{18.55}$$

The viscosity analogy has been pushed a bit further. The drop in $|\eta^*|$ with frequency resembles the variation in steady-flow viscosity η with shear rate. On a purely empirical basis, Cox and Merz[11] suggested that $|\eta^*|$ and η are the same when compared at equal values of frequency and shear rate:

$$|\eta^*|(\omega) \approx \eta(\dot{\gamma}) \qquad \text{at } \omega = \dot{\gamma} \tag{18.56}$$

While not exact, (18.56) appears to be at least a reasonable approximation.

For a generalized Maxwell model consisting of n elements (Fig. 18.10) the results obtained for a single Maxwell element in Example 8 are readily generalized to

$$G' = \sum_{i=1}^{n} \frac{G_i(\omega\lambda_i)^2}{1 + (\omega\lambda_i)^2} \tag{18.57}$$

and

$$G'' = \sum_{i=1}^{n} \frac{G_i\omega\lambda_i}{1 + (\omega\lambda_i)^2} \tag{18.58}$$

(the other dynamic properties may be obtained from the two above, as in Example 8). Presumably, the G_i and λ_i determined from, for example, stress-relaxation data can be used in (18.57) and (18.58) to predict dynamic response, and vice versa. The relations (18.27) and (18.28) should be equally applicable to both. Also, the continuous distributions $G(\lambda)$ and $J(\lambda)$ as obtained from stress-relaxation and creep measurements are at least approximately interconvertible with $G'(\omega)$ and $G''(\omega)$.[2, 4, 8, 9]

Basically, three methods are available for determining dynamic properties: free oscillation, forced oscillation, and steady-state rotation. Experimental and analytical details for the first two are reviewed extensively by Ferry.[2]

Free-oscillation measurements are made with a torsion pendulum (Fig. 18.17). The sample is given an initial torsional displacement, and the frequency and amplitude decay of the oscillations are observed on release. G' is determined from the sample geometry, moment of inertia of the oscillating mechanism, and the observed period of oscillation. For example, with a cylindrical specimen of length L and radius R,

$$G' = \frac{8\pi LI}{R^4 P^2} \tag{18.59}$$

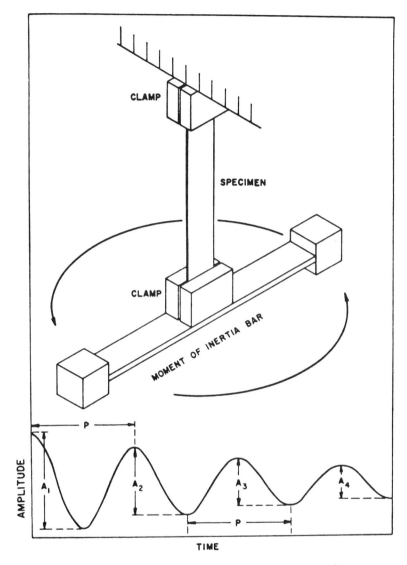

Figure 18.17   Torsion pendulum with its output.[19]

where I is the moment of inertia of the oscillating mechanism and P is the observed period of oscillation ($P = 2\pi/\omega$). The damping, or logarithmic decrement Δ is calculated from the amplitude decay of the oscillations (with a perfectly elastic material, there would be no damping of the oscillations):

$$\Delta = \ln\frac{A_1}{A_2} = \ln\frac{A_2}{A_3} = \ln\frac{A_i}{A_{i+1}} = \frac{1}{n}\ln\frac{A_i}{A_{i+n}} \qquad (18.60)$$

and for $\Delta < 1$

$$\Delta \approx \pi \tan \delta \tag{18.61}$$

Expressions for other specimen geometries and higher damping are reviewed by Nielsen.[5] Solid or rubbery samples are twisted as illustrated in the form of rods, tubes, strips, etc. Liquids or soft solids may be contained in one of the geometries described for rotational viscometry in Chapter XVI (Couette, cone-and-plate, etc.).

In *torsional braid analysis*,[12] a flexible, braided fiber, usually glass, is impregnated with the material to be studied. The impregnated fiber then becomes the torsion member in the pendulum. This type of device is useful for following the cure of a material that starts out as a liquid and cures to a solid (e.g., an epoxy). In analyzing the data from such a device, care must be exercised in separating interactions between the sample and supporting fiber.

Although the frequency can be varied somewhat by changing the moment of inertia of the oscillating portion of the mechanism, torsion pendula are usually intended to study only the temperature dependence of dynamic properties at a constant, relatively low frequency (≈ 1 cps). On the other hand, they are inexpensive and rather simple to construct.

Forced-oscillation devices ("jiggle machines") apply a sinusoidal stress or strain of known amplitude and frequency and measure the resulting strain or stress. The dynamic properties are calculated from the relation between the two. For liquid samples, the geometries discussed in conjunction with rotational viscometry in Chapter XVI are often used with the drive system modified to produce sinusoidal rather than steady rotational deformation. Flexible samples such as fibers, films, and rubber are pre-loaded in tension and oscillated about a positive tensile strain so that they don't go slack at the "bottom" of the sine wave. Such tests give dynamic *tensile* properties, E', E'', etc., which are related to the corresponding shear properties by

$$|E^*| = 2|G^*|\,(1 + v) \tag{18.62}$$

where v is Poisson's ratio ($v = 1/2$ for an incompressible material). Another type of forced-oscillation device applies a sinusoidal shear or compression wave to one end of a sample and monitors the attenuation of the wave as it progresses through the sample.

Forced-oscillation devices are generally intended to study dynamic properties as a function of frequency as well as temperature. Drive and detection systems for such devices may be strictly mechanical, but the newer and more sophisticated ones make use of piezoelectric crystals, inorganic crystals (e.g., barium titanate) that change dimension in proportion to an applied voltage and, conversely, generate an output voltage proportional to an imposed deformation. Thus, the amplitude and especially the frequency of the applied strain can be conveniently controlled over a wide range in the form of electrical signals (the response of the crystals extends to very high frequencies, and electrons have very little inertia, unlike mechanical linkages). Input and output signals may be fed through A/D converters to a computer for on-line control and computations.

The expense and complexity of oscillating drives and associated detection systems has led to the increased popularity of instruments that determine dynamic properties, but are driven in steady rotation and detect steady (non-oscillating) forces. This sounds like a contradiction in terms, but they really do work! We will attempt to describe the operation of one of the more popular of them. Mathematical details are provided in elegant fashion by Walters.[13]

In the Maxwell *orthogonal rheometer*[14] (Fig. 18.18), material is sheared between two parallel disks of radius R, separated from one another by a distance h, each rotating at the same steady angular frequency ω, but *with their axes of rotation displaced by a distance d*. Transducers are set up to measure three steady, orthogonal force components (hence the name) on one of the disks, F_x, F_y, and F_z.

How, you ask, does this result in oscillating deformation? Well, the easiest way to see is to follow the motion of a point on the upper disk (point 2), relative to one on the lower disk (point 1) as the disks rotate. Here, we choose point 1 on the axis of rotation of the lower disk, and point 2 directly above it at the start of the analysis, $\omega t = 0$. This is the simplest choice, because point 1 remains

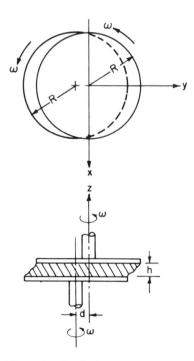

Figure 18.18 Orthogonal rheometer.

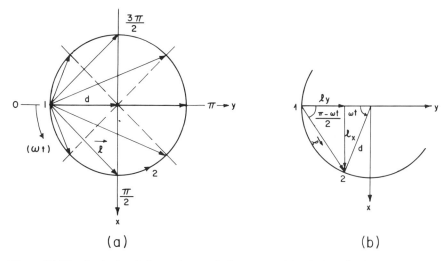

Figure 18.19 Analysis of the orthogonal rheometer: (a) relative displacement vector between point 1 on center of lower disk and point 2 immediately above it at $\omega t = 0$; (b) geometry of the displacement vector and its components.

stationary, but the same result will be obtained for any pair of points initially at the same values of x and y (you can prove this with a compass, ruler, and protractor). Figure 18.19a shows the relative displacement vector in the xy plane, l, and how it varies with the angle of rotation ωt. From Fig. 18.19b, the magnitude of the relative displacement is then

$$|\mathbf{l}| = 2d \, \sin \frac{\omega t}{2} \tag{18.63}$$

and its x and y components are

$$l_x = d \, \sin(\omega t) \tag{18.64}$$

and

$$l_y = |\mathbf{l}| \cos\left(\frac{\pi}{2} - \frac{\omega t}{2}\right) = d[1 - \cos(\omega t)] \tag{18.65}$$

If we assume that the strain is simply the relative displacement divided by the disk separation h, the x and y components of strain are*

---

* This requires that each fluid element move at fixed z in a circle that is centered on a line connecting the centers of the disks. There has been some controversy about this, but it is probably at least a good approximation for $d/h < 0.5$.

$$\gamma_x = \frac{d}{h} \sin(\omega t) \tag{18.66}$$

$$\gamma_y = \frac{d}{h}[1 - \cos(\omega t)] \tag{18.67}$$

Thus, both components of strain undergo simple harmonic oscillation as a result of the steady angular rotation.

It is important to note that the x component of strain is symmetrical about 0, while the y component is not (it varies between 0 and $+2d/h$). For linear materials, because stress is directly proportional to strain, any shear stresses, and therefore forces parallel to the xy plane produced by γ_x's in the region $0 < \omega t < \pi$, will be cancelled by those arising from the equal-and-opposite x-component strains in the region ($\pi < \omega t < 2\pi$). As a result, the measured forces F_x and F_y can depend only on γ_y.

Looking at it another way, consider a purely elastic material in the rheometer to be represented by rubber bands stretched between points 1 and 2. Viewed from above, the rubber bands coincide with 1. It is obvious that the x components of the tug of the rubber bands will cancel, leaving only a net force (and therefore only a stress component τ_{zy}) in the y direction. Thus, the purely elastic stress is in phase with γ_y, so Eq. 18.67 represents the variation of γ' with location in a disk at constant z; that is,

$$\gamma'(\omega t) = \gamma_y(\omega t) = \frac{d}{h}[1 - \cos(\omega t)] \tag{18.68}$$

and

$$\tau_{zy} = \tau' = \tau'(\gamma') \tag{18.69}$$

Any measured value of F_x, therefore, must arise from a viscous component, whose stress τ_{zx} will be 90° out of phase with γ_y; that is,

$$\tau_{zx} = \tau''(\gamma') \tag{18.70}$$

To determine the total force component on a disk, the stress–area product must be integrated over the surface of the disk:

$$F_y = \int_{\text{surface of disk}} \tau' \, dA \tag{18.71}$$

From (18.35), $\tau' = G'\gamma'$, and in polar coordinates, $dA = r \, d(\omega t) \, dr$, so that

$$F_y = G'\left(\frac{d}{h}\right)\int_0^R \int_0^{2\pi} [1 - \cos(\omega t)] \, d(\omega t) \, r \, dr = \pi R^2 \left(\frac{d}{h}\right)G' \tag{18.72}$$

or

$$G' = \frac{h/d}{\pi R^2} F_y \tag{18.73}$$

Similarly,

$$F_x = \int_{\text{surface of disk}} \tau'' \, dA \tag{18.74}$$

and since $\tau'' = G''\gamma'$ (18.36), it follows that

$$G'' = \frac{h/d}{\pi R^2} F_x \tag{18.75}$$

The third force component F_z is believed to be related to the second normal stress difference N_2 (17.4b).

Equations 18.73 and 18.75 show how dynamic properties can be obtained from a device in steady rotation by measuring steady forces, which mechanically and analytically represents a great simplification over forced-oscillation techniques. It must be pointed out that there have been some questions as to whether a nonviscometric flow such as this can be used to determine quantities such as G', G'', η', η'', N_2, etc., which are really defined in terms of viscometric deformations, but it is now pretty generally agreed that the technique is valid, at least in the limit of small strains, d/h.[15, 16]

Also, analyses are based on the assumption that both disks rotate at the same ω. In the usual instrument, one disk is driven and the other goes along for the ride, so with bearing friction and hydrodynamic effects, the assumption might not be strictly true. An analysis by Davis and Macosko[17] shows it to be a pretty good assumption under most conditions of interest.

Other devices have been developed that operate along similar lines. The so-called balance rheometer confines a test fluid between two concentric hemispheres that rotate at the same rate but whose axes of rotation are inclined at an angle to one another. Similarly, cone-and-plate geometry, in which the axis of the cone is not perpendicular to the plate, and Couette (cup-and-bob) geometry, with the bob not centered in the cup can also be used to obtain dynamic data. These devices are analyzed by Walters.[13] Dynamic tensile properties can be obtained for relatively rigid materials by subjecting a rotating cylindrical rod to a cantilever deflection.[18]

Figure 18.20 illustrates G' and damping $\approx \pi \tan \delta$ vs. T for polymethyl methacrylate, an amorphous, linear polymer. The data were obtained with a torsion pendulum at about 1 cycle/s. At low temperatures, the typical glassy modulus 10^{10} to 10^{11} dyn/cm^2 (10^9 to 10^{10} N/m^2 = Pa) is observed. In the vicinity of 110 to 130°C, G' drops precipitously, ultimately reaching a plateau of 10^6 to 10^7 dyn/cm^2 (10^5 to 10^6 N/m^2), the typical rubbery modulus. Also,

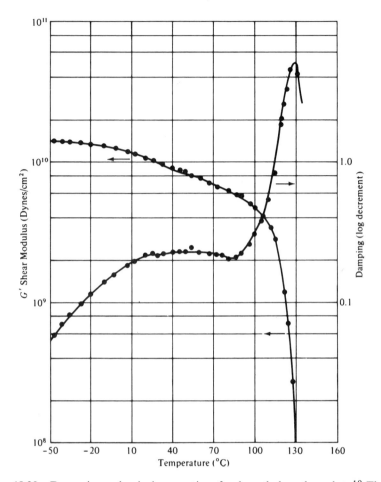

Figure 18.20 Dynamic mechanical properties of polymethyl methacrylate.[19] The data were obtained with a torsion pendulum at about 1 cycle/s.

a sharp peak in the damping is observed in this region. Although the temperature at which this peak and drop are observed is frequency dependent, at low frequencies (such as are obtained with the usual torsion pendulum), they are identified with the material's glass transition temperature, and the drop in G' is indicative of the decrease in "stiffness" at T_g, going from the straining of bond angles and lengths to coiling and uncoiling as the dominant response mechanism. The damping peak represents the onset of cooperative motion of 40 to 50 main-chain carbon atoms at T_g (Chapter VIII).

In terms of the four-parameter model, below T_g, only spring 1 is operative, and the material is almost completely elastic (low damping). In the vicinity of T_g, the viscosity of dashpot 2 drops to the point where it can deform and dissipate energy, giving the damping peak. At higher temperatures, its viscosity drops to the point where it dissipates little energy, and the material is again highly elastic, mainly through spring 2. At still higher temperatures, the modulus drops off rapidly due to viscous flow, dashpot 1. The broad damping "hill" centered at about 40°C (and the accompanying gradual drop in G') in Fig. 18.20 has been shown to arise from the motion of the $-CO-O-CH_3$ side groups on the molecules.

Note that the dimensions of the angular frequency ω are time^{-1}. The angular frequency thus corresponds to a *reciprocal* time scale. For dynamic (oscillating) deformations, then, the Deborah number is

$$De = \lambda_c \omega \tag{18.76}$$

In dynamic tests, a viscoelastic material becomes more solid-like as the frequency is increased and the time-dependent response mechanisms (coiling and uncoiling, slip) are less able to follow the rapidly reversing stress. In the limit of very high frequencies, the straining of bond angles and lengths will be the only operative response mechanism, and the polymer will exhibit the typical glassy modulus even though it may be well above its glass-transition temperature.

18.8 TIME–TEMPERATURE SUPERPOSITION

Anyone who has ever wrestled with a cheap garden hose in cold weather appreciates the fact that polymers become stiffer and more rigid at lower temperatures, while at high temperatures they are softer and more flexible. In preceding examples, we have seen that the time scale (or frequency) of the application of stress has a similar influence on mechanical properties, short times (or high frequencies) corresponding to low temperatures and long times (low frequencies) corresponding to high temperatures. The quantitative application of this idea, *time–temperature superposition*, is one of the most important principles of polymer physics. It is based on the fact that the Deborah number determines quantitatively just how a viscoelastic material will behave mechanically. Changing either t_s (or ω) or λ_c can change De. The nature of the applied deformation determines t_s (or ω), while a polymer's characteristic time is a function of temperature. The higher the temperature, the more thermal energy the chain segments possess, and the more rapidly they are able to respond, lowering λ_c. Thus, for example, De can be doubled by halving t_s (or doubling ω in a dynamc test) or by lowering the temperature enough to double λ_c. The change in mechanical response will be the same either way, according to the time–temperature superposition principle.

Although time–temperature superposition is applicable to any viscoelastic response test (creep, dynamic, etc.), its application will be illustrated here with stress relaxation. Figure 18.21 shows tensile stress relaxation data at various temperatures for polyisobutylene, plotted in the form of a time-dependent

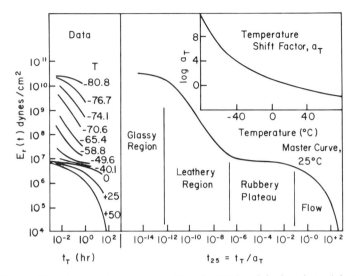

Figure 18.21   Time–temperature superposition for NBS polyisobutylene. Adapted from Tobolsky and Catsiff.[20]

tensile (Young's) modulus $E_r(t)$ vs. time on a log–log scale:

$$E_r(t) = \frac{\sigma(t)}{\varepsilon_0} = \frac{f(t)/A}{\Delta l/l} \tag{18.77}$$

where $f(t)$ is the measured tensile force in the sample held at a constant strain $\varepsilon_0 = \Delta l/l$ and A is its cross-sectional area. The technique is, of course, equally applicable to shear deformation. In stress relaxation, the lower measurement time limit is set by the assumption that the constant strain is applied instantaneously. In practice, inertia and other mechanical limitations make this impossible, so data are valid only at times an order of magnitude or so longer than it actually takes to apply the constant strain. The upper limit is set by the dedication of the experimenter and the long-term stability of the sample and equipment. These data were obtained over a range from seconds to a couple of days. As might be expected, the modulus drops with time at a given temperature, and at a given time, it drops with increasing temperature.

Staring at the curves for a while indicates that they appear to be sections of one continuous curve, chopped up, with the sections displaced along the log–time axis. That this is indeed so is shown in Fig. 18.21. Here, 25°C has arbitrarily been chosen as a reference temperature T_0 and the curves for other temperatures shifted along the log time axis to line up with it. The data below 25°C are shifted to the left (shorter times) and those above 25°C are shifted to the right (longer times), giving a *master curve* at 25°C.

Sometimes, the relaxation moduli at each temperature T are corrected to the reference temperature T_0 by multiplying by the ratio T_0/T before superposing.

This correction is based on the theory of ideal rubber elasticity (Chapter XIV), which states that the modulus is proportional to the absolute temperature. This procedure is open to question, however. While ideal rubber elasticity might be a reasonable approximation when the major response mechanism is chain coiling and uncoiling, it certainly isn't where response is dominated by straining of bond angles and lengths (glassy region) or molecular slippage (viscous flow region). In any event, such corrections are of minor practical significance.

Shifting a constant-temperature curve along the log–time axis corresponds to dividing every value of its abscissa by a constant factor (it is immaterial what kind of scale is used for the ordinate). This constant factor, which brings a curve at a particular temperature T into alignment with the one at the reference temperature T_0, is known as the *temperature shift factor* a_T:

$$a_T = \frac{t_T}{t_{T_0}} \qquad \text{(for same response)} \qquad (18.78)$$

where t_T is the time required to reach a particular response (E_r in this case) at temperature T and t_{T_0} is the time required to reach the *same response* at the reference temperature T_0. The abscissa of the master curve from (18.78) is $t_{T_0} = t_T/a_T$.

For temperatures above the reference temperature, it takes less time to reach a particular response (the material responds faster, i.e., has a shorter relaxation time), so a_T is less than one, and vice versa. The logarithm of the experimentally determined temperature shift factor is plotted as a function of temperature in Fig. 18.21.

The master curve now represents stress relaxation at 25°C *over 17 decades of time*. Since $t_T = a_T t_{T_0}$, multiplication by the appropriate value of a_T (shifting along the log–time axis) establishes the master curve at any other temperature T, and can thus be used to predict response at that temperature over 17 decades of time. With such a wide time scale, more terms can be established through "Procedure X" and the functions $G(\lambda)$ and $J(\lambda)$ can be determined and interconverted more accurately, as noted in Section 18.5.

Two additional aspects enhance the utility of the time–temperature superposition concept. First, the same temperature shift factors apply to a particular polymer regardless of the nature of the mechanical response; that is, the shift factors as determined in stress relaxation are applicable to the prediction of the time–temperature behavior in creep or dynamic testing. Second, *if the polymer's glass-transition temperature is chosen as the reference temperature*, the shift factors are given by the Williams–Landel–Ferry (WLF) equation in the range $T_g < T < (T_g + 100°C)$:

$$\log a_T = \frac{-C_1(T - T^*)}{C_2 + (T - T^*)} \qquad \text{(for } T_o = T_g\text{)} \qquad (18.79)$$

With $T^* = T_g$, the "universal" constants $C_1 = 17.44$ and $C_2 = 51.6$ (with T's in

K) give a rough fit for a wide variety of polymers. The WLF equation is most useful in this form because T_g is extensively tabulated.[21] Better fits can be obtained by using constants C_1, C_2, and T^* specific to the polymer, but these constants are not readily available. It has also been suggested that fit can be improved by using $C_1 = 8.86$ and $C_2 = 101.6$, with T^* adjusted to fit specific data, if available. When this is done, T^* generally turns out to be $T_g + (50 \pm 5)°C$.[22]

Example 8. The damping peak for polymethyl methacrylate in Fig. 18.20 is located at 130°C. Assuming the data were obtained at a frequency of 1 cycle/s, at what temperature would the peak be located if measurements were made at 1000 cycles/s? For polymethyl methacrylate, $T_g = 105°C$.

Solution. Keep in mind that frequency is a *reciprocal* time scale, and therefore $a_T = \omega_{T_0}/\omega_T$. Applying the WLF equation with the "universal" constants gives

$$\log a_T = \log \frac{\omega_{T_g}}{\omega_T} = \frac{-17.44(T - T_g)}{51.6 + (T - T_g)}$$

Shifting the measurements at 1 cycle/s to T_g (i.e., finding the frequency at which the peak would be located at T_g) gives

$$\log \frac{\omega_{105°C}}{\omega_{130°C}} = \frac{-17.44(130 - 105)}{51.6 + (130 - 105)} = -5.69$$

$$\omega_{105°C}/\omega_{130°C} = 2.03 \times 10^{-6}$$

and

$$\omega_{105°C} = (1 \text{ cps})(2.03 \times 10^{-6}) = 2.03 \times 10^{-6} \text{ cps}$$

Now, shifting from T_g to T,

$$\log \frac{2.03 \times 10^{-6}}{1000} = -8.69 = \frac{-17.44(T - 105)}{51.6 + T - 105}$$

$$T = 156°C$$

This illustrates quantitatively the statements made at the end of Section 18.7.

Example 9. The master curve for the polyisobutylene in Fig. 18.21 indicates that stress relaxes to a modulus of 10^6 dyn/cm² in about 10 h at 25°C. Using the WLF equation, estimate the time it will take to reach the same modulus at a temperature of $-20°C$. For PIB, $T_g = -70°C$.

Solution. To use the WLF equation, the reference temperature must be $T_g = -70°C$:

$$\log \frac{t_T}{t_{T_g}} = \log \frac{t_{25°C}}{t_{-70°C}} = \frac{-17.44(25 + 70)}{51.6 + 25 + 70} = -11.3$$

$$\frac{t_{25°C}}{t_{-70°C}} = 5.01 \times 10^{-12} \quad \text{and} \quad t_{-70°C} = \frac{10}{5.01 \times 10^{-12}} = 2 \times 10^{12}\,\text{h}$$

$$\log \frac{t_{-20°C}}{t_{-70°C}} = \frac{-17.44(-20 + 70)}{51.6 - 20 + 70} = -8.59$$

$$t_{-20°C}/t_{-70°C} = 2.57 \times 10^{-9}$$

$$t_{-20°C} = (2 \times 10^{12})(2.57 \times 10^{-9}) = 5140\,\text{h}$$

This shows how lowering the temperature maintains mechanical "stiffness" for much longer periods of time.

Let's take a closer look at the stress-relaxation master curve. The one shown in Fig. 18.21 is typical of linear, amorphous polymers, and illustrates the *five regions of viscoelastic behavior.*[4] At low temperatures or short times (large De), only bond angles and lengths can respond, and so the typical glassy modulus of 10^{10} to $10^{11}\,\text{dyn/cm}^2$ (10^9 to 10^{10} Pa) is observed. This is the so-called *glassy region.* At longer times or higher temperatures, response is governed by the uncoiling of the chains, with the characteristic modulus 10^6 to $10^7\,\text{dyn/cm}^2$ (10^5 to 10^6 Pa) in the *rubbery plateau.* The intermediate region, where the modulus drops from glassy to rubbery, is sometimes known as the *leathery region* from the leather-like feel of materials with moduli in this range. At still longer times or higher temperatures (low De), the modulus drops from the rubbery plateau into the *rubbery-flow region*, where the material is still quite elastic but has a significant flow component and then falls off rapidly as a result of molecular slippage in the *viscous-flow region* (this last distinction is somewhat artificial).

Figure 18.22 illustrates the effects of molecular weight and crosslinking on the stress-relaxation master curve. Molecular weight should have no significant influence on straining of bond angles and lengths or on uncoiling, so the glassy and rubbery moduli are unchanged. Flow, however, is severely retarded by increasing molecular weight, which extends the rubbery plateau. For very low molecular weight polymers, the rubbery plateau isn't even seen. In the limit of infinite molecular weight (light crosslinking), flow is eliminated entirely, and the curve levels off with the rubbery modulus. Higher degrees of crosslinking restrict uncoiling, ultimately leading to a material that responds only by straining of bond angles and lengths.

The effects of crystallinity on properties are similar to those of crosslinking. However, the applicability of time–temperature superposition in and across the

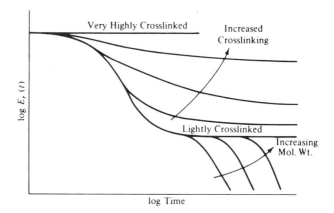

Figure 18.22 The effects of molecular weight and crosslinking on stress-relaxation master curves.

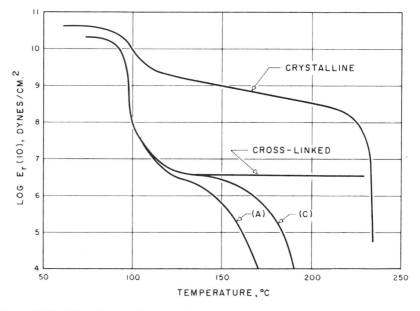

Figure 18.23 The 10-s tensile relaxation modulus $E_r(10)$ vs. temperature for several polystyrenes.[4] Samples (A) and (C) are linear, amorphous (atactic) materials with narrow molelcular weight distributions. $\bar{M}_n(A) = 140\,000$, $\bar{M}_n(C) = 217\,000$. The crosslinked sample is also amorphous. The crystalline material is isotactic.

region of T_m is open to question. In this region, the degree of crystallinity and crystalline morphology may change, and one would be, in effect, superposing data for different materials. A second, vertical shift has been suggested to help superpose data for crystalline polymers.

To illustrate the effect of temperature on mechanical properties, it is sometimes preferable to plot the property vs. temperature for constant values of time. For example, data of the type shown in Fig. 18.21 may be cross-plotted as $E_r(10)$ (the 10-second relaxation modulus) vs. T. Such a plot is given in Fig. 18.23 for several polystyrene samples.[4] The five regions of viscoelastic behavior are evident in the linear, amorphous (atactic) samples (A) and (C) along with the effect of molecular weight in the flow region. The drop in modulus in the vicinity of T_g ($100°C$) is clearly seen. The crystalline (isotactic) sample maintains a fairly high modulus all the way up to T_m ($\approx 235°C$). Given values of a_T, one can convert data in the form E_r vs. t at constant T (a master curve) to E_r vs. T at constant t and vice versa.

REFERENCES

1. Reiner, M., *Phys. Today*, Jan. 1964, p. 62.

2. Ferry, J. D., *Viscoelastic Properties of Polymers*, 3rd ed., Wiley, New York, 1980.

3. Eirich, F. R. (ed.), *Rheology*, Vols. 1–4, Academic, New York, 1956–1964.

4. Tobolsky, A. V., *Properties and Structure of Polymers*, Wiley, New York, 1960.

5. Nielsen, L. E., *Mechanical Properties of Polymers and Composites*, Vol. 1, Dekker, New York, 1974.

6. Spriggs, T. W., *Chem. Eng. Sci.* **20**, 931 (1965).

7. Rouse, P. E., Jr., *J. Chem. Phys.* **24**, 269 (1956).

8. Schwarzl, F. R., *Pure and Appl. Chem.* **23**, 219 (1970).

9. Schwarzl, F. R., *Rheol. Acta* **8**, 6 (1969); **9**, 382 (1970); **10**, 166 (1971); **14**, 581 (1975).

10. McKelvey, J. M., *Polymer Processing*, Wiley, New York, 1962.

11. Cox, W. P., and E. H. Merz, *J. Polym. Sci.* **28**, 619 (1958).

12. Gillham, J. K., *Am. Inst. Chem. Eng. J.* **20**, 1066 (1974).

13. Walters, K., *Rheometry*, Halsted, New York, 1975.

14. Maxwell, B., and R. P. Chartoff, *Trans. Soc. Rheol.* **9**, 41 (1965).

15. Bird, R. B., and E. K. Harris, *Am. Inst. Chem. Eng. J.* **14**, 758 (1968).

16. Macosko, C. W., and W. M. Davis, *Rheol. Acta* **13**, 814 (1974).

17. Davis, W. M., and C. W. Macosko, *Am. Inst. Chem. Eng. J.* **20**, 600 (1974).

18. Maxwell, B., *J. Polym. Sci.* **20**, 551 (1956).

19. Nielsen, L. E., *SPE J.* **16**, 525 (1960).

20. Tobolsky, A. V., and E. Catsiff, *J. Polym. Sci.* **19**, 111 (1956).

21. Brandrup, J., and E. H. Immergut (eds.), *Polymer Handbook*, 3d ed., Wiley–Interscience, New York, 1989.

22. Bird, R. B., R. C. Armstrong, and O. A. Hassager, *Dynamics of Polymeric Liquids*, Vol. 1, Fluid Mechanics, Wiley, New York, 1977.

PROBLEMS

1. Given enough springs and dashpots, it is possible in principle to fit any linear response to any desired degree of accuracy. For each of the individual

springs, it is true by definition that $G_i = 1/J_i$ (the modulus of each individual spring is the reciprocal of its compliance).

Does it follow that $G_r(t) = 1/J_c(t)$? That is, is a material's stress-relaxation modulus always the reciprocal of its creep compliance? *Hint:* Examine this question for the simplest of materials, a Maxwell element.

2. Isothermal tensile creep data on polymers can sometimes be fit by an empirical equation of the form

$$\varepsilon = A(B + t^C)\sigma_0$$

where ε is the tensile strain, σ_0 is the (constant) applied tensile stress, and t is the time. A, B, and C are positive constants.

   a. Is a material that follows this equation linear?
   b. What is the instantaneous elastic compliance?
   c. Is a material that follows this equation a liquid or solid?
   d. Can this equation ever describe equilibrium viscous flow? If so, under what conditions, and what is the equilibrium tensile viscosity under those conditions?

3. Describe how to apply "Procedure X" to recoverable elastic compliance data, $J_c^\dagger(t)$ to obtain the discrete parameters $J_i^\dagger$ and λ_i in a generalized Voigt–Kelvin model.

4. Further digging by archeologists has unearthed creep-recovery data on the material in Problem 1, Chapter XV. The data are in the form of recoverable (elastic) shear strain, $\gamma^\dagger(t)$ (obtained from creep *recovery*).

t(s)	$\gamma^\dagger(t)$	t(s)	$\gamma^\dagger(t)$
0	0	7.33	0.781
1.69	0.500	7.89	0.793
2.26	0.563	8.45	0.803
2.82	0.610	9.02	0.813
3.38	0.650	9.58	0.821
3.95	0.676	10.13	0.831
4.51	0.701	10.71	0.840
5.07	0.719	11.83	0.853
5.64	0.739	12.40	0.859
6.20	0.754	12.97	0.865
6.76	0.768	∞	0.977

In this experiment, $\tau_0 = 9.0 \times 10^4$ dyn/cm^2.

   a. Convert the data to recoverable (elastic) creep compliance, $J_c^\dagger(t)$.
   b. Estimate a characteristic time for this material.

c. Apply the solution to Problem 3 above to these data. Plot $J_c^\dagger(t)$ predicted by your generalized Voigt–Kelvin model along with the experimental data to see how good a fit is obtained.

5. Example 18.1 analyzes the response of a Maxwell element in an engineering stress–strain test. Do the same thing for a Voigt–Kelvin element. Illustrate the effect of strain rate with a sketch. Don't expect great realism.

6. The four-parameter model discussed in the text consists of a series combination of Maxwell and Voigt–Kelvin elements. Here, consider a four-parameter model that consists of a *parallel* combination of Maxwell (G_1, η_1) and Voigt–Kelvin (G_2, η_2) elements.

   a. Does this model represent a fluid of a solid?

   b. Is the model suited for representing creep, stress relaxation, neither, or both?

   c. Write in terms of model parameters the equilibrium ($t = \infty$) strain in response to an applied stress τ_0.

   d. Which of the following is the model capable of representing?

   (1) Instantaneous elastic deformation

   (2) Retarded elastic deformation

   (3) Equilibrium viscous flow

   (4) Instantaneous elastic recovery

   (5) Retarded elastic recovery

   (6) Permanent set

7. Analyze the dynamic properties of a Voigt–Kelvin element; i.e., obtain G' and G'' in terms of model parameters G and η and the frequency ω. *Hint:* Unless you're a confirmed masochist, do not use the Boltzmann principle here. Just examine the response to a sinusoidal strain.

8. Consider the three-parameter model (Fig. 18.1e).

   a. Write an expression for $J(t)$ in terms of model parameters.

   b. Write an expression for $J^\dagger(t)$ in terms of model parameters.

   c. Write an expression for the characteristic time λ_c in terms of model parameters.

   d. A three-parameter model initially free of stress and strain is subjected to the following stress history:

$$0 \le t \le \lambda \quad \tau = \tau_0$$
$$\lambda \le t \le 2\lambda \quad \tau = 2\tau_0$$
$$2\lambda \le t \quad \tau = 0$$

Sketch $\gamma(t)$ and determine the equilibrium strain $\gamma(\infty)$.

9. Examine the response of a two-element generalized Maxwell model (Fig. 18.10) with the following parameters:

$$G_1 = 10^{10} \, \text{dyn/cm}^2 \qquad \eta_1 = 10^6 \, \text{P}$$

$$G_2 = 10^6 \, \text{dyn/cm}^2 \qquad \eta_2 = 10^6, \, 10^8, \, \infty \, \text{P}$$

a. Plot $\log G_r(t)$ vs. $\log t$.
b. Plot $\log G'$ and $\log G''$ vs. $\log \omega$ with $\eta_2 = 10^8 \, \text{P}$.
c. Assume that the above model parameters apply at $T_g + 20°\text{C}$, and further that the model's temperature shift factors a_T are given by the WLF equation with the "universal" constants. Plot the log of the 10-s relaxation modulus, $G_r(10)$ vs. $T - T_g$.
Cover a wide enough range of the independent variable so that the G's range between 10^4 and $10^{10} \, \text{dyn/cm}^2$.

10. The text defines the Deborah number for transient and dynamic tests. How would you define De for an engineering stress–strain test?

11. A Maxwell element has $G = 10^8 \, \text{dyn/cm}^2$ and $\lambda = 1$ s at a temperature of $T = T_g + 20°\text{C}$. Assume that the element's shift factors are given by the WLF equation with the "universal" constants. Calculate G' and G'' at a frequency of $\omega = 1 \, \text{s}^{-1}$ and a temperature of $T = T_g + 40°\text{C}$.

12. A three-parameter model, the Zener element or standard linear solid (SLS), has been used to represent viscoelastic behavior in certain solids. Two equivalent forms of the SLS are:
a. A spring (G_2) in series with a Voigt–Kelvin element (G_1, η_1).
b. A spring (G_3) in parallel with a Maxwell element (G_4, η_2).
Obtain the equations necessary to relate the spring constants in one form of the SLS to the spring constants in the other; i.e., given G_1 and G_2, how could you determine G_3 and G_4, or vice versa?

13. The shear creep compliance of a thermoplastic at 25°C is described by

$$J_{c25}(t) = 1.2 \times 10^{-3} \, t^{0.10} \qquad (\text{m}^2/\text{N with } t \text{ in s})$$

a. This material has $T_g = 0°\text{C}$. Assume that its temperature shift factors are given by the WLF equation with the "universal" constants. Obtain an equation that gives its shear creep compliance at 35°C, $J_{c35}(t)$.
b. The material is subjected to the following stress history at 25°C:

$$t \leq 0 \, \text{s} \qquad \tau = 0 \, \text{Pa}$$

$$0 \leq t \leq 1000 \, \text{s} \qquad \tau = 1000 \, \text{Pa}$$

$$1000 \leq t \leq 2000 \, \text{s} \qquad \tau = 1500 \, \text{Pa}$$

$$2000 \leq t \, \text{s} \qquad \tau = 0 \, \text{Pa}$$

Calculate the shear strain at 2500 s.

14. Consider the four-parameter model (Fig. 18.1f). The model is subjected to a creep test. Identify the following quantities in terms of model parameters:

    a. η_0

    b. $J_c(0)$

    c. $J_c(\infty)$

    d. $J_c^{\dagger}(0)$

    e. $J_c^{\dagger}(\infty)$

    f. λ_c

15. The four-parameter model above is subjected to the following stress history, in which τ_0 is a constant stress:

$$t \leq 0 \qquad \tau = 0$$
$$0 \leq t \leq t_s \qquad \tau = -\tau_0$$
$$t_s \leq t \leq 2t_s \qquad \tau = +2\tau_0$$
$$2t_s \leq t \qquad \tau = 0$$

    Write an expression for the permanent set $\gamma(\infty)$ that results from this stress history.

16. Dynamic measurements taken on a polymer solution at a frequency 1 radian/s give values of $G' = 5.0 \times 10^4$ dyn/cm^2 and $G'' = 2.0 \times 10^4$ dyn/cm^2. On the basis of these data, what quantitative statement (if any) can you make about the steady-state viscosity of this material at the same temperature?

17. A particular polystyrene sample shows a 10-s relaxation modulus $E_r(10)$ of 1.0×10^8 dyn/cm^2 at 100°C. Estimate how long it would take for this same sample to reach the same modulus at a temperature of 115°C. $T_g = 100$°C.

18. Consider the four-parameter model (Fig. 18.1f). In terms of model and test parameters, answer the following:

    a. What is $G_r(0)$, the initial modulus in stress relaxation?

    b. What is the equilibrium modulus $G_r(\infty)$ in stress relaxation?

    c. Repeat (*b*) when the model is used to represent a lightly crosslinked polymer.

    d. In the shear analog of an engineering stress–strain test ($\dot{\gamma}_0$ constant), what is the limiting high strain rate ($\dot{\gamma}_0 \to \infty$) modulus?

    e. Repeat (*d*) for the limiting low strain rate ($\dot{\gamma}_0 \to 0$) modulus.

    f. In a dynamic test, what is the limiting low-frequency storage modulus, $G'(\omega \to 0)$?

    g. In a dynamic test, what is the limiting high-frequency storage modulus, $G'(\omega \to \infty)$?

    h. Repeat (*f*) for G''.

    i. Repeat (*g*) for G''.

19. Consider a four-parameter model made up of a Maxwell element (G_1, η_1) *in parallel* with a Voigt–Kelvin element (G_2, η_2). Obtain expressions for $G'(\omega)$ and $G''(\omega)$ for this model in terms of the model parameters.

20. Creep data for a particular material are fit well with a standard four-parameter model (Fig. 18.1f), with the following parameters:

$$G_1 = 1250\,\text{dyn/cm}^2 \qquad \eta_1 = 1.0 \times 10^6\,\text{P}$$
$$G_2 = 500\,\text{dyn/cm}^2 \qquad \eta_2 = 2.5 \times 10^5\,\text{P}$$

For the creep response of this material, calculate
a. $J_c^\dagger(\infty)$.
b. λ_c.
c. $\gamma(1000\,\text{s})$ for the applied stress history.

$$t \le 0\,\text{s} \qquad \tau = 0\,\text{dyn/cm}^2$$
$$0 \le t \le 500\,\text{s} \qquad \tau = -1000\,\text{dyn/cm}^2$$
$$500 \le t\,\text{s} \qquad \tau = +2000\,\text{dyn/cm}^2$$

21. Construct a master curve log $E_r(t)$ vs. log t for the linear polystyrene sample (C) in Fig. 18.23. Assume that the shift factors are given by the WLF equation with the "universal" constants. $T_g = 100°\text{C}$.

22. The Kepes balance rheometer is another device that determines dynamic properties in steady rotation. Material is confined between concentric hemispheres whose axes of rotation form an angle θ. Both hemispheres rotate with the same angular velocity ω. $R_o - R_i = \delta \ll R_i$ and θ is small.
a. Obtain an expression for the shear strain in this device.
b. A polymer melt is tested in such a device. Do you think the forces that arise (if any) will tend to increase, decrease, or not influence θ?

23. Dynamic measurements on a sample of poly(ethyl procrastinate) (PEP) give $G' = 1.0 \times 10^8\,\text{dyn/cm}^2$ at $T_g + 20°\text{C}$ and a frequency of 1 cycle/s. At what frequency would this sample have the same G' at a temperature of $T_g + 40°\text{C}$?

24. Crud Chemicals is enviously noticing the market for Tygon® flexible tubing. They have developed a linear, amorphous polymer with $T_g = -10°\text{C}$ which they think may be a good competitor in this market. Their polymer has excellent clarity and chemical resistance, but they're not too sure about its mechanical suitability.
    To check it out mechanically, they extrude some tubing and subject it to a burst tests at 20°C. These tests consist of taking a fixed length of tubing, pressurizing it to 50 psi, and determining the average burst time for a large

number of samples. The average life of their specimens turns out to be 40 h at 20°C.

    a. To what common laboratory test of viscoelasticity does this test most closely correspond?

    b. Estimate the average life of the same tubing in the same burst test at 30°C.

25. Figure 18.6 crudely illustrates the effect of strain rate on the engineering stress–strain test of a viscoelastic material at constant temperature.

    a. Make a similar qualitative sketch that illustrates the effect of temperature at constant strain rate.

    b. A stress–strain test is run on a sample of polyvinyl acetate, $T_g = 24°C$ at $\dot{\varepsilon}_0 = 1\,\mathrm{s}^{-1}$. The initial slope (Young's modulus) is 5000 psi. At a temperature of 30°C, what would $\dot{\varepsilon}_0$ have to be to give again an initial modulus of 5000 psi?

26. The 25°C master curve in Fig. 18.21 extends over the time range from about 10^{-14} to 10^{+2} h. What approximate time range would the master curve at $T_0 = -40°C$ cover?

27. Items are sometimes fastened to plastic parts by screws driven into smooth holes molded in the parts. Over time, however, the screws loosen. To what common laboratory test of viscoelasticity does this correspond?

28. Given the statement that a polymer's mechanical behavior depends on the Deborah number, how is the material's characteristic time λ_c related to its temperature shift factor a_T?

29. Sections of large-diameter pipe are commonly joined by bolting together flanges with a polymer-based gasket in between. A pipeline carrying waste acid at atmospheric pressure is assembled using gaskets made from a linear, amorphous polymer that has a T_g of 0°C. During the winter, when the average temperature is 0°C, the joints last about 1000 h before they begin to leak and have to be retorqued.

    a. To what common laboratory test does this situation correspond?

    b. Estimate the joint life during summer, when the average temperature is 25°C.

PART IV

Polymer Technology

CHAPTER XIX

Processing

19.1 INTRODUCTION

The term *processing* is used here to describe the technology of converting raw polymer, or compounds containing raw polymer, to articles of a desired shape. Considering the wide variety of polymer types and the even wider variety of articles made from them, a complete description and analysis of the myriad of processing techniques that have sprung up through the years would be impossible here. There are references that provide detailed qualitative descriptions of polymer processing operations.[1-4] Similarly, quantitative treatment of processing operations has been the subject of a number of books.[5-10] Here, we outline the common processing techniques, introduce relevant terminology, and consider how the various techniques are based on the fundamentals previously discussed. The reader wishing additional detail may consult these references and the more specialized ones cited in the sections to follow.

19.2 MOLDING

Molding consists of confining a material in the fluid state in a mold where it solidifies, taking the shape of the mold cavity. *Injection molding*[11,12] is the most common means of fabricating thermoplastic articles. Figure 19.1 illustrates a typical injection press. The molding compound, usually in the form of pellets of approximately $\frac{1}{8}$-in. (3-mm) cubes or $\frac{1}{8}$-in.-diam $\times \frac{1}{8}$-in.- long cylinders (molding "powder") is fed from a hopper (which may be heated with circulating hot air to dry the material) to an electrically heated barrel. The pellets are conveyed forward through the barrel by a rotating screw. The material is melted as it goes by a combination of heat from the barrel and the shearing (viscous energy dissipation) of the screw.

Older machines had a reciprocating plunger in place of the screw. All the heat for melting had to be supplied by conduction from the barrel, and mixing was

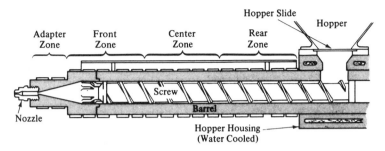

Figure 19.1 Injection molding machine. From *Modern Plastics Encyclopedia*, McGraw-Hill, New York, 1969–1970 ed.

very poor. This resulted in low plasticizing (melting) rates, and a thermally nonuniform melt. The screw generates heat *within* the material and provides some mixing, thereby increasing plasticizing rates and giving a more uniform melt temperature.

Molten material passes through a check valve at the front of the screw, and as it is deposited ahead of the screw, it pushes the screw backward while material from the previous shot is cooling in the mold. The cooled parts are ejected from the mold, the mold closes, and the screw is pushed forward hydraulically, injecting a new shot into the mold.

Some machines have a two-stage injection unit. A rotating-screw preplasticizer feeds molten polymer into an injection cylinder, from which it is injected into the mold by a hydraulically driven plunger. This configuration is claimed to provide better control of shot size.

The molten polymer flows through a *nozzle* into the water-cooled mold, where it travels in turn through a *sprue, runners,* and a narrow *gate* into the *cavity* (Fig. 19.2). When the part has cooled sufficiently, the mold opens and knockout pins eject the parts, with material cooled in the sprues and runners attached. This material is removed by hand or by robots. With thermoplastics, it is chopped and recycled back to the hopper as *regrind*. Depending on the extent of material degradation in the cycle and the property requirements of the finished part, injection-molding operations may tolerate up to 25% regrind in the hopper feed. *Hot-runner molds* contain heaters that maintain the material in the sprue and runners in the molten state, and often a shutoff valve ("valve gate") at the gate. This eliminates regrind.

The molds are opened and closed by hydraulic cylinders, toggle mechanisms, or combinations of both. These have to be pretty hefty, because pressures in the mold can reach several thousand psi and the projected area of some parts may be hundreds of square inches.

The molds themselves often involve much intricate hand labor and can be quite complex mechanically. Surfaces are usually chrome-plated for wear resistance. Molds can thus be quite expensive, but when amortized over a production

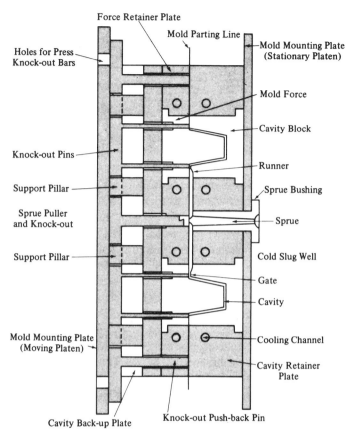

Figure 19.2 Two-cavity injection mold. From *Modern Plastics Encyclopedia*, McGraw-Hill, New York, 1969–1970 ed.

run of many thousand parts, the contribution of mold cost to the cost of the finished item may be insignificant.

Injection-molding presses are rated in terms of tons of mold-clamping capacity and in ounces (of general-purpose polystyrene) of shot size. They range from 2-ton, 0.25-oz laboratory units to 6600-ton, 2400-oz monsters that are used to mold garbage cans, TV cabinets, dishwasher tubs, lawn furniture, etc. They are usually completely automated. Cycle times (and therefore production rates) sometimes depend on plasticizing capacity (the rate at which the material can be melted), but more often than not are limited by the cooling time in the mold, which is in turn established by the thickness of the part and the (usually very low) thermal diffusivity of the material. Typical cycle times are between $\frac{1}{4}$ and 2 min. As usual, there are compromises involved. It is often tempting to try to reduce cooling time by lowering the mold temperature, and thereby increase the cooling rate, and/or lowering the temperature at which the material is injected into the mold, reducing the amount it must be cooled before solidifying enough to allow removal from the mold without distortion. The higher material viscosity that results, however, can give rise to short shots (incomplete mold filling), poor surface finish, and part distortion from frozen-in strains (see below).

Injection-molding machines now are commonly equipped with feedback control systems that monitor cavity pressure, screw position, and/or screw velocity and control these variables through the hydraulic system to provide high-quality, uniform parts, using a minimum amount of material and minimizing the number of rejects. A new line of machines has appeared in which the hydraulic systems are replaced with ac servomotors. They are claimed to provide precise control and extreme cleanliness.

Example 1. Use the four-parameter model (Fig. 18.2*f*), to explain the presence of frozen-in strains in injection-molded parts.

Solution. When a polymer melt is squirted into a mold, it is subjected to a combination of volumetric (from the high pressures), shear, and elongational (from flow-induced orientation) strains. On a molecular level, these all arise because of deviations from the most-random, highest-entropy configuration, and therefore can be represented by extension of the Voigt–Kelvin element, spring 2 and dashpot 2 of the four-parameter model. When the mold has filled and the gate freezes, these strains begin to relax (the Voigt–Kelvin element begins to contract). However, before retraction is complete, the rapidly dropping temperature in the mold raises η_2 to the point where the rate of retraction becomes very low, and the part is ejected from the mold with the strain frozen in. Subsequent recovery outside the mold causes distortion of the part. The problems are compounded by the fact that temperatures and flow fields are not uniform within the mold, so the frozen-in strains vary from point to point in the part, giving rise to nonuniform distortions (warping). (It is for this reason that the thermally uniform melt from a screw machine in beneficial.) These problems can, of course, be minimized by using higher melt temperatures and/or mold temperatures, but both lengthen cycle time. It might also be pointed out here that part failure can result if these frozen-in strains are suddenly released, as, for example, by the concentration of stress by a scratch or sharp corner on the part, or by plasticization by a liquid (stress cracking).

Foamed plastic articles are molded by incorporating a blowing agent (Chapter XX). These blowing agents may be inert but volatile liquids, compounds that decompose chemically at elevated temperatures to liberate a gas such as nitrogen or carbon dioxide, or simply nitrogen dissolved in the polymer. In any case, the blowing gas is kept in solution by the high pressures ahead of the screw. The mold is partially filled with a shot from the cylinder. The reduced pressure within the mold allows the blowing agent to vaporize, expanding the shot to fill the cavity and giving a foamed part.

Until fairly recently, one drawback of injection molding was its inability to produce parts with complex internal passages. To be sure, simple holes could be incorporated by using core pins which are retracted mechanically either parallel or perpendicular to the mold-opening direction. To permit retraction, however, these pins (and the resulting holes) had to be straight and if not of constant cross section, tapered appropriately with regard to the direction of retraction.

The *lost-core* process allows the molding of parts with curved internal passages of varying cross section. The core is cast in a low-melting bismuth–tin alloy. This cast core is placed in the injection mold and the plastic is molded around it. After the part is ejected from the mold, the metal core is melted out. The high thermal diffusivity of the metal core prevents its melting during the molding step. This process is becoming increasingly popular for molding, e.g., automotive intake manifolds, where it provides the necessary curved passages with a nice, smooth interior finish.

Injection molding was at one time confined exclusively to thermoplastics. It is now also used for thermosets, which are injected into a heated mold, where they solidify through the curing (crosslinking) reaction. This is a tricky operation. The compound must be heated just enough in the barrel to achieve fluidity and injected into the mold before it begins to cure appreciably. This requires precise control of temperatures and cycle timing to prevent premature cure in the barrel and the resulting shutdown-and-cleanout operation. Thermoset sprues and runners cannot be recycled, of course, and must be minimized to cut waste.

Thermosetting compounds are traditionally *compression* molded. The molds are mounted in hydraulic presses on steam-, electric-, or oil-heated platens. The molding compound is fed to the heated mold, which closes, maintaining the material under pressure until cured. The part is then ejected from the mold.

Molding compound in the form of granules or powder may be fed to the mold automatically in weighed shots, or as pre-formed (by cold pressing) tablets. The charge is often preheated to reduce heating time in the mold and thereby cycle time.

Mold temperature is a critical variable in compression molding. The higher it is, the faster the material cures, but if it cures too fast, it will not have enough time to fill thin sections and far corners of the mold, so a compromise must be reached. Material suppliers attempt to optimize the cure characteristics to provide minimum cycle times.

Transfer molding is a variation of compression molding. Here, the material is melted in a separate transfer pot, from which it is squirted into the mold. This can give faster cycle times and since no solid material is pushed around in the mold cavity itself, damage to delicate inserts (e.g., metal electrical contacts molded into a part) is minimized, mold wear is reduced and greater ease in filling intricate molds and more uniform cures are obtained. Material cured in the transfer pot and sprue is wasted, however.

Reaction injection molding (RIM)[13] (Fig. 19.3) is a process developed to mold large polyurethane (Chapter II, Example 4Q) or polyurea (Chapter II, Problem

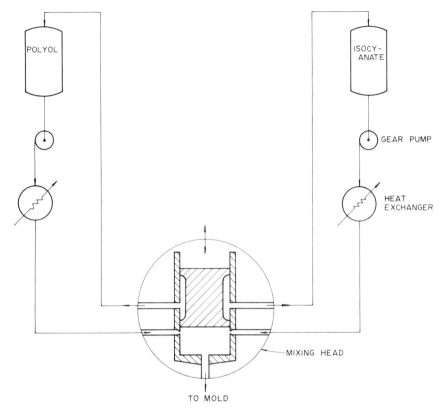

Figure 19.3   Reaction injection molding (RIM). The Kraus–Maffei mixing head is shown in the shot position. At the end of the shot, the plunger moves down, packing the mold, cleaning the mixing chamber, and recirculating the reactant streams back to their storage tanks.

4T) articles directly from the starting chemicals with cycle times comparable to injection molding. Between shots, the highly reactive liquid polyol (for polyurethanes) or polyamine (for polyureas) and isocyanate components are circulated from separate tanks through heat exchangers which regulate their temperatures. During the shot, the streams are pumped at about 2500 psi through nozzles in a mixing head, where they are impinged at high velocity and thoroughly mixed by the resulting turbulence. The mixed stream begins to react as it flows into the mold, where the polymerization reaction is completed. The mold is maintained at about 150°F, initially heating the reacting system, but later removing the exothermic heat of polymerization as the reaction approaches completion. Since the unreacted stream is relatively low in viscosity (about like pancake syrup), unlike the molten polymer in an injection-molding process, it flows easily to fill large molds with narrow clearances at relatively low pressures (100 psi, or so, in contrast to the thousands of psi in injection molding).

These low pressures allow the use of relatively inexpensive molds and low-force mold clamping systems (compared to injection molding).

The polyurethanes and polyureas may be formulated to be flexible or rigid, solid or foamed. Fillers and/or reinforcing agents (for example, glass fibers or flake) may be added to one or both components (in which case the process is sometimes known as RRIM—reinforced RIM). A major application of RIM is to produce energy-absorbing front and rear ends and body panels for automobiles. Cycle times of two minutes or less are feasible for such large parts.

RIM has achieved considerable success, particularly in the automotive industry, where the drive to cut weight is intense. RIM panels have already replaced stamped steel in a variety of automotive applications. It is now facing stiff competition from other plastics and processes, injection-molded thermoplastics, for example, for this large and lucrative market.

A somewhat similar process known as *resin-transfer molding* (RTM) is used to produce highly reinforced parts from low-viscosity, reactive starting materials. Reinforcing preforms, usually consisting of layers of fiberglass cloth or mat, are placed in the mold. When the mold is closed, they pretty much extend throughout the cavity and therefore the finished part. The resin and catalyst components are forced by positive-displacement piston pumps through a static or motionless mixer (Fig. 19.12) into the closed mold, where they cure. To insure rapid and complete impregnation of the reinforcing material, care must be taken to vent the mold properly to allow the escape of air as the resin is injected. In some cases, the air is sucked out of the mold with a vacuum pump prior to resin injection.

RTM has been used mainly with unsaturated polyesters (Chapter II, Example 3), but it is also applicable to epoxies, polyurethanes, etc. Parts weighing hundreds of pounds have been produced (truck air deflectors, for example). As with RIM, because the viscosity of the flowing material is low, relatively modest pressures and inexpensive molds are required.

19.3 EXTRUSION[14-17]

Thermoplastic items with a uniform cross section are formed by extrusion. This includes many familiar items such as pipe, hose and tubing, gaskets, wire and cable insulation, sheeting, window-frame moldings, and house siding. Molding powder is conveyed down an electrically or oil-heated barrel by a rotating screw. It melts as it proceeds down the barrel and is forced through a *die* which gives it its final shape (Fig. 19.4). Vented extruders incorporate a section in which a vacuum is applied to the melt to remove volatiles such as traces of unreacted monomer, moisture, solvent from the polymerization process, or degradation products.

The design of extruder screws is an interesting and complex technical problem and has received considerable study. Screws are optimized for the particular polymer being extruded. Basically, a screw consists of three sections: melting,

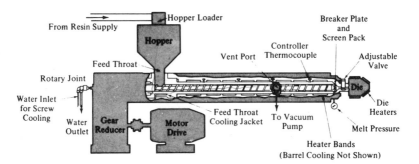

Figure 19.4 Vented extruder. From *Modern Plastics Encyclopedia*, McGraw-Hill, New York, 1969–1970 ed.

compression, and metering. The function of the melting section is to convey the solid pellets forward from the hopper and convert them into molten polymer. Its analysis involves a combination of fluid and solid mechanics and heat transfer. The compression section, in which the depth of the screw flight decreases, is designed to compact and mix the molten polymer to provide a more-or-less homogeneous melt to the metering section, the function of which is to pump the molten polymer out through the die.

This last section is well understood. Analysis of the metering section is an interesting application of the rheological principles discussed previously, as is much of die design. The determination of the die cross section needed to produce a desired product cross section (other than circular) is still pretty much a trial-and-error process, however. Viscoelastic polymer melts swell upon emerging from the die (recovering stored elastic energy), and the degree of die swell cannot be reliably predicted.

In addition to die swell, as the extrusion rate is increased, the extrudate begins to exhibit roughness, and then an irregular, severely distorted profile. This phenomenon is known as *melt fracture*. It is generally attributed to melt elasticity, but there is currently no way of quantitatively predicting its onset or severity. It can be minimized by increasing die length, smoothly tapering the entrance to the die, and raising the die temperature.

Extruders are normally specified by screw diameter and length-to-diameter ratio. Diameters range from 1 in. (2.5 cm) in laboratory or small production machines to 1 ft (30 cm) for machines used in the final pelletizing step of production operations. Typical L/D ratios seem to grow each year or so, with values now in the 20/1 to 36/1 range.

Single-screw extruders depend for their pumping ability on the drag flow of material between the rotating screw and stationary barrel. As a result, they are not positive-displacement pumps and tend to give a rather broad residence time distribution. Moreover, they are not particularly good mixing devices. Counter-rotating *twin-screw* extruders are true positive-displacement pumps, capable of generating the high pressures needed in certain profile extrusion applications.

Co-rotating twin-screw extruders, though not positive displacement pumps, with proper screw design can give excellent mixing and a narrow residence time distribution (subjecting all the material to essentially the same shear and temperature history). They are, therefore, used extensively in polymer compounding (mixing) operations, and to a certain extent as continuous polymerization reactors.

In steady-state extruder operation, most or all of the energy needed to plasticize the polymer is supplied by the drive motor through viscous energy dissipation. The heaters on extruders are needed mainly for startup and because enough heat can't always get from where it's generated (the compression and metering zones) to where it's needed (the melting zone). In fact, cooling through the barrel walls and/or screw center is sometimes necessary. A steady-state energy balance on an operating extruder is shown below[17]:

Extruder: 3.5-in. (8.9-cm) barrel, 32/1 L/D

Material: high-impact polystyrene

Operating conditions: screw speed 121 rpm
throughput 566 lb/h (257 kg/h)

Energy inputs

Shaft work (viscous energy dissipation)	+ 90%
Heaters	+ 10%

Energy outputs

Polymer enthalpy increase	− 65%
Losses to ambient	− 15%
Cooling water	− 20%

The complex interplay between the rate of viscous energy dissipation, temperature, and material viscosity in an extruder sometimes results in a variation in output (*surging*) even at constant screw speed. Where it is necessary to minimize surging, positive-displacement *gear pumps* are often added after the extruder.

Most packaging film is produced by *blow extrusion* (Fig. 19.5). A thin-walled, hollow cylinder in extruded vertically upward. Air is introduced to the interior of the cylinder, expanding it to a tube of film (less than 0.01-in. thickness). The tube is grasped between rolls at the top, preventing the escape of air and flattening it for subsequent slitting and windup on rolls. The expanded tube is rapidly chilled by a blast of air from a chill ring as it proceeds upward. With crystalline polymers such as polyethylene and polypropylene, this rapid chilling produces smaller crystallites and enhances film clarity.

"Cast" film is produced by extruding the polymer from a slot die onto polished metal chill rolls. The rolls maintain a more uniform thickness (gage) and the more rapid cooling and smoother surface finish provide superior film clarity. On the other hand, cast film is oriented (stretched) only in the machine

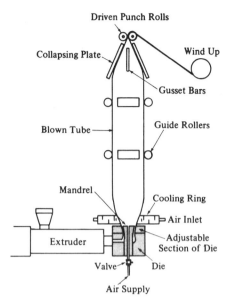

Figure 19.5 Blow extrusion film line. From *Modern Plastics Encyclopedia*, McGraw-Hill, New York, 1969–1970 ed.

direction, whereas blown film is biaxially oriented, and therefore has superior mechanical properties in equivalent thickness (or equivalent properties in a thinner film).

19.4 BLOW MOLDING[18]

A quick walk through a supermarket provides convincing proof of the economic importance of plastic bottles. They're nearly all made by blow molding. In one form of this process, *extrusion blow molding* (Fig. 19.6), a hollow cylindrical tube or *parison* is extruded downward. The parison is then clamped between halves of a water-cooled mold. The mold pinches off the bottom of the parison and forms the threads on the neck of the bottle. Compressed air expands the parison against the inner mold surfaces, and when the part has cooled sufficiently, the mold opens and the part is ejected. In one form of machine, the mold "shuttles" aside for blowing and ejection as a new parison is being formed beneath the die. High-production rotary machines may have two parison heads and index as many as twenty molds past them on a rotating table.

The rheological properties of the parison are important. If it sags too much before being grasped by the mold, the walls of the bottle will be too thin in places. Sag is minimized by using high molecular weight compounds with high viscosities (high "melt strength").

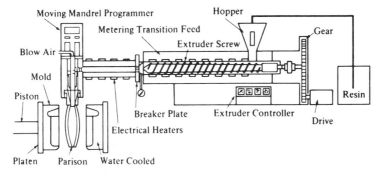

Figure 19.6   Extrusion blow molding. From *Modern Plastics Encyclopedia*, McGraw-Hill, New York, 1969–1970 ed.

Many blow-molding extruders are equipped with *parison programmers*, which vary the orifice diameter as the parison is being extruded, minimizing variations in wall thickness in the blown product. The parisons may be "ovalized" to reduce variations in wall thickness in objects of noncircular cross section.

Extrusion blow molding has passed well beyond the bottle-production stage. It is now being used to produce large items such as drums (to replace the familiar 55-gallon steel drum) and truck, automobile, and recreational-vehicle gasoline tanks. In such metal-replacement applications, blow-molded containers offer light weight and great design flexibility.

In *injection blow molding*, the parison is formed by injection molding rather than extrusion. A variation known as *stretch* (or *orientation*) *blow molding* is responsible for the now ubiquitous plastic soda-pop bottle. In this process, parisons (often called *preforms* in this case) are injection molded with the bottom end closed and the threads and neck molded on the open top. They are allowed to cool to room temperature. Prior to blowing, they are reheated in a radiant-

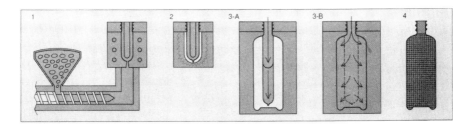

Figure 19.7   Stretch blow molding: 1, parison injection molded; 2, parison is reheated; 3-A, a rod stretches the parison, imparting axial orientation; 3-B, air expands parison against mold walls, imparting tangential orientation; 4, finished bottle ejected from mold. From *Modern Plastics* **55**(10), 22 (1978).

heat oven in which close control is exercised over the temperature profile of the parison. When introduced to the blow mold, the parison is normally a fraction of the length of the mold (Fig. 19.7), but before blowing, a rod rapidly stretches the parison to nearly the full length of the mold, orienting the polymer molecules in the axial direction. This is followed by blowing with compressed air, which imparts tangential (hoop) orientation. The resulting *biaxial* orientation improves the toughness, creep resistence, clarity, and barrier properties (resistance to permeation) of the bottle material. In this case, the frozen-in strains that are normally detrimental to thick, injection-molded parts (Example 1) are deliberately introduced to thin-walled blown bottles with highly beneficial results. The biaxial orientation provided by stretch blow molding is absolutely essential to the success of the poly(ethylene terephthalate) soda-pop bottles.

19.5   ROTATIONAL, FLUIDIZED-BED AND SLUSH MOLDING

Molding techniques have been developed to take advantage of the availability of finely powdered plastics, mainly polyethylene and nylons. In *rotational molding*, a charge of powder is introduced to a heated mold, which is then rotated about two mutually perpendicular axes. This distributes the powder over the inner mold surfaces, where it fuses. The mold is then cooled by compressed air or water sprays, opened, and the part is ejected.

Rotational molding is capable of producing extremely irregular hollow objects. The molds are inexpensive, often simply sheet metal, because no elevated pressures are involved, and they can be heated in simple hot-air ovens. Thus, capital outlay is relatively low. The process is in many ways competitive with extrusion blow molding for the production of large, hollow items such as drums and gasoline tanks. Blow molding requires a much larger initial investment, but is capable of higher production rates.

These two processes also provide a good example of how the processing operation, polymer, and finished part properties are often intimately connected. Blow molding can handle very high molecular weight linear polyethylenes, and in fact usually *requires* high molecular weight material to prevent excessive parison sag. Rotational molding, on the other hand, requires a low molecular weight resin because a low viscosity is needed to permit fusion of the powder under the influence of the low forces in a rotational mold.

When subjected to stresses for long periods of time, particularly in the presence of certain liquids, linear polyethylene has a tendency to fail through stress cracking. It turns out that high molecular weight resins are much more resistant to stress cracking. Thus, although the parts might appear similar, those produced by extrusion blow molding will ordinarily have superior resistance to stress cracking. The difference can be narrowed or eliminated by using more material (thereby lowering stress for a given load) or by using a crosslinkable polyethylene in rotational molding. Either of these solutions increases material cost, however.

Polymer powders are also used in a process known as *fluidized-bed coating.* When a gas is passed up through a bed of particles, the bed expands and behaves much like a boiling liquid. When a heated object is dipped into a bed of fluidized particles, those that contact it fuse and coat its surface. Such *100%-solids* coating processes are increasing in importance because they eliminate the pollution often caused by solvent evaporation when ordinary paints are used.

Similar procedures have been in use for years with plastisols (Chapter VII). Liquid plastisol is poured into a heated female mold. The plastisol in contact with the mold surface fuses and the remainder is poured out for reuse. This slush-molding process is used to produce objects such as doll's heads. Dipping a heated object into a liquid plastisol coats it with plasticized polymer. Vinyl-coated wire dishracks are familiar products of this process.

19.6 CALENDERING

Polymer sheet (greater than 0.01-in. thickness) may be produced either by extrusion through thin, flat dies or by *calendering* (Fig. 19.8). Basically, a calender consists of a series of rotating, heated rolls, between which the polymer compound (most often plasticized PVC) is squeezed into sheet form. The thickness of the sheet is determined by the clearance between the rolls. Commercial calenders may be very large (rolls up to 3 ft (1 m) in diameter by 8 ft (3 m) long) and represent a big capital outlay, but are capable of tremendous production rates (up to 100 yards/min). The polymer may be laminated to a layer of fabric between two rolls to give a supported sheeting. The final rolls may also be

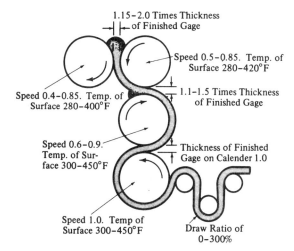

Figure 19.8 Inverted "L" calender, illustrating process variables for the production of PVC sheet. From *Modern Plastics Encyclopedia,* McGraw-Hill, New York, 1969–1970 ed.

embossed to impart a pattern to the sheet. Shower curtains, vinyl upholstery materials (Naugahyde®) and vinyl floor tile are produced by calendering.

19.7 SHEET FORMING (THERMOFORMING)[19, 20]

Thermoplastic sheet is converted to a wide variety of finished articles by processes known generically as *sheet forming* or *thermoforming*. Although the details vary considerably, these processes all involve heating the sheet above its softening point and forcing it to conform to a cooled mold (Fig. 19.9).

In *vacuum forming*, for example, the heat-plasticized sheet (either directly from an extruder or preheated in an oven) is drawn against the mold surface by the application of a vacuum from beneath the surface. Similarly, a positive pressure may be used to force the sheet against the mold surface, and where very deep draws are required, mechanical assists—*plug forming*—are used. In *drape forming*, the plasticized sheet is draped over a male mold, perhaps with a vacuum assist.

From a material standpoint, polymers used for sheet forming should have high "melt strengths," that is, high melt viscosities so that they do not draw down or thin out excessively or perhaps even tear in the forming operation. Thus, high molecular weight resins are preferred. One of the great advantages of sheet forming is that relatively inexpensive molds are required since no high pressures are involved. Epoxy molds are often used because they can be easily cast to shape from a hand-made pattern.

Among the many familiar items made by sheet forming are drinking cups, meat trays, cigarette packs, aircraft canopies, advertising displays, and lighting globes.

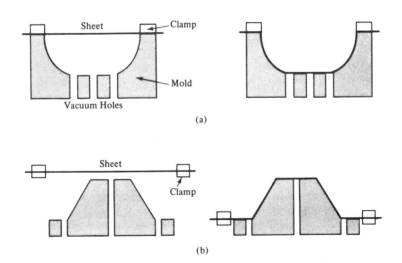

Figure 19.9 Sheet-forming process: (*a*) vacuum forming; (*b*) drape forming.

19.8 STAMPING

The plastics industry has long envied the ability to stamp sheet-metal parts with cycle times of a few seconds. Because of the high production rates possible with a stamping process (this is one important reason why most auto bodies are still largely stamped steel) and the large existing investment in stamping presses, the stamping process has been adapted to form thin-walled (0.1-in.) parts from glass-reinforced thermoplastic sheet (Azdel®). The sheet is heated in an infrared oven and then stamped in modified metal-forming equipment, with a dwell time of about 8 s as the material cools in the mold. Because of the thin sections, cooling is rapid and cycle times as low as 15 to 20 s can be achieved with automated feeding and part-removal systems.[21,22] So far, the process has been limited to rather simple shapes and shallow draws (crankcase oil pans, battery trays, bucket seat bottoms and backs, etc.), but because of its economic potential, it should achieve more widespread use in the future.

19.9 SOLUTION CASTING

Plastic sheet or film may be produced by dissolving the polymer in an appropriate solvent, spreading the viscous solution onto a polished surface, and evaporating the solvent. In the manufacture of photographic film base, the solution is spread with a "knife" onto a slowly rotating wheel about 2 ft wide and 20 ft in diameter. As the wheel revolves, heated air evaporates the solvent (which is recovered for reuse), and the dried film is stripped from the wheel before the casting point is again reached.

19.10 CASTING

The raw materials for many thermosetting polymers are available as low molecular weight liquids which are converted to solid or rubbery materials by the crosslinking reaction. Similarly, liquid vinyl monomers can be converted to solid, linear polymers through an addition reaction. These materials are easily cast to shape at atmospheric pressure in inexpensive molds. A good example of this process is the reproduction of carved wood furniture panels. Only the original is actually carved (which is the expensive step). A mold is made by pouring a room-temperature-curing silicone rubber over the original. When cured, the rubber mold is stripped from the original and used to cast multiple copies from a liquid, unsaturated polyester (Example 2.3). A free-radical initiator causes crosslinking to the final solid object, which reproduces faithfully the detail of the carved original at a fraction of the cost.

Such furniture components provide a good example of how the processing method of choice often depends on the production volume. They can also be injection molded from, for example, high-impact polystyrene. Because the panels are usually large, they require a large and very expensive molding machine. Similarly, the mold, with its intricate carving and wood-grain detail in hardened steel, is very expensive to produce. Thus, the capital outlay for injection-molded panels is huge, while the reverse is true for cast panels. Nevertheless, parts can be injection molded with cycle times under a minute and very little hand labor involved, and when the capital investment is amortized over a large number of panels, it becomes a small contribution to the total panel cost. On the other hand, the long cure times and hand labor required for the casting process make it prohibitively expensive for large production runs, but it is the preferred process for small numbers of parts.

Delicate electrical components are often encapsulated or "potted" by casting a thermosetting liquid resin (usually an epoxy) around them. The casting of acrylic sheet from monomer or syrup is described in Chapter XIII.

Example 2. A polymer of divinyl benzene $H_2C=CH$ ⟨◯⟩ $HC=CH_2$ is to be used to make rotameter bobs (1-in.-long × $\frac{1}{2}$-in.-diam cylinders). What processing technique should be used?

Solution. The monomer is tetrafunctional (Chapter II) and hence will be highly crosslinked in the polymerization reaction, and therefore cannot be softened by heat. The only suitable technique is casting directly from the monomer, either to final shape, or casting rods that can then be cut and machined to the final shape.

19.11   REINFORCED THERMOSET MOLDING

Many plastics, when used by themselves, do not possess enough mechanical strength for structural applications. When reinforced with high-strength fibers, however, the composites have high strength-to-weight ratios and can be fabricated into a wide variety of complex shapes. Glass and other fibers are used extensively to reinforce thermosetting plastics, mostly polyesters and epoxies. FRPs (fiberglass-reinforced plastics) are now used for just about all boats under 40 ft in length, truck cabs, automobile body panels, structural panels, aircraft components, and bathtubs.

Many such objects are fabricated by a process known as *hand layup*. The mold surface is often first sprayed with a pigmented but nonreinforced *gel coat* of the liquid resin to provide a smooth surface finish. The gel coat is followed up by successive layers of fiberglass, either in the form of woven cloth or random matting, impregnated with the liquid resin, which is then cured (crosslinked) to give the finished product.

This process is tremendously versatile. The molds are relatively inexpensive, because no high pressure is required. Objects may be selectively reinforced by adding extra layers of material where desired. The major drawback of the process is the expense of the hand labor involved, which makes it uneconomical for large production volumes. For this reason, it is giving way to a *sprayup* process. A special gun chops continuous fibers into approximately one-inch lengths. The chopped fibers are combined with a stream of liquid resin and sprayed directly onto the mold surfaces. Although the random chopped fibers do not reinforce quite as well as woven cloth or random mat, the labor savings are substantial. Sprayed-on polyester–fiberglass backings are also applied to vacuum-formed polymethyl methacrylate sheets, thus combining a smooth, hard, strain- and UV-resistant acrylic surface skin with the lightweight strength of a fiberglass-reinforced plastic. Such composites are used for sinks, bathtubs, and recreational vehicle bodies.

In many objects, stresses are not uniformly distributed, so the reinforcing fibers may be arranged to support the stress most efficiently. In "fiberglass" or "graphite" fishing rods, vaulting poles, golf club shafts, etc., the fibers are arranged along the long axis to resist the bending stresses applied. The process of *filament winding* extends this principle to more complex structures. Continuous filaments of the reinforcing fiber are impregnated with liquid resin and then wound on a rotating *mandrel*. The winding pattern is designed to resist most efficiently the anticipated stress distribution. The range of the Polaris submarine ballistic missile was increased by several hundred miles by replacing the metallic rocket casing with a filament-wound reinforced plastic system. This technique is also used to produce tanks and pipe for the chemical process industries, and even gun barrels, by filament winding about a thin metal core which provides the necessary heat and abrasion resistance.

Pultrusion[23] is used to produce continuous lengths of objects with a constant cross section, structural beams, for example. Continuous fibers (roving) and/or mat are impregnated by passing them through a tank of liquid resin and pulled slowly through a heated die of the desired cross section. The resin cures to a solid as it passes through the die.

19.12   FIBER SPINNING

The polymers used for synthetic fibers are similar, and in many cases identical to those used as plastics, but for fibers, the processing operation must produce an essentially infinite length-to-diameter ratio. In all cases, this is accomplished by forcing the plasticized polymer through a *spinnerette*, a plate in which a multiplicity of holes has been formed to produce the individual fibers, which are then twisted together to form a thread for subsequent weaving operations. The cross section of the spinnerette holes obviously has a lot to do with the fiber cross section, which in turn greatly influences the properties of the fiber. The three basic types of spinning operations differ mainly in the method of plasticizing and deplasticizing the polymer.

Melt spinning is basically an extrusion process. The polymer is plasticized by melting and pumped through the spinnerette. The fibers are usually solidified by a cross-current blast of air as they proceed to the drawing rolls. The drawing step stretches the fibers, orienting the molecules in the direction of stretch and inducing the high degrees of crystallinity necessary for good fiber properties. Nylons are commonly melt spun.

In *dry spinning*, a solution of the polymer is forced through the spinnerette. As the fibers proceed downward to the drawing rolls, a countercurrent stream of warm air evaporates the solvent. In this process, the cross section of the fiber is determined not only by the shape of the spinnerette holes, but also by the complex nature of the diffusion-controlled solvent evaporation process, because there is considerable shrinkage as the solvent evaporates. The acrylic fibers (Orlon®, Creslan,® Acrilan®, etc.), mainly polyacrylonitrile, are produced by dry spinning.

Wet spinning is similar to dry spinning in that a polymer solution is forced through the spinnerette. Here, however, the solution strands pass directly into a liquid bath. The liquid might be a nonsolvent for the polymer, precipitating it from solution as the solvent diffuses outward into the nonsolvent. The bath might also contain a substance that precipitates a polymer fiber by chemically reacting with the dissolved material. Here again, the process as well as the spinnerette influences the fiber cross section. Kevlar® is an example of a wet-spun fiber (Section 5.7). The polymer is dissolved in concentrated sulfuric acid and the solution is spun into a water bath. The water leaches out the acid, causing the polymer to precipate in fiber form.

Example 3. Suggest processing techniques for the manufacture of the following:

a. 100 000 ft of plasticized PVC garden hose.

b. 50 000 polystyrene pocket combs.

c. 100 000 polyethylene detergent bottles.

d. 5000 phenolic (phenol-formaldehyde) TV knobs.

e. 6 souvenir paperweights of polymethyl methacrylate containing a coin.

f. A strip of chlorinated rubber (a linear, amorphous polymer) and white pigment (60:40), roughly 0.001 in. thick, 4 in. wide, running 20 miles down the center of a highway.

g. 1 000 000 polystyrene meat trays, 0.005 in. thick.

Solution.

a. Extrusion.

b. Injection molding.

c. Blow molding.

d. Compression, transfer, or thermoset injection molding.

e. Although these objects could be injection molded, the small number of articles required would make injection molding uneconomical. They can easily be cast from the monomer.

f. Dissolve rubber in solvent, add pigment, and spray on highway. Evaporation of solvent leaves the desired strip.

g. Extrude sheet, then vacuum form trays.

19.13 COMPOUNDING

Polymers are almost always used in combination with other ingredients. These ingredients are discussed in subsequent chapters, but they must be combined with the polymer in a *compounding* operation.

Occasionally, if they don't interfere with the polymerization reaction, such ingredients may be incorporated at the monomer or low molecular weight polymer (e.g., unsaturated polyester, Chapter II, Example 3) stage and carried through the polymerization and/or crosslinking reaction. In such cases, viscosities are low enough to permit the use of standard mixing equipment. Similarly, powdered PVC and thermosets are compounded with other ingredients in the usual tumbling-type of blending equipment.

Because of their extremely high melt viscosities, specialized equipment is usually needed to compound ingredients with high molecular weight thermoplastics, however. In general, high shear rates and large power inputs per unit volume of material are required to achieve a uniform and intimate dispersion of

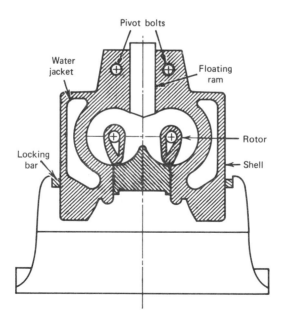

Figure 19.10 Banbury mixer. From L. K. Arnold, *Introduction to Plastics*, Iowa State UP, Ames (1968).

Figure 19.11   Two-roll mill.

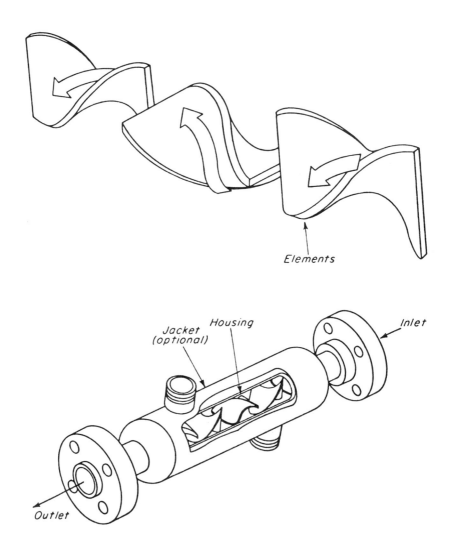

Figure 19.12   The Kenics Static Mixer. From W. L. McCabe, J. C. Smith, and P. Harriott, *Unit Operations of Chemical Engineering*, 4th. ed., McGraw-Hill, New York, 1985.

ingredients in the melt. Single- and twin-screw extruders (Section 19.3) are used extensively for continuous compounding. The latter, with screws often modified to incorporate special mixing sections, are better mixers and provide a narrower, more uniform distribution of residence times, while the former offer lower cost and greater mechanical simplicity. *Intensive mixers*, such as the Banbury (Fig. 19.10), subject the material to high shear rates and large power inputs in a closed, heated chamber containing rotating, intermeshing blades.

A *two-roll mill* (Fig. 19.11) generates high shear rates in a narrow nip between two heated rolls which counter-rotate with slightly different velocities. In commercial mills, the rolls are about 1 ft in diameter by 3 ft long. Once the polymer has banded on one of the rolls, ingredients are added to the bank between the rolls. The band is cut off the roll with a knife, rolled up, and fed back to the nip at right angles to its former direction. This is done several times to improve mixing. Despite extensive safety precautions, operators of two-roll mills often have n fingers ($n < 10$). To the author's knowledge, no systematic studies of the effect of the additional ($10-n$) ingredients on the properties of the compounded polymers have been reported.

Motionless or *static mixers* are a relatively recent development for continuous compounding.[24-26] One example of this type of device, the Kenics mixer (Fig. 19.12), consists of a series of alternating right- and left-handed helices which continuously divide the melt and cause it to rotate around its own hydraulic axis. Each element in a Kenics mixer breaks the stream into two parts. Therefore, an n-element mixer generates 2^n layers, so n does not have to be very large to achieve intimate mixing. The obvious advantage of this type of device is that it contains no moving parts (although the material must be pumped through it). It also requires relatively low power input per unit of material processed. Such mixers provide good radial mixing and relatively narrow residence-time distributions, despite the fact that flow is invariably laminar in the processing of polymer melts. They are also incorporated in extruders, between the screw and die, and in injection molding machines ahead of the nozzle to improve thermal homogeniety of the melt.

REFERENCES

1. Kroschwitz, J. (ed.), *Encyclopedia of Polymer Science and Engineering*, 2d. ed., Wiley, New York, 1985.

2. *Modern Plastics Encyclopedia*, McGraw-Hill, New York (yearly).

3. Rosato, D. V., and D. V. Rosato, *Plastics Processing Data Handbook*, Van Nostrand Reinhold, New York, 1989.

4. Morton-Jones, D. H., *Polymer Processing*, Chapman & Hall, London, 1989.

5. Han, C. D., *Rheology in Polymer Processing*, Academic, New York, 1976.

6. McKelvey, J. M., *Polymer Processing*, Wiley, New York, 1962.

7. Middleman, S., *Fundamentals of Polymer Processing*, McGraw-Hill, New York, 1977.

8. Pearson, J. R. A., *Mechanics of Polymer Processing*, Elsevier, New York, 1985.

9. Pearson, J. R. A., and S. M. Richardson (eds.), *Computational Analysis of Polymer Processing*, Applied Science, New York, 1983.

10. Tadmor, Z., and C. G. Gogos, *Principles of Polymer Processing*, Wiley-Interscience, New York, 1979.

11. Rosato, D. V., and D. V. Rosato, *Injection Molding Handbook*, Van Nostrand Reinhold, New York, 1985.

12. Rubin, I. I., *Injection Molding. Theory and Practice*, Wiley-Interscience, New York, 1973.

13. Macosko, C. W., *Fundamentals of Reaction Injection Molding*, Hanser, Munich, 1988.

14. Stevens, M. J., *Extruder Principles and Operation*, Elsevier, New York, 1985.

15. Tadmor, Z. and I. Klein, *Engineering Principles of Plasticating Extrusion*, Van Nostrand Reinhold, New York, 1970.

16. Rauwendaal, C., *Polymer Extrusion*, Hanser, Munich, 1986.

17. Private communication, NRM Corp.

18. Rosato, D. V., and D. V. Rosato (eds.), *Blow Molding Handbook*, Hanser, Munich, 1989.

19. Gruenwald, G., *Thermoforming: A Plastics Processing Guide*, Technomic, Lancaster, PA 1987.

20. Throne, J. L., *Thermoforming*, Hanser, Munich, 1987.

21. Sikes, S., *Mod. Plast.* **51**(1), 70 (1974).

22. Ward, L. G., *Plast. Eng.* **35**(4), 47 (1979).

23. Meyer, R. W., *Handbook of Pultrusion Technology*, Chapman & Hall, London, 1985.

24. Skoblar, S. M., *Plast. Technol.* Oct. 1974, p. 37.

25. Schott, N. R., B. Weinstein, and D. LaBombard, *Chem. Eng. Prog.* **71**(1), 54 (1975).

26. Chen, S. J., *Chem. Eng. Prog.* **71**(8), 80 (1975).

CHAPTER XX

Plastics

20.1 INTRODUCTION

As was mentioned in the introduction to this book, there are five major applications for polymers: plastics, rubbers, synthetic fibers, surface finishes, and adhesives. Previous sections have dealt with the properties of the polymers themselves. While these properties are undoubtedly the most important in determining the ultimate application, polymers are rarely used in a chemically pure form, so in a discussion of the technology of polymers, it is necessasry to mention in addition to the properties of the polymers required for a given application, the nature and reasons for use of the many other materials often associated with the polymers. The following brief chapters do this for the five major applications.

Plastics are normally thought of as being polymer compounds possessing a degree of structural rigidity—in terms of the usual stress–strain test, a modulus on the order of 10^8 Pa (N/m^2) (15 000 psi) or greater. The *molecular requirements* for a polymer to be used in a plastic compound are as follows: (1) If linear or branched, the polymer must be below its glass transition temperature (if amorphous) and/or below its crystalline melting point (if crystallizable) at use temperature; otherwise (2) it must be crosslinked sufficiently to restrict molecular response essentially to straining of bond angles and lengths (e.g., ebonite or "hard" rubber).

20.2 MECHANICAL PROPERTIES OF PLASTICS

The engineering properties of commercial plastics vary considerably within the broad definition given above. Figure 20.1 sketches some representative stress–strain curves for three types of plastics. As discussed previously, for any given material, the quantitative nature of the curves depends markedly on the

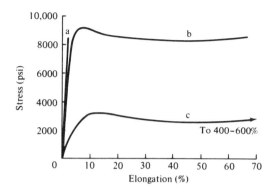

Figure 20.1 Typical stress–strain curves for plastics: (*a*) rigid and brittle; (*b*) rigid and tough; (*c*) flexible and tough.

rate of strain and the temperature. In general, faster straining and lower temperatures give higher moduli (slopes) and smaller ultimate elongations.

Plastics with stress–strain curves of the type in Fig. 20.1a are *rigid* and *brittle*. The former term refers to the high initial modulus. The latter refers to the area under the stress strain curve, which represents the energy per unit volume required to cause failure. These materials usually fail by catastrophic crack propagation at strains on the order of 2%. Since hardness correlates well with tensile modulus, it is another valuable property of this type of plastic. Examples of this class are polystyrene, polymethyl methacrylate, and most thermosets.

Curve *b* represents *rigid* and *tough* materials, sometimes known as *engineering plastics*. In addition to high modulus, tensile strength and hardness, these materials undergo *ductile deformation* or drawing beyond the yield point, evidence of considerable molecular orientation before failure. The latter confers toughness or *impact resistance*, that is, the ability to withstand shock loading without brittle failure. Examples of this class of materials are polycarbonates, cellulose esters, and nylons. Although the reasons for their toughness are not entirely clear, such plastics generally exhibit a secondary dynamic damping peak (Chapter XVIII) well below their use temperature.

Curve *c* represents flexible and tough plastics, as typified by low- and medium-density polyethylenes. Here, the ductile deformation leading to very high ultimate elongations results from the conversion of low-crystallinity material to high-crystallinity material. The tensile samples "neck down," with the density, percent crystallinity, and modulus of the material in the neck being appreciably greater than those of the parent material. Despite their good toughness, these materials are limited in their structural applications by their low moduli and tensile strengths.

20.3 CONTENTS OF PLASTIC COMPOUNDS

In addition to the polymer itself, which is seldom used alone, plastics usually contain at least small amounts of one or more of the following additives.[1,2]

1. Reinforcing Agents.[3] The function of these is to enhance the structural properties of the compound, in particular properties such as modulus (stiffness), strength, and the retention of these properties at higher temperatures. The use of long fibers to reinforce epoxy and polyester thermosets was mentioned in the previous chapter. While glass is by far the most common reinforcing fiber, largely due to its strength, ready availability, and relatively low cost, other, more exotic fibers are used where very high strength-to-weight ratios are required and cost is of secondary concern. Carbon or "graphite" fibers, made by the pyrolysis of pitch or polyacrylonitrile fibers, offer superior performance at a cost premium over glass. They are used in many aerospace applications and in more down-to-earth objects such as golf-club shafts, tennis-racket frames, racing-car bodies, and automotive springs and driveshafts. Aramid (*aromatic poly*amid) fibers (Kevlar®) are also used to reinforce plastics for premium-property applications, sometimes in combination with other fibers.

Short ($\frac{1}{8}$- to $\frac{1}{2}$-in.) (3- to 13-mm) glass and other fibers are used extensively to reinforce thermoplastics.[4] Addition of up to 40 wt% short glass fibers to nylon, polypropylene, polybutylene terephthalate, and so on, provides a relatively inexpensive way of greatly improving the structural strength of the plastic, as illustrated in Table 20.1. The glass-reinforced materials also have less mold shrinkage, and therefore give parts with better dimensional control and stability. The down side is that they are more difficult to process and often have a rough surface finish.

Note that major gains in heat-distortion temperature are obtained with the crystalline polymers nylon 6/6, polybutylene terephthalate, and polypropylene, but not with the amorphous polystyrene and polycarbonate. This is a general result. The crystallites "lock on" to the fibers, maintaining mechanical integrity well above T_g.

Another example of reinforcement is the use of up to 20% dispersed rubber particles, on the order of 1 μm in diameter, to increase the *toughness* or *impact*

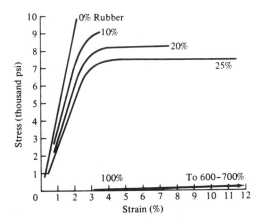

Figure 20.2 The influence of discrete, micrometer-sized rubber particles on the high-speed (133 in./in. min) stress–strain curves of polystyrene.[6]

Table 20.1 Some Properties^a of Short Glass-Fiber-Reinforced Thermoplastics

Plastic^b	PP		Nylon 6/6^c		PC		PBT		PS	
wt % glass	0	40	0	30	0	30	0	30	0	30
Ultimate tensile strength, psi ASTM D638	5300	11 700	12 400	25 000	9500	19 500	8200	16 500	6400	12 000
Ultimate elongation % ASTM D638	350	2.8	300	4	110	3.5	175	3	1.9	1.1
Tensile modulus, 10^3 psi ASTM D638	195	1300	390	1300	345	1300	360	1380	400	1250
Izod impact, $\frac{1}{8}$-in. ft-lb/in. ASTM D256A	0.8	1.7	1.3	3.0	16	2.3	0.85	1.3	0.4	2.0
Rockwell hardness ASTM D758	R91	R106	R120	R110	M70	M92	M73	M90	M67	M90

Linear coefficient of expansion, $10^{-6}/°C$ ASTM D696	90	30	80	35	68	22	67	25	77	20
264 psi DTUL,[d] °F ASTM D648	130	315	179	371	270	297	154	430	186	217
Density, g/cm^3 ASTM D792	0.91	1.22	1.14	1.30	1.20	1.41	1.34	1.51	1.05	1.29

[a] Where a range is given, the average is listed here.
[b] PP, polypropylene; PC, polycarbonate; PBT, polybutylene terephthalate; PS, polystyrene.
[c] Nylon properties are very sensitive to moisture content.
[d] Deflection temperature under load.

Source. *Modern Plastics Enyclopedia*, McGraw-Hill, New York, 1990.

strength of a normally brittle plastic. Figure 20.2 illustrates the effect of adding increasing levels of a rubber to polystyrene in a high rate-of-strain (to simulate shock loading) tensile test. The addition of the rubber particles decreases the modulus (slope) and ultimate tensile strength, but introduces considerable ductile deformation (elongation beyond the "knee"), thereby increasing the area under the stress–strain curve. It also imparts a secondary dynamic damping peak below use temperature. In effect, it converts a type *a* material (Fig. 20.1) to a type *b* material. For many applications, the improvement in impact strength considerably outweighs the slight decreases in modulus and tensile strength produced by the addition of rubber. *Impact modifiers* are rubbery additives that impart such toughness when blended with plastics. Proper compatibility (good adhesion but not solubility) between the phases is essential in these applications. This is often achieved with graft and block copolymers.

Commercially important examples of these *rubber-toughened* plastics include the high-impact polystyrenes (HIPS), in which polystyrene is toughened with a polybutadiene rubber, and the ABS (acrylonitrile-butadiene-styrene) plastics, in which a polybutadiene or poly(butadiene-*co*-acrylonitrile) rubber toughens a poly(styrene-*co*-acrylonitrile) glassy phase. More complete discussions of these two-phase polymer systems are available elsewhere.[6, 7]

2. Fillers.[8] Fillers are particulate materials whose major function often is simply to extend the polymer and thereby reduce the cost of the plastic compound. One of the earliest examples of a filler is the wood flour (fine sawdust) long used in phenolics and other thermosets. Calcium carbonate is used in a variety of plastics and polypropylene is often filled with talc. Even water has been used to extend polyester casting compounds.

Fillers may also be used to improve certain properties of the compound. They almost all reduce mold shrinkage and thermal expansion coefficient. They reduce warpage in molded parts. Mica and asbestos increase heat resistance. Hollow glass or phenolic microspheres reduce the density of the composite. *Conductive fillers*, such as certain carbon blacks and aluminum flake, impart electrical conductivity which bleeds off static charge and provides plastics with EMI (electromagnetic interference) shielding capability. Just about any filler whose hardness and modulus are greater than those of the polymer will increase the hardness and *initial* modulus of a mixture with that polymer, but generally, elongation, ultimate strength, toughness, and processability suffer.

3. Coupling Agents. For a reinforcing fiber such as glass to be of maximum benefit (or a filler not to be too detrimental to mechanical properties), stress must be efficiently transferred from the polymer to the reinforcing agent or filler. Unfortunately, most inorganics have hydrophilic surfaces, while the polymers are hydrophobic, so interfacial adhesion is often poor. This problem is exacerbated by the tendency of the surface of many inorganics (particularly glass) to adsorb water, which further degrades adhesion. Coupling agents are intended to help overcome these difficulties.

Currently, the most common coupling agents are silanes, with the general formula YRSi(OR′)₃. The (OR′) group reacts with the inorganic substrate, and the Y group reacts (or at least forms strong secondary bonds) with the polymer, thereby enhancing interfacial adhesion. The reaction sequence by which vinyl-triethoxysilane is believed to couple to a substrate containing hydroxyl surface functionality (e.g., glass) is shown below.

Step 1. Hydrolysis (most likely with water adsorbed on the substrate):

$$H_2C{=}CH{-}Si{-}(OC_2H_5)_3 + 3H_2O \rightarrow H_2C{=}CH{-}Si{-}(OH)_3 + 3C_2H_5OH$$

Step 2. Reaction with substrate

$$H_2C{=}CH{-}Si{-}(OH)_3 + HO{-}\Big\langle \rightarrow H_2C{=}CH{-}Si(OH)_2{-}O{-}\Big\langle + H_2O$$

The "dangling double bonds" so introduced can then participate in the cure of an unsaturated polyester resin (Example 2.3), covalently bonding the polymer to the surface.

A wide variety of silane coupling agents is available, with the functional groups optimized for use with various polymers and substrates. Titanate coupling agents perform in a similar fashion. Fillers and reinforcing fibers are supplied pretreated with coupling agent, or the agent may be added during compounding (in which case, presumably most of it migrates to the interface).

In the case of fillers, coupling agents have an added benefit. By converting an otherwise hydrophilic surface to a hydrophobic surface, compatibility with the polymer is improved. This results in easier and more uniform dispersion of the filler in the polymer, and at a given filler loading can reduce the viscosity substantially, thereby improving processability.

Graft and block copolymers often perform a coupling function in the case of the rubber-toughened plastics mentioned in the discussion of reinforcing agents. In the manufacture of high-impact polystyrene, for example, the polybutadiene is dissolved in the polymerizing styrene monomer. As a result, some of the styrene is grafted to the rubber. The grafted polystyrene chains are physically compatible with the surrounding homopolystyrene and are chemically bound to the rubber, thereby enhancing adhesion between the dispersed polybutadiene particles and the polystyrene matrix. The graft copolymer is thought to be essential for effective impact enhancement in this particular system, because simple mechanical blends of polystyrene and polybutadiene of similar composition do not show much impact enhancement. However, minor amounts of styrene-butadiene *block* copolymers will improve the compatibility and therefore the impact strength of such blends, much as a soap emulsifies oil in water.

4. Stabilizers. Most polymers are susceptible to one or more forms of degradation, usually as a result of environmental exposure to oxygen or UV radiation, or to high temperatures during processing operations. Stabilizers inhibit degradation of the polymer. In the case of polyvinyl chloride, for example, the major product of thermal degradation during processing is HCl, which catalyzes further degradation. Compounds such as metal oxides, which react with the HCl to form stable products, are used as stabilizers. Oxidative degradation of polymers is thought to take place by a free-radical mechanism involving crosslinking and/or chain scission initiated by free radicals from peroxides formed in the initial oxidation step. Similarly, these reactions can be initiated by free radicals produced by ultraviolet radiation. Stabilizers against such reactions are generally quinone-type organics which are effective free-radical scavengers.

5. Pigments. Plastics are often colored by the addition of pigments, which are finely divided solids. If the polymer is itself transparent, a pigment imparts opacity. Titanium dioxide is a common pigment where a brilliant, opaque white is desired. Sometimes, pigments perform other functions. For example, calcium carbonate acts as both a filler and a pigment in many plastics, and carbon black is often a stabilizer as well as a pigment. It prevents degradation by absorbing and preventing the penetration of UV light beyond the surface of the article.

6. Dyes. Dyes are colored organic chemicals that dissolve in the polymer to produce a transparent compound (assuming the polymer is transparent to begin with). Some thermoplastics, though transparent, develop a slight yellow tinge from minor degradation during processing which causes selective absorption of light toward the blue end of the spectrum. The yellow can largely be "canceled out" by the addition of a blue dye which reduces the transmission of yellowish wavelengths. This technique unavoidably results in a slight lowering of the total light transmission.

7. Plasticizers. The addition of a relatively low molecular weight organic plasticizer to a normally glassy polymer will progressively reduce its modulus by bringing the compound's glass transition temperature down closer to use temperature. The level at which the modulus is reduced to the point where the compound is considered an elastomer rather than a plastic is somewhat arbitrary, but plastics sometimes contain small amounts of a plasticizer to reduce brittleness.

8. Lubricants. *External* lubricants are low molecular weight organics that are relatively insoluble in the polymer and "plate out" or migrate to the surface of the compound and form a slippery coating during processing operations. They produce smoother extrudates and molded articles, and minimize sticking in the mold by acting as mold release agents. Where subsequent painting or printing of the surface is required, they can reduce adhesion, and so must be used with care. Stearic acid and its metal salts are common external lubricants. *Internal* lubricants are soluble in the polymer, and ease processing by lowering the compound's viscosity.

9. Processing Aids. Various compounds (often a second polymer) are used to modify the rheological properties of a polymer during processing to provide higher outputs, better surface finish, and easier handling in general. Low molecular weight polyethylene is added to PVC to improve extrusion behavior.

10. Curing Agents. These are chemicals whose function is to produce a cross-linked, thermosetting plastic from an initially linear or branched polymer. A vinyl monomer, such as styrene, a free-radical initiator, and sometimes a *promoter* (e.g., cobalt naphthenate—to speed up the reaction) are dissolved in a low molecular weight unsaturated polyester resin and crosslink it by an ordinary addition mechanism involving the double bonds in the polyester Chapter II, Example 3).

The polymer for a *two-stage* thermosetting compound is deliberately produced with a stoichiometric shortage of one of the reagents to give a highly branched, but not yet crosslinked structure, which is later cured in the mold with a curing agent. An "A-stage" phenolic polymer is produced by reacting phenol and formaldehyde in about a 1.25/1 mol ratio (a molal excess of formaldehyde is required for crosslinking). This still-thermoplastic product (a *novolac*) is compounded with fillers, pigments, reinforcing agents, etc., and a curing agent, hexamethylenetetramine, to give a "B-stage" resin. The curing agent decomposes in the presence of moisture (a product of the condensation reaction) upon heating in the mold, giving the additional formaldehyde required for crosslinking, and ammonia, which acts as a basic catalyst for the reaction:

$$\text{Hexamethylenetetramine} + 6H_2O \longrightarrow 6CH_2{=}O + 4NH_3$$

Hexamethylenetetramine $\qquad$ Water $\qquad$ Formaldehyde $\qquad$ Ammonia

Single-stage resins (*resoles*) are made using the final desired reactant ratio, about 1/1.5 phenol/formaldehyde, but the reaction is stopped short of crosslinking. Crosslinking is completed in the mold without the necessity of a curing agent.

Free-radical initiators (usually organic peroxides) are used as curing agents for saturated thermoplastic polymers. The curing agent must be compounded with the polymer at temperatures low enough to prevent appreciable decomposition to free radicals. The compound is then heated in the mold, producing free radicals which abstract protons from the polymer, leaving unshared electrons on the chains, which combine to form crosslinks:

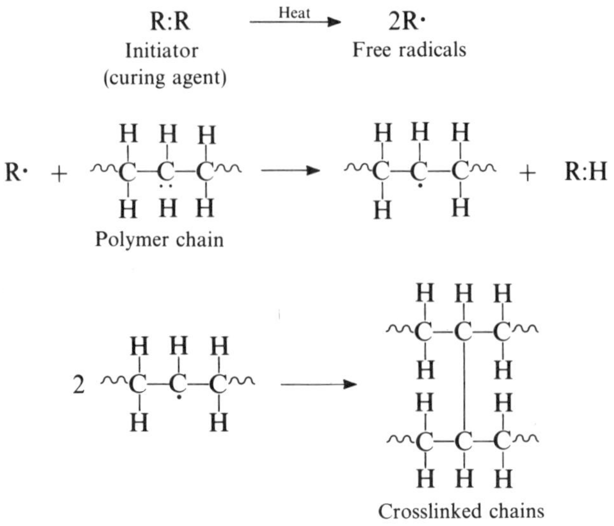

$$R{:}R \xrightarrow{\text{Heat}} 2R\cdot$$

Initiator   Free radicals
(curing agent)

Polymer chain

Crosslinked chains

Sometimes, sulfur or multifunctional vinyl monomers are added to improve crosslinking efficiency by helping to "bridge the gap" (i.e., produce longer crosslinks) between the chains. In this manner, polyethylene is crosslinked with dicumyl peroxide, converting it from a thermoplastic to a thermoset with much greater heat resistance and resistance to stress cracking and abrasion. It can also tolerate much higher levels of carbon black filler without becoming excessively brittle.

11. Blowing Agents. Foamed plastics must contain a material that generates a gas to produce foaming. There are *chemical blowing agents* (CBAs) which generate gas through a chemical reaction, and physical blowing agents, inert but volatile chemicals that are dissolved in the polymer or its precursors and simply vaporize upon heating or a reduction in pressure. As an example, polyurethanes (Chapter II, Example 4Q) may be "water blown" by the reaction of excess diisocyanate with water, generating CO_2,

$$O{=}C{=}N{-}R{-}N{=}C{=}O + 2H_2O \rightarrow R(NH_2)_2 + 2CO_2$$

Diisocyanate   Diamine

This is not often done in practice because of the expense of the diisocyanate and the difficulty in controlling foam structure. Alternatively, they can be foamed with a physical blowing agent, traditionally CCl_3F. The latter gives excellent control of foam structure, and because of its low thermal conductivity, makes good insulating foams. Unfortunately, it poses a threat to the atmospheric ozone layer and is rapidly being phased out. The search is on for adequate replacements.

CBAs must be compounded with a polymer below their decomposition temperatures, and then decompose to generate the gas at temperatures low enough to prevent degradation of the polymer. A variety of CBAs is available, covering a range of decomposition temperatures, for use with most common plastics.[9, 10]

In the manufacture of polystyrene foam molding beads, pentane is added to the monomer in a suspension polymerization or absorbed by the beads afterward. When the beads are placed in a mold and heated (usually with steam), the pentane volatilizes, expanding the beads against each other and the mold walls. These beads are used in many familiar applications, such as drinking cups, picnic coolers, and packaging supports.

12. Flame Retardants.[11-13] Most synthetic polymers, being composed largely of carbon and hydrogen, are flammable (although no more so than natural materials such as wood and cotton, which they often replace—a fact sometimes overlooked). Plastics are increasingly being compounded with flame retardants to reduce their flammability. The most common flame-retarding additives for plastics contain large proportions of chlorine or bromine. These elements are believed to quench the free-radical flame propagation reactions. They may be compounds which are simply mixed with the plastic, for example, decabromodiphenyl ether, or they may be reactive monomers which become part of the polymer. Examples of the latter include tetrabromobisphenol-A substituted for some of the normal bisphenol-A in epoxies (Chapter II, Example 4O) or polycarbonates (Chapter II, Example 4N) and tetrabromophthalic anhydride in polyesters (Chapter II, Example 3). Antimony trioxide is often used in synergistic combination with halogenated compounds. Organic phosphates are thought to function as flame retardants by forming a char which acts as a barrier to the flame. A compound such as tris(2,3-dibromopropyl) phosphate combines halogen and phosphate. Polyvinyl chloride, because of its chlorine content, is inherently flame resistant, but if it is to maintain this valuable property in plasticized form, it must be compounded with halogenated or phosphate plasticizers.

The safety of halogen and antimony flame-retardant systems has been questioned, both the compounds themselves during processing operations and their combustion products. This has lead to a recent swing toward phosphorus-based systems.[14] Hydrated alumina is a particulate filler which contains 35% water of hydration, the evaporation of which absorbs energy and inhibits flame spread.

13. Miscellaneous. Various other materials are added to plastics to provide certain end-use properties. For example, where a polymer will be subjected to a warm, humid environment (vinyl shower curtains, silicone-rubber bathtub caulks, etc.), it often contains a *biocide* to inhibit the growth of mold, mildew, and fungus. These biocides are generally organic copper, mercury, or tin compounds. Fatty-acid amines are added to compounds for use in bottles and phonograph records as anti-static agents. These chemicals, because of their limited compatibility with the polymer, migrate to the surface of the article, and because of their polarity, attract moisture from the atmosphere. The moisture bleeds off static electricity charges, which otherwise would attract considerable dust over a period of time. These same fatty-acid amines also function as *slip* or *anti-block* agents, which prevent layers of plastic film and sheet from sticking to each other. *Nucleating agents*, which promote more but smaller crystallites, were discussed in Chapter V.

20.4   SHEET MOLDING COMPOUND

The subject of sheet molding compoud (SMC) is introduced here because (1) they are commercially important, (2) they make use of many of the additives discussed in the previous section, and (3) they don't seem to fit logically anywhere else. SMCs consist mainly of unsaturated polyester resin (Chapter II, Example 3), filler (usually $CaCO_3$), and chopped-glass reinforcing fiber. Early SMCs contained roughly a third of each, but newer materials are closer to 25% resin, 30% glass, and 45% filler.[15] The liquid polyester and filler are compounded in a high-shear mixer along with minor amounts of a curing agent (usually an organic peroxide), a thickener (MgO or CaO), an external lubricant as a mold-release agent (zinc stearate), and sometimes a low-shrink additive.

The material leaves the mixer looking (but not smelling) like pancake batter, and is spread continuously onto layers of polyethylene film (Fig. 20.3). Chopped

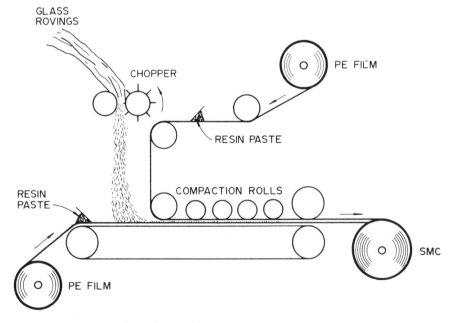

Figure 20.3   Sheet molding compound line.

glass fibers (about 1 in. long) are sprinkled on, the layers are combined, and the sheet is kneaded by rollers to wet out the glass. The sheet is then rolled up and allowed to age, during which time it thickens to the consistency of cardboard. This thickening is believed to involve interaction of the MgO or CaO with the carboxylic acid groups on the polyester, with water playing a role also, but it is not a true polymerization.

Before use, the sheets are cut to size and the polyethylene stripped off. They are then placed in a heated compression-type mold (several layers may be stacked for greater thicknesses) where true cure (crosslinking) of the polyester takes place.

The choice of a curing agent for an SMC involves a compromise between an active initiator, which will give quick cures, but conversely, imparts a short shelf life to the stored SMC, and a less active initiator with a longer shelf life but slower cure. A drawback of early SMCs was the wavy surface and sometimes visible glass-fiber texture produced by shrinkage of the polyester during cure. This necessitated hand finishing operations, which are not at all popular in the automotive industry. The low-shrink additives that counteract this are usually a second, linear polymer dissolved in the polyester, which precipitate out as a micrometer-sized dispersed phase as the polyester cures. Why this should reduce shrinkage is not at all clear, but if the low-shrink additive is a rubber, it may act as a toughening agent as well (see Reinforcing Agents).

Sheet molding compounds provide a high strength-to-weight ratio, with low cost and easy, rapid processing. They are used for automotive body parts such as front ends and fender extensions and for other structural parts such as business machine and air-conditioner housings. They (along with RIM polyurethanes and polyureas and injection-molded thermoplastics) are in the thick of the competition for increased use in auto body panels (doors, hoods, fenders, etc.). They have recently been adopted for the body panels in the GM minivans.

Bulk molding compounds (BMCs) are similar in composition to SMCs, with somewhat shorter fibers ($\frac{1}{4}$–$\frac{1}{2}$ in.), and perhaps less glass (15%) and more filler. They are produced in chunk form and compression molded like other thermosetting compounds.

REFERENCES

1. Seymour, R. B. (ed.), *Additives for Plastics*, Vol. 1, *State of the Art*, Vol. 2, *New Developments*, Academic, New York, 1978.
2. Greek, B. F., *Chem. Eng. News*, June 13, 1988, p. 35.
3. Milewski, J. V., and H. S. Katz (eds.), *Handbook of Reinforcements for Plastics*, Elsevier, New York, 1987.
4. Clegg, D. W., and A. A. Collyer (eds.), *Mechanical Properties of Reinforced Thermoplastics*, Elsevier, New York, 1986.
5. *Modern Plastics Encyclopedia*, McGraw-Hill, New York, 1990.
6. Rosen, S. L., *Trans. N.Y. Acad. Sci.* **35**(6), 480 (1973).

7. Riew, C. K. (ed.), *Rubber-Toughened Plastics*, American Chemical Society, Washington, DC, 1989.

8. Katz, H. S., and J. V. Milewski (eds.), *Handbook of Fillers for Plastics*, Elsevier, New York, 1987.

9. Heck, R. L., III, *Plast. Compd.* Nov./Dec. 1978, p. 52.

10. Hallans, R. S., *Plast. Eng.* Dec. 1977, p. 17.

11. Schongar, L. H., *Plast. Compd.* May/June 1978, p. 44.

12. Blake, W. P., *Plast. Compd.*, July/Aug. 1978, p. 26.

13. Nametz, R. C., *Plast. Compd.*, Jan./Feb. 1979, p. 31.

14. Wood, S. A., *Mod. Plast.* **67**(5), 40 (1990).

15. Wigotsky, V., *Plast. Eng.* **47**(6), 8 (1991).

CHAPTER XXI

Rubber

21.1 INTRODUCTION

A rubber is generally defined as a material that can be stretched to at least twice its original length, and that will retract rapidly and forcibly to substantially its original dimensions on release of the force. An elastomer is a rubber-like material from the standpoint of modulus but one that has limited extensibility and incomplete retraction. The most common example is highly plasticized polyvinyl chloride.

Rubber has the following characteristics: (1) From molecular standpoint, a rubber must be a high polymer, since rubber elasticity is due mainly to the coiling and uncoiling of long chain segments (Chapter XIV). (2) For the molecules to be able to coil and uncoil freely, the polymer must be above its glass-transition temperature at use temperature. (3) Furthermore, the polymer must be amorphous in its unstretched state, because crystallinity would hinder coiling and uncoiling. (4) There used to be an additional requirement that the polymer be crosslinked. If it were not, the chains would slip past one another (undergo viscous flow) under stress and recovery would be incomplete.

21.2 THERMOPLASTIC ELASTOMERS[1,2]

The introduction of so-called thermoplastic elastomers (actually rubbers) has gotten around the requirement for crosslinking in the strictest sense—that of covalent bonding between chains. The most common thermoplastic elastomers are styrene-*b*-butadiene-*b*-styrene (SBS) block copolymers, produced in solution by anionic polymerization (Example 1, Chapter XI). Since most polymer pairs are mutually insoluble, the polystyrene chain ends aggregate together in microscopic domains. These polystyrene domains, being below their glass transition temperature (100°C) at normal use temperatures, are rigid, and act to tie together the long, flexible polybutadiene segments (above their T_g) as do ordinary crosslinks. Unlike the usual covalent crosslinks, however, the polystyrene domains soften above the T_g of polystyrene, and the polymer behaves as a true thermoplastic.

All thermoplstic elastomers have in common at use temperature a continuous rubbery phase of "soft segments" tied together by glassy or crystalline "hard segments" which soften at elevated temperatures. Thus, thermoplastic elastomers don't have to be vulcanized, and can be processed by the usual economical thermoplastic techniques.

21.3 CONTENTS OF RUBBER COMPOUNDS

Natural rubber, and with a few exceptions the many synthetic rubber polymers commercially available, are unsaturated; that is, they contain double bonds which provide sites for the vulcanization (crosslinking) reaction. The polymers are mostly linear or branched, but some also contain (either intentionally or unintentionally) minor amounts of gel (crosslinked particles) prior to vulcanization. This can have a profound effect on processing properties.[3,4] The following substances are also included in rubber compounds.

1. Reinforcing Fillers. Most rubber compounds that are intended to develop a reasonable tensile strength, abrasion, and tear resistance will contain up to 50 phr (parts per hundred parts rubber by weight) or more of a reinforcing filler, nearly always a carbon black. The use of carbon black in rubbers is quite different than in plastics, where it is strictly a pigment and limited to much lower loadings.

Carbon blacks are not simply carbon. Basic blacks have hydroxyl groups at the particle surfaces and acid blacks have carboxylic acid functionality. It has been shown that the rubber polymer forms strong secondary and primary covalent bonds with the carbon black, which accounts for its reinforcing ability. A wide variety of carbon blacks is available.[5] In addition to chemical functionality, they differ in such factors as particle size, degree of aggregation, and surface area, and different types of rubber polymers require particular kinds of black for optimum reinforcement. Silicone rubbers are sometimes reinforced with finely divided silica (SiO_2).

Natural rubber (cis-1,4-polyisoprene) and its synthetic counterpart and butyl rubber are among the few rubber polymers that can develop reasonable mechanical strength without reinforcement. This is because they crystallize with molecular orientation at high elongations, and the crystallites function as a reinforcing agent, causing the sharp upturn in the stress–strain curves at high elongations shown in Fig. 21.1.[6]

2. Fillers. As with plastics, the function of fillers in rubber is mainly to reduce the cost of the compound. Most rubber fillers are finely divided inorganics such

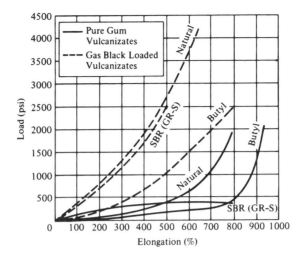

Figure 21.1 Stress–strain curves for filled and unfilled rubber vulcanizates.[6] From A. X. Schmidt and C. A. Marlies, *Principles of High-Polymer Theory and Practice*, Copyright 1948 by the McGraw-Hill Book Company, Inc., New York. Used with permission of McGraw-Hill Book Company.

as $CaCO_3$. The addition of such high-modulus fillers raises the modulus (stiffens the compound), and too much will cause a loss of rubbery properties, all other things being equal. Carbon blacks sometimes perform as fillers as well as reinforcing agents in rubber compounds.

3. Extending Oils. Hydrocarbon oils are often used in rubber compounds. Their function is twofold. First, they plasticize the polymer, making it softer and easier to process. This is particularly important with very high molecular weight polymers. Second, since they are usually cheaper than the rubber polymer, they act like fillers in reducing the cost of the compound. Extending oils are available in various degrees of aromaticity, and the properties of the compound depend on the type of oil in relation to the polymer as well as oil level.

Carbon black and extending oils have opposite effects on the modulus of a rubber compound. One trick for producing low-cost compounds from synthetic polymers is to polymerize to higher-than-normal molecular weight. The modulus is cut down by the addition of an extending oil and then brought back up to the desired level by the addition of large amounts of black. From projecting figures showing the increasing levels of black and oil in rubber compounds, it appears that there soon won't be any polymer left. Seriously, there is a limit to this technique, because other properties are soon degraded excessively.

4. Vulcanizing or Curing Systems. The function of these is to crosslink the polymer. The most common curing systems are based on sulfur. While sulfur alone will cure unsaturated rubbers with heating, the process is slow and inefficient in its use of the sulfur. The mechanisms of sulfur curing are not well understood, but are thought to include (among other things) the formation of sulfide or disulfide links between chains and the abstraction of protons from adjacent chains to form H_2S, with the chains crosslinking at the remaining unshared electrons.

To speed up the vulcanization process, *accelerators* are generally used. These are usually complex sulfur-containing organic compounds, often of proprietary composition. *Promotors* or activators are used to improve the cure still further. Zinc oxide is a common example, particularly in conjunction with stearic acid.

Nonsulfur cures are used occasionally. Free-radical initiators will provide crosslinks, as discussed in the previous chapter, and zinc oxide will crosslink polymers containing chlorine.

5. Antioxidants or Stabilizers. These are particularly important with natural rubber and the many synthetic rubbers that contain a major proportion of butadiene or isoprene. These polymers are highly unsaturated and the double bonds are extremely susceptible to attack by oxygen and ozone, resulting in

Table 21.1   Radial Passenger-Tire Tread Formulation

Ingredient	phr[a]	Function
SBR 1712[b]		
Polymer	45	Rubber polymer
Aromatic oil	37.5	Extending oil
cis-Polybutadiene 1252	55	Rubber polymer
Carbon black N-234	70	Reinforcing agent
Oil-soluble sulfonic acid	1	Processing aid
Sulfur	1.75	Vulcanizing agent
Stearic acid	2	Promotor
Zinc oxide	3	Promotor
N-Cyclohexyl-2-benzothiazole sulfenamid	1	Accelerator
Polymerized 1,2-dihydrotrimethylquinoline	2	Antioxidant
N-(1,3-dimethylbutyl)-N'-Phenyl-p-phenylene diamine	1	Antiozonant
Blended petroleum wax	3	Crack inhibitor and antiozonant
Total	222.25	

[a] Parts per hundred parts rubber, by weight.
[b] Oil-extended master batch. Cold emulsion polymer $\approx 75\%$ butadiene, 25% styrene extended with highly aromatic oil.

Source. Stephens.[7]

embrittlement, cracking, and general degradation. Since the degradation reactions take place by free-radical mechanisms, antioxidants are compounds that scavenge free radicals.

6. Pigments. Where great mechanical strength is not required, and thus carbon black not used, rubbers can be colored with pigments as can plastics.

A typical radial passenger-tire tread formulation is shown in Table 21.1. Notice that the final compound is less than 45% polymer.

21.4 RUBBER COMPOUNDING[7,8]

The ingredients above must be *compounded* with the rubber polymer to produce the final rubber compound for molding, extrusion, etc. This is usually done in *two-roll* mills or *Banbury* mills (Section 19.13). Both devices are driven by relatively large electric motors and put a lot of energy per unit time and volume into the polymer. This energy input serves two purposes. First, it breaks down the polymer and reduces its "nerve" to the point where it can be easily compounded and processed. Nerve is a term used to describe the difficult-to-handle highly elastic response caused by too high a molecular weight. This mastication step mechanically degrades the polymer, lowering its molecular weight and therefore increasing viscous response. Premastication is particularly important with natural rubber, since its molecular weight can't be controlled during polymerization. Second, the high energy input breaks up and disperses the compounding ingredients evenly throughout the polymer.

REFERENCES

1. Legge, N. R., G. Holden, and H. E. Schroeder (eds.), *Thermoplastic Elastomers*, Hanser, Munich, 1987.
2. Walker, B. M., and C. P. Rader (eds.), *Handbook of Thermoplastic Elastomers*, 2d ed., Van Nostrand Reinhold, New York, 1988.
3. Rosen, S. L., and F. Rodriguez, *J. Appl. Polym. Sci.* **9**, 1601 (1965).
4. Rosen, S. L., and F. Rodriguez, *J. Appl. Polym. Sci.* **9**, 1615 (1965).
5. Byers, J. T., Fillers, Part I: carbon black, Chapter 3 in *Rubber Technology*, 3d ed., M. Morton (ed.), Van Nostrand-Reinhold, New York, 1987.
6. Schmidt, A. X., and C. A. Marlies, *Principles of High Polymer Theory and Practice*, McGraw-Hill, New York, 1948, p. 537.
7. Stephens, H. L., The compounding and vulcanization of rubber, Chapter 2 in *Rubber Technology*, 3d. ed., M. Morton (ed.), Van Nostrand-Reinhold, New York. 1987.
8. Barlow, F. W., *Rubber Compounding, Principles, Materials and Techniques*, Dekker, New York, 1988.

CHAPTER XXII

Synthetic Fibers

22.1 INTRODUCTION

While many of the polymers used for synthetic fibers are identical to those in plastics, the two industries grew up separately, with completely different terminologies, testing procedures, and so on. Many of the requirements for fabrics are stated in nonquantitative terms such as "hand" and "drape" which are difficult to relate to normal physical property measurements, but which can be critical from the standpoint of consumer acceptance, and therefore the commercial success, of a fiber.

A fiber is often defined as an object with a length-to-diameter ratio of at least 100. Synthetic fibers are spun (Section 19.12) in the form of continuous *filaments*, but may be chopped to much shorter *staple*, which is then twisted into thread before weaving. Natural fibers, with the exception of silk, are initially in staple form. The thickness of a fiber is most commonly expressed in terms of *denier*, which is the weight in grams of a 9000-m length of the fiber. Stresses and tensile strengths are reported in terms of *tenacity*, with units of grams/denier.

22.2 FIBER PROCESSING

The polymer molecules in synthetic fibers are only slightly oriented by flow as they emerge from the spinnerette. To develop the tensile strengths and moduli necessary for textile fibers, the fibers must be drawn (stretched) to orient the molecules along the fiber axis and develop high degrees of crystallinity. All successful fiber-forming polymers are crystallizable, and so from a molecular standoint, the polymer must have polar groups, between which strong hydrogen bonding holds the chains in a crystal lattice (e.g., polyacrylonitrile, nylons), or be sufficiently regular to pack closely in a lattice held together by dispersion forces (e.g., isotactic polypropylene).

In Section 19.12, it was pointed out that the cross section of fibers is determined by the cross section of the spinnerette holes and the nature of the spinning process. This plays an important role in establishing the properties of the fiber. For certain applications, the spun fibers are textured after spinning. Carpet fibers, for example, are often given a heat-set twist and/or are crimped by passing them through a pair of gear-like rollers.

22.3   DYEING

The dyeing of fibers is a complex art in itself. A successful dye must either form strong secondary bonds to polar groups on the polar, or react to form covalent bonds with functional groups on the polymer. Furthermore, since the fibers are dyed after spinning, the dye must penetrate the fiber, diffusing into it from the dye bath. The dye molecules can't penetrate the crystalline areas of the polymer, so it is mainly the amorphous regions that are dyed. This often conflicts with the requirement of high crystallinity. The chains of polyacrylonitrile, for example, while possessing the necessary polar sites for dye attachment in abundance, are so strongly bound to each other that it's difficult for the dye to penetrate. For this reason, acrylic fibers usually contain minor amounts of plasticizing comonomers to enhance dye penetration. Nonpolar, nonreactive fibers such as polypropylene, on the other hand, have no sites to which the dye can bond even if it could penetrate. This was long a problem with polypropylene fibers, and was initially overcome by incorporating a finely divided solid pigment in the polymer before melt spinning. Copolymers of propylene and a monomer with dye-accepting sites are now available.

22.4   ADDITIVES AND TREATMENTS

Static electricity can be a big problem with carpets. Many carpet fibers therefore incorporate an anti-static agent (Section 20.3.13) to bleed off static charge.

The same polar bonding sites that make fibers dyeable also make them stainable. In the past, the finished item, a carpet, for example, would be treated with an anti-staining agent such as a fluorocarbon telomer. Nowadays, these coatings are often applied to the fibers before weaving. Another approach is to make use of bicomponent fibers. For example, a core fiber of nylon 6/6 could be coated with a sheath of highly stain-resistant polypropylene.

22.5   EFFECTS OF HEAT AND MOISTURE

The polarity of the polymer also directly influences its degree of water absorption. Other things being equal, the more polar the polymer, the higher its equilibrium moisture content under any given conditions of humidity. As with dyes, however, moisture content is reduced by strong interchain bonding. The moisture content exerts a strong influence on the feel and comfort of fibers. Hydrophobic fibers tend to have a "clammy" feel in clothing, and can build up static electricity charges.

Perhaps the most important effect of moisture on polar polymers is as a plasticizer. Since fiber-forming polymers are linear, heat is also a plasticizer. This explains why suits wrinkle on hot, humid days, and why the wrinkles can be removed by steam pressing. "Wash-and-wear" and "permanent-press" fabrics are produced by operations that cross-link the fibers by reacting with functional groups on the chains, such as the hydroxyls on cellulose. The more hydrophobic polymers are inherently more wrinkle resistant because they are not plasticized by water. Wash-and-wear shirts, therefore, usually are made of blends of polyethylene terephthalate and cotton, about 65%/35%

Surface Finishes

23.1 INTRODUCTION

Nearly all surface finishes and coatings, with the exception of ceramic types for high-temperature applications, are based on a polymer film of some sort. They account for the use of a lot of polymer, but determining just how much and which polymers is not easy, because most formulations are proprietary, and production figures do not always separate polymer and nonpolymer components. The next section describes the five traditional types of surface finishes and the role played by polymers in each. Subsequent sections discuss more recent developments in the field.

23.2 TRADITIONAL TYPES OF SURFACE FINISHES

1. Lacquers. A lacquer consists of a polymer solution to which a pigment has been added. The film is formed simply by evaporation of the solvent, leaving the pigment trapped in the polymer film. Since no chemical change occurs in the polymer, it retains its original solubility characteristics. Hence, a major drawback of lacquers is their poor solvent resistance.

Much of the technology of lacquers involves the development of polymers that form tougher, more adhesive, and more stable films, and the choice of solvent systems that provide the optimum application viscosity and minimum cost, and meet environmental constraints. The volatility of the solvent system is important, too. If volatility is too high, the solvent will evaporate before the film has had a chance to "level," leaving brush marks or an "orange-peel" or rough surface when sprayed. If too low, the coating will "sag" excessively after application.

A wide variety of polymers is used for lacquers. Newer systems favor acrylic polymers for their superior chemical stability. Acrylic lacquers may also incorporate a curing agent[2] such as hexamethoxymelamine, which crosslinks the film in a subsequent baking operation. Crosslinking toughens the film and improves solvent resistance.

The term "spirit varnish" is an old and imprecise name for a lacquer in which the solvent is alcohol. Shellac is an example. Since the polymers are soluble in a highly polar solvent, spirit varnishes have rather poor water resistance.

2. Oil Paints. These popular and widely used finishes typically consist of a suspension of pigment in a *drying oil*, an ester of glycerine and unsaturated fatty acid such as linseed oil. The film is formed by a free-radical reaction involving atmospheric oxygen which polymerizes and crosslinks the drying oil through its double bonds. Sometimes an inert solvent (mineral spirits, turpentine) is added to control viscosity, and catalysts (cobalt naphthenate, for example) are used to promote the crosslinking reaction. Oil paints, once cured, are no longer soluble, although they may be swelled and softened considerably by appropriate solvents (as used in paint removers).

3. Oil Varnish or Varnish. These coatings consist of a polymer, either natural or synthetic, dissolved in a drying oil, with perhaps an inert solvent to control viscosity, and a catalyst to promote the crosslinking reaction with oxygen. When cured, they produce a clear, tough, solvent-resistant film.

4. Enamel. An enamel is a pigmented oil varnish. It is much like an oil paint except that the added polymer provides a tougher, glossier film.

5. Latex Paints. These versatile finishes have just about replaced oil paints and enamels for home use because of their quick drying, low odor, and water cleanup properties. Basically, they are polymer latexes, produced by emulsion polymerization (Chapter XIII) to which pigments and rheological-control agents have been added.

The film is formed by coalescence of the polymer particles upon evaporation of the water. The polymer itself is not water soluble. Although they are sometimes termed "water-soluble" paints, this is a serious misnomer, as is well known to anyone who has tried to clean a brush after the water has evaporated. As long as the individual latex particles have not coalesced, they are water-dispersable.

In order that the particles coalesce to form a film when the water evaporates, the polymer must be deformable under the action of surface-tension forces. Thus, polymers for latex paints must be near or above their glass transition temperatures at use temperature. This formerly resulted in a rather soft paint film, which inherently lacked "scrubability." Reasonable resistance to abrasion can be achieved by the incorporation of fairly large particle size inorganic filler-pigments such as $CaCO_3$. The rough paint film that results scatters reflected light and therefore has a "flat" finish, a characteristic of all early latex paints. By incorporating a low-volatility leveling solvent, latex "enamels" (they are not true enamels in the traditional sense) are produced. The polymer is formulated to have a T_g above use temperature. It is plasticized by the solvent so that its T_g is below application temperature, allowing the particles to coalesce to form a film. Over a period of a day or so after application, the solvent evaporates, leaving the hard, scrub-resistant, high-T_g polymer film, with a reasonable gloss if desired.

Early latex paints were based on styrene–butadiene copolymers or polyvinyl acetate. These have been largely supplanted by paints based on acrylic latices (acrylates, $ROOCH=CH_2$ or methacrylates, $ROOCCH_3=CH_2$), because of the acrylics' superior chemical stability and therefore resistance to color change and degradation.

Much work has gone into adjusting the formulations of latex paints to allow high pigment contents and thick films for one-coat coverage, together with reasonable ease of application. The main rheological property desired is *thixotropy* (Chapter XV), which gives a high viscosity to prevent the settling of pigments and sagging and dripping of the applied film, together with a lower viscosity under the shearing of brushing or rolling for easy application. These properties are achieved by the addition of finely divided inorganics and water-soluble polymers.

23.3 SOLVENTLESS COATINGS

Traditionally, the application of most surface finishes has been accompanied by the evaporation of an organic solvent. These evaporated solvents are now recognized as a significant source of air pollution. Moreover, since the hydrocarbon crunch, it has become economically undesirable simply to lose them to the atmosphere. While the water-based latex paints have gone a long way toward eliminating these problems, the coatings produced are sometimes lacking in thickness, uniformity, and durability. As a result, there is considerable interest in solventless or *100% solids* coatings.

Two techniques for producing surface coatings without solvent evaporation, plastisol coating and fluidized-bed coating, were discussed in Section 19.5. Like the latter, *electrostatic spraying* makes use of a polymer powder. The powder is sprayed past an electrode, charging the particles, and onto a heated substrate with the opposite charge. The powder contacting the substrate melts to form a continuous film. Thermosets may be further oven-cured, if necessary. The resulting film is usually quite uniform. Another advantage of this process over traditional spraying of either latex- or solvent-based paints is that the material that is sprayed but not actually deposited on the surface (*overspray*) can be recovered for reuse. Since overspray may account for much of the material sprayed, this can be economically significant.

The other approach to 100%-solids coatings is to carry out the polymerization reaction right on the surface. Reactive liquid monomer or oligomer (low molecular weight polymer) is deposited on the surface and polymerized there. The reactants may be two components of a condensation pair (e.g., epoxies, Chapter II, Example 4O, or polyurethanes, Chapter II, Example 4Q), which are mixed just prior to application and heat-cured on the surface, or they may be unsaturated materials that undergo addition polymerization (acrylates, $ROOC-CH=CH_2$, because of their high reactivity, are favored here).

In *radiation curing* processes, addition reactions are activated by various forms of radiation: infrared, microwave, radio frequency, ultraviolet, and electron beam.[2,3] Basically, the energies of the first three are such that they simply thermally activate the system, that is, heat it, and are used in conjunction with the usual free-radical initiators.

Ultraviolet-cured systems have achieved significant commercial importance, not only as coatings, but also as printing links, as a result of their low energy requirements (as compared to heat-cured systems) and freedom from pollution. The polymeric portion of a typical UV-cured coating has three main ingredients: (1) a reactive oligomer, (2) a multifunctional acrylate monomer, and (3) a photoinitiator. The reactive oligomer can be just about any low molecular weight polymer containing at least a couple of double bonds, preferably acrylate double bonds because of their high reactivity. A simple example would be the system formed by reacting an excess of 1,6-hexanediol, $HO(CH_2)_6OH$, with adipic acid, $HOOC(CH_2)_4COOH$. The resulting hydroxyl-capped polyester is then condensed with acrylic acid, $HOOC-CH=CH_2$, to give the diacrylate polyester

$$H_2C=\overset{\overset{\displaystyle H}{|}}{C}-\overset{\overset{\displaystyle O}{\|}}{C}-O(CH_2)_6[O-\overset{\overset{\displaystyle O}{\|}}{C}(CH_2)_4\overset{\overset{\displaystyle O}{\|}}{C}-O(CH_2)_6]_xO-\overset{\overset{\displaystyle O}{\|}}{C}-\overset{\overset{\displaystyle H}{|}}{C}=CH_2$$

This material, being tetrafunctional (two double bonds), will cure in an addition reaction, but the cross-link density tends to be too low and, even though x is kept low, the viscosity is too high for most applications, so it is diluted with a multifunctional acrylate monomer such as trimethylol propane triacrylate:

$$H_3C-CH_2-C(CH_2-OOC-CH=CH_2)_3$$

This reduces the viscosity of the mix for easier application and increases the crosslink density in the cured film.

The film is cured by free radicals generated by the UV-induced decomposition of a photoinitiator, for example, benzoin methyl ether:

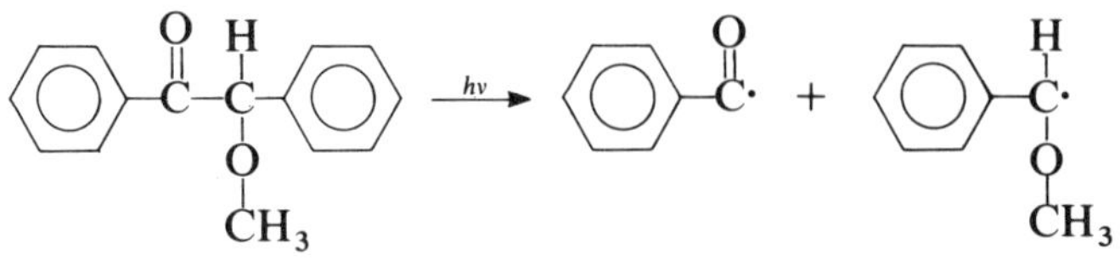

Electron-beam curing systems are similar in composition. Although they have been around a long time, they have not yet achieved widespread commercial application, probably because of the high capital investment required to produce the electrons and the need to shield the equipment (stray x-rays are generated). Nevertheless, for highly pigmented systems, where UV radiation tends to be absorbed preferentially near the surface, they offer more uniform cures. The economics appear reasonable in high-volume applications, so as equipment continues to improve, they should increase in importance.

The chemistry of polymers for high-solids coatings has been extensively reviewed.[4]

23.4 ELECTRODEPOSITION[5,6]

The process of electrodeposition has been developed to provide uniform, highly adhesive, corrosion-resistant primer coats, particularly for the automotive industry. (The ultimate solution to the corrosion problem is to replace the metal with plastic, but that's another story.) In the most common form of the process, polymers that contain carboxylic acid functionality are produced. These polymers are then "solubilized" in an aqueous medium by partial neutralization with a base to give macroanions (P represents a polymer backbone)

$$P(\overset{O}{\overset{\|}{C}}-OH)_n + mBOH \rightarrow (HO-\overset{O}{\overset{\|}{C}})_{n-m}P(\overset{O}{\overset{\|}{C}}-O^-)_m + mB^+ + mH_2O$$

| Macrocarboxylic acid | Base | Macroanion |

The macroanions are believed to exist in solution as micelles.

The part to be coated is made an anode (+) and immersed in a tank containing the macroanion solution plus pigment, plasticizer, curing agent, and other ingredients. The applied electric field draws the macroanions to the surface where they lose their charge and are deposited as a film. From the standpoint of forming a continuous, uniform, and therefore highly corrosion-resistant film, this process has significant advantages. The electric field draws the particles into all the nooks and crannies of the part (good "throwing power"). Sharp exterior corners, which in other processes tend to be thinly coated because surface tension draws the material away from the corner, produce a stronger electric field, and thus attract more polymer. Because the deposited coating is a dielectric, the field is stronger at holes in the film, preferentially attracting polymer to seal the hole. Furthermore, pollution and fire hazards are minimized, and the process wastes very little material (unlike spraying) and so offers economic advantages.

A cationic deposition process is also used. Theoretically, it offers the best potential for rust resistance. This process is based on amine-functional polymers treated with an acid:

$$P(NR_2)_n + mHX \rightarrow (R_2N)_{n-m}P(NR_2H^+)_m + mX^-$$

| Acid | Macrocation |

The macrocations are then deposited on a cathodic substrate.

Following electrodeposition, objects are usually rinsed to remove any non-adhering material and then baked to dry and fuse the film, and perhaps cross-link it.

REFERENCES

1. Morgans, W. M., *Outlines of Paint Technology*, 3d ed., Wiley, New York, 1990.

2. McGinnis, V. D., and G. W. Gruber, Curing methods for coatings, Chapter 35 in *Applied Polymer Science*, 2d ed., R. W. Tess and G. W. Poehlein (eds.), American Chemical Society, Washington, DC, 1985.

3. Randell, D. R. (ed.), *Radiation Curing of Polymers*, The Royal Society of Chemistry, London, 1987.

4. Athey, R. D., Jr., *Prog. Org. Coatings* **7**, 289 (1979).

5. Beck, F., *Prog. Org. Coatings* **4**, 1 (1976).

6. Brewer, G. E. F., Electrodeposition of paint, Chapter 34 in *Applied Polymer Science*, 2d ed., R. W. Tess and G. W. Poehlein (eds.), American Chemical Society, Washington, DC, 1985.

CHAPTER XXIV

Adhesives

Adhesives have been a technologically important application of polymers for thousands of years. Many of the early natural adhesives are still used. These include starch and protein-based formulations such as hydrolyzed collagen from animal hides, hooves, and bones and casein from milk. As new adhesive formulations based on synthetic polymers (often the same polymers used in other applications) continue to be developed, the range of applications for adhesives has expanded dramatically.[1-7]

An adhesive has been defined as a substance capable of holding materials (adherends) together by surface attachment. Adhesives offer a number of significant advantages as a means of bonding: (1) They are often the only practical means available, particularly in the case of small adherends. For example, it's hard to imagine welding abrasive grains to a paper backing to make sandpaper, or bolting the grains together to make a grinding wheel. (2) In the adhesive joining of large adherends, forces are fairly uniformly distributed over large areas of the adherend, resulting in low stresses, and holes (necessary for riveting or bolting), which invariably act as stress concentrators in the adherends, are eliminated, thus lowering the possibility of adherend failure. (3) In addition to joining, adhesives may also act as seals against the penetration of fluids. In the case of corrosive fluids, this, coupled with the absence of holes, where corrosion usually gains an initial foothold, can minimize corrosion problems. (4) In terms of weight, it doesn't take much adhesive to join much larger adherends. Hence, it is not surprising that many of the newer high-performance adhesives were originally developed for aerospace applications. (5) Adhesive joining may offer economic advantages, often by reducing the hand labor necessary for other bonding techniques.

A detailed treatment of the science of adhesion is beyond the scope of this chapter. Nevertheless, some important generalizations will be drawn.[8,9] Adhesion results from (1) mechanical bonding between the adhesive and adherend, and (2) chemical forces—either primary covalent bonds or polar secondary forces between the two. The latter are thought to be the more important, and this explains in part why inert, nonpolar polymeric substrates such as polyethylene and polytetrafluoroethylene are very difficult to adhesive bond: They must first be chemically treated to introduce polar sites on the surface. To promote mechanical bonding, adherend surfaces are often roughened before joining, but this is sometimes counterproductive. It can trap air bubbles at the bottom of crevices which act as stress concentrators to promote failure in rigid adhesives.

With good bonding between adhesive and adherend, joint failure is cohesive (the adhesive itself or the substrate fails). Where the adhesive is weaker than the substrate, to a good approximation, the properties of the adhesive polymer determine the properties of the adhesive joint; that is, the bond can be no stronger than the glue line. Brittle polymers give brittle joints, polymers with high shear strengths give bonds of high shear strength, heat-resistant polymers produce bonds with good heat resistance, and so on.

To form a successful joint, the adhesive must intimately contact the adherend surface. This requires first that it wet the surface. The subject of wetting is considered in detail in treatises on surface chemistry.[10,11] In general, wetting is promoted by polar secondary forces between adhesive and substrate. This is another reason why low-polarity polymeric adherends such as polyethylene and polytetrafluoroethylene are difficult to bond with adhesives. To insure proper wetting and interfacial bonding, it is often necessary to clean the adherend surfaces carefully before joining.[12] Good contact also requires a viscosity low enough under conditions of application to allow the adhesive to flow over the surface and into its nooks and crannies. Once contact has been established, the adhesive must harden to provide the necessary joint strength. There are five general categories of organic adhesive that accomplish these objectives in different ways.

1. Solvent-Based Adhesives. Here the adhesive polymer is made to flow by dissolving it in an appropriate solvent to form a cement. The adhesive hardens by evaporation of the solvent. Thus, the polymers used must be linear or branched to allow solution, and the joints formed will not be resistant to solvents of the type used initially to dissolve the polymer. To get a good bond, it helps if the solvent attacks the adherend also. In fact, solvent alone is often used to "solvent weld" polymers, dissolving some of the adherend to form an adhesive on application.

One of the drawbacks to solvent-based adhesives based on rigid polymers is the shrinkage that results when the solvent evaporates. This can set up stresses that weaken the joint. An example of this type of adhesive is the familiar model airplane cement, basically a cellulose nitrate solution, with perhaps some plasticizer. Rubber cements, of course, maintain their flexibility, but cannot support as great a stress. Commercial rubber cements are based on natural, SBR (poly(butadiene-co-styrene)), nitrile (poly(butadiene-co-acrylonitrile)), chloroprene (poly(2-chlorobutadiene)), and reclaimed (devulcanized) rubbers. Examples are household rubber cement and Pliobond®. Rubber cements may also incorporate a curing agent to crosslink the polymer after application and evaporation of the solvent. This greatly increases solvent resistance and strength.

2. Latex Adhesives. These materials are based on polymer latexes made by emulsion polymerization. They flow easily while the continuous water phase is present and dry by evaporation of the water, leaving behind a layer of polymer. In order that the polymer particles coalesce to form a continuous joint and be able to flow to contact the adherend surfaces, the polymers used must be above their glass transition temperature at use temperature. These requirements are similar to those for latex paints, so it is not surprising that some of the same polymers are used in both applications, for example, styrene–butadiene copolymers and polyvinyl acetate. Nitrile and neoprene rubbers are used for increased polarity. A familiar example of a latex adhesive is "white glue," basically a plasticized polyvinyl acetate latex. Latex adhesives are displacing solvent-based adhesives in many applications because of their reduced pollution and fire hazards. They are used extensively for bonding pile to backing in carpets.

3. Pressure-Sensitive Adhesives.[13] These are really viscous polymer melts at room temperature, so the polymers used must be above their glass transitions. They are caused to flow and contact the adherends by applied pressure, and when the pressure is released, the viscosity is high enough to withstand the stresses produced by the adherends, which obviously cannot be very great. The key property for a polymer used in this application is tack, which basically is a viscosity low enough to permit good surface contact, yet high enough to resist separation under stress, something on the order of 10^4–10^6 cP,[14] although elasticity probably plays a role, also. Natural, SBR, and reclaimed rubbers are common in this application. The many varieties of pressure-sensitive tape are faced with this type of adhesive.

Contact cements are a variation in which the rubbery polymer is applied to each adherend surface in the form of a solution, or increasingly, a latex. Evaporation of the solvent or water leaves a polymer film with the tack necessary to grab and hold the adherends when they are pressed together.

4. Hot-Melt Adhesives. Thermoplastics often form good adhesives simply by being melted to cause flow and then solidifying on cooling after contacting the surfaces under moderate pressure. Polyamides and poly(ethylene-*co*-vinyl acetate) are used frequently as hot-melt adhesives. Electric "glue guns" have been introduced to the consumer market which operate on this principle.

5. Reactive Adhesives. These compounds are either monomers or low molecular weight polymers which solidify by a polymerization and/or cross-linking reaction after application. They can develop tremendous bond strengths and have good solvent resistance and good (for polymers, anyhow) high-temperature properties. The most familiar example of reactive adhesives are the epoxies (Chapter II, Example 4O) generally cured by multifunctional amines. Polyurethanes (Chapter II, Example 4Q) also make excellent reactive adhesives.

The α-alkyl cyanoacrylate "super glues" ("one drop holds 5000 lbs") are now a familiar part of the consumer market. Originally, the monomers had extremely low viscosities, and so could crawl into narrow crevices and wet the adherend surfaces rapidly. On the other hand, they wouldn't fill gaps, and were absorbed into porous adherends, giving poor bonds. Newer versions are available with higher viscosities to overcome these drawbacks. Cyanoacrylates can polymerize in seconds by an anionic addition reaction believed initiated by hydroxyl ions from water adsorbed on the adherend surfaces:

$$x\ H_2C{=}\underset{\underset{\displaystyle CN}{|}}{C}{-}\underset{\displaystyle O}{\overset{\displaystyle O}{\|}}{C}{-}OR \quad \xrightarrow{\ OH^-\ } \quad HO{-}\!\!\left[\underset{\underset{\displaystyle H}{|}}{\overset{\overset{\displaystyle H}{|}}{C}}{-}\underset{\underset{\displaystyle H}{|}}{\overset{\overset{\displaystyle C{=}O}{\overset{\displaystyle |}{\displaystyle OR}}}{C}}\right]_{\!x}\!\!{-}H$$

Unfortunately (or fortunately, if you stick your fingers together), being linear and polar, they have poor resistance to polar solvents (acetone is a good solvent), and they are subject to hydrolysis, and so have poor environmental stability.

Phenolic and other formaldehyde condensation polymers are also important reactive adhesives. Powdered phenolic resin is mixed with abrasive grains and the mixture is compression molded to form grinding wheels. A B-stage phenolic (Chapter XX) in a solvent is used to impregnate tissue paper. The solvent is evaporated, and the dry sheets are placed between layers of wood in a heated press, where the resin first melts and then cures, bonding the wood to form plywood. Similarly, sheets of paper impregnated with a B-stage melamine–formaldehyde resin are laminated and cured to form the familiar Formica® counter tops.

Unlike the previous examples of reactive adhesives, the phenolics and other formaldehyde condensation polymers evolve water as they cure. If trapped in the joint, this can result in serious weakness, which limits their adhesive applications.

Note that all these examples of reactive adhesives are highly polar polymers. It is largely this polarity that accounts for their good bonding capabilities.

REFERENCES

1. Temin, S. C., Adhesive compositions, p. 547 in *Encyclopedia of Polymer Science and Engineering,* 2d ed., Vol. 1, J. Kroschwitz (ed.), Wiley, New York, 1985.

2. Patrick, R. L., Chemistry and technology of adhesives, Chapter 34 in *Applied Polymer Science,* J. K. Craver and R. W. Tess (eds.), American Chemical Society, Washington, DC, 1975.

3. Shields, J., *Adhesives Handbook*, 3d ed., Butterworths, London, 1984.

4. Hartshorne, S. R. (ed.), *Structural Adhesives: Chemistry and Technology*, Plenum, New York, 1986.

5. Wake, W. C., *Adhesion and the Formulation of Adhesives*, 2d ed., Applied Science, London, 1982.

6. Wake, W. C. (ed.), *Synthetic Adhesives and Sealants*, Wiley, New York, 1987.

7. Landrock, A. H., *Adhesives Technology Handbook*, Noyes, Park Ridge, NJ, 1985.

8. Gent, A. N., and G. R. Hamed, Adhesion, p. 476 in *Encyclopedia of Polymer Science and Engineering*, 2d ed., Vol. 1, J. Kroschwitz (ed.), Wiley, New York, 1985.

9. Meyer, F. J., Bonding, p. 518 in *Encyclopedia of Polymer Science and Engineering*, 2d ed., Vol. 1, J. Kroschwitz (ed.), Wiley, New York, 1985.

10. Adamson, A. W., *Physical Chemistry of Surfaces*, 5th ed., Wiley, New York, 1990.

11. Hiemenz, P. C., *Principles of Colloid and Surface Chemistry*, 2d ed., Dekker, New York, 1986.

12. Wegman, R. F., *Surface Preparation Techniques for Adhesive Bonding*, Noyes, Park Ridge, NJ, 1989.

13. Sates, D. (ed.), *Handbook of Pressure-Sensitive Adhesives*, 2d ed., Van Nostrand Reinhold, New York, 1987.

14. Skeist, I., Adhesive compositions, p. 482 in *Encyclopedia of Polymer Science and Technology*, Vol. 1, N. Bikales (ed.), Wiley, New York, 1971.

Index